Advanced Plasma Technology

Edited by
Riccardo d'Agostino, Pietro Favia,
Yoshinobu Kawai, Hideo Ikegami,
Noriyoshi Sato, and
Farzaneh Arefi-Khonsari

Related Titles

Hippler, R., Kersten, H, Schmidt, M., Schoenbach, K H. (eds.)

Low Temperature Plasmas

Fundamentals, Technologies and Techniques

approx. 1110 pages in 2 volumes
exp. 2007
Hardcover
ISBN: 978-3-527-40673-9
ISBN: 3-527-40673-5

d'Agostino, R., Favia, P., Oehr, C, Wertheimer, M. R. (eds.)

Plasma Processes and Polymers

16th International Symposium on Plasma Chemistry Taormina/Italy June 22-27, 2003

545 pages with 275 figures
2005
Hardcover
ISBN: 978-3-527-40487-2
ISBN: 3-527-10487-2

Woods, L C.

Physics of Plasmas

226 pages with 69 figures
2004
Softcover
ISBN: 978-3-527-40461-2
ISBN: 3-527-40461-9

Marcus, R. K, Broekaert, J. A. C. (eds.)

Glow Discharge Plasmas in Analytical Spectroscopy

498 pages
2002
Hardcover
ISBN: 978-0-471-60699-4
ISBN: 0-471-60699-5

Diver, D.

A Plasma Formulary for Physics, Technology and Astrophysics

220 pages with 19 figures and 23 tables
2001
Hardcover
ISBN: 978-3-527-0294-6
ISBN: 3-527-10294-2

Advanced Plasma Technology

Edited by
Riccardo d'Agostino, Pietro Favia, Yoshinobu Kawai,
Hideo Ikegami, Noriyoshi Sato, and
Farzaneh Arefi-Khonsari

WILEY-VCH

WILEY-VCH Verlag GmbH & Co. KGaA

The Editors

Prof. Riccardo d'Agostino
Department of Chemistry
University of Bari
Bari, Italy

Prof. Pietro Favia
Department of Chemistry
University of Bari
Bari, Italy

Prof. Yoshinobu Kawai
Kyushu University
Engineering Sciences
Fukuoka, Japan

Prof. Hideo Ikegami
Nagoya, Japan

Prof. Noriyoshi Sato
Graduate School of Engineering
Tohoku University
Sendai, Japan

Prof. Farzaneh Arefi-Khonsari
Laboratoire de Genie des
Procedes Plasmas
Paris, Frankreich

Library of Congress Card No.:
applied for

British Library Cataloguing-in-Publication Data
A catalogue record for this book is available from the British Library.

Bibliographic information published by the Deutsche Nationalbibliothek
The Deutsche Nationalbibliothek lists this publication in the Deutsche Nationalbibliografie; detailed bibliographic data are available in the Internet at <http://dnb.d-nb.de>.

© 2008 WILEY-VCH Verlag GmbH & Co. KGaA, Weinheim

Typesetting   Thomson Digital Noida, India
Printing   Strauss GmbH, Mörlenbach
Binding   Litges & Dopf GmbH, Buchbinderei Heppenheim

Printed in the Federal Republic of Germany

Printed on acid-free paper

ISBN: 978-3-527-40591-6

Contents

Advanced Plasma Technology. Edited by Riccardo d'Agostino, Pietro Favia, Yoshinobu Kawai, Hideo Ikegami,
Noriyoshi Sato, and Farzaneh Arefi-Khonsari
Copyright © 2008 WILEY-VCH Verlag GmbH & Co. KGaA, Weinheim
ISBN: 978-3-527-40591-6

Preface

Plasma processes started to be applied for surface modification of materials in the 70's, in the fields of microelectronics (dry etching processes for fabricating integrated circuits) and semiconductors (deposition processes of semiconductor thin films for solar cells). Since then, enormous advancements in the basic, diagnostic and experimental aspects of plasma sciences have been made, so that many other science areas and industrial fields have been permeated by plasma processes: polymers, textiles, biomaterials, microfluidics, composite materials, paper, packaging, automobile, waste treatment, cultural heritage and corrosion protection, to mention but a few.

The idea of organizing this book was developed during the second International School of Industrial Plasma Application, held at Villa Monastero in Varenna, Italy, in October 2004, where approximately one hundred attendees from all over the world were assembled. The aim of the School was to describe, in a tutorial way, the numerous modern industrial applications of plasmas.

Now, three years later, this book is issued with the same tutorial purpose: to describe advances of low and atmospheric pressure plasmas in technological fields, such as polymers, semiconductors, solar cells, biomaterials, displays, water treatment, and space, with the introduction of some fundamental chapters on diagnostics, reactor design, modeling and process control.

Advanced Plasma Technology is a collection of 25 chapters on various aspects of plasma processes authored by well known plasma scientists. We are convinced that this book will be of help to both students and researchers, in academia as well as in the industry.

To all the authors, to the referees and to our publishers at Wiley-VCH we would like to extend our warmest "thank you" for the creation of this book. We hope that you, reader, will enjoy reading this book as much as we enjoyed editing it.

October 2007

Riccardo d'Agostino
Pietro Favia
Yoshinobu Kawai
Hideo Ikegami
Noriyoshi Sato
Farzaneh Arefi-Khonsari

List of Contributors

E. Amanatides
Plasma Technology Lab
Department Chemical Engineering
University of Patras
P.O. Box 1407
26504 Patras
Greece

Akira Ando
Department of Electrical Engineering
Tohoku University
6-6-05 Aoba-yama
Aoba
Sendai 980-8579
Japan

Farzaneh Arefi-Khonsari
Laboratoire de Génie des Procédés
Plasma et Traitements de Surface
Université Pierre et Marie Curie
ENSCP
11 rue Pierre et Marie Curie
75231 Paris cedex 05
France

Natalia Babaeva
Department of Electrical and Computer
Engineering
Iowa State University
Ames, IA 50011
USA

Jen-Shih Chang
McMaster University (Professor
Emeritus)
NRB 118 Hamilton
Ontario L8S 4M1
Canada

Francis F. Chen
Electrical Engineering Department
University of California
Los Angeles, CA 90095-1594
USA

Vittorio Colombo
Department of Mechanical Engineering
(DIEM) and Research Center for
Applied Mathematics (CIRAM)
University of Bologna
Via Saragozza 8
40123 Bologna
Italy

Pascal Colpo
European Commission
Joint Research Centre
Institute for Health and Consumer
Protection
21020 Ispra (VA)
Italy

Advanced Plasma Technology. Edited by Riccardo d'Agostino, Pietro Favia, Yoshinobu Kawai, Hideo Ikegami,
Noriyoshi Sato, and Farzaneh Arefi-Khonsari
Copyright © 2008 WILEY-VCH Verlag GmbH & Co. KGaA, Weinheim
ISBN: 978-3-527-40591-6

Riccardo d'Agostino
Department of Chemistry
University of Bari
via Orabona 4
70126 Bari
Italy

Sarah E. Dickson
Department of Civil Engineering
McMaster University
Hamilton
Ontario L8S 4B2
Canada

Monica B. Emelko
Department of Civil and
Environmental Engineering
University of Waterloo
200 University Avenue West
Waterloo
Ontario N2L 3G1
Canada

R. Engeln
Department of Applied Physics
Eindhoven University of Technology
P.O. Box 513
5600 MB Eindhoven
The Netherlands

F. Fanelli
Department of Chemistry
University of Bari
via Orabona 4
70126 Bari
Italy

Pietro Favia
Department of Chemistry
University of Bari
via Orabona 4
70126 Bari
Italy

F. Fracassi
Department of Chemistry
University of Bari
via Orabona 4
70126 Bari
Italy

Emanuele Ghedini
Department of Mechanical Engineering
(DIEM) and Research Center for
Applied Mathematics (CIRAM)
University of Bologna
Via Saragozza 8
40123 Bologna
Italy

Roberto Gristina
Department of Chemistry
University of Bari
via Orabona 4
70126 Bari
Italy

Yiping Guo
Department of Civil Engineering
McMaster University
Hamilton
Ontario L8S 4B2
Canada

Marina Hasiwa
European Commission
Joint Research Centre
Institute for Health and Consumer
Protection
21020 Ispra (VA)
Italy

Kunihiko Hattori
Nippon Institute of Technology
Miyashiro-machi
Minami-saitama-gun
Saitama 345-8501
Japan

H. Huang
Department of Materials Engineering
The University of Tokyo
7-3-1
Hongo
Bunkyo-Ku
Tokyo 113-8656
Japan

Hideo Ikegami
National Institute for Fusion Science
Toki
Gifu 509-5292
Japan

Masaaki Inutake
Department of Electrical Engineering
Tohoku University
6-6-05 Aoba-yama
Aoba
Sendai 980-8579
Japan

Ryohei Itatani
601-1311 Daigo-Ohtakacho 11-18
Husimi-ku
Kyoto
Japan

Felipe Iza
Department of Electronic and Electrical
Engineering
Pohang University of Science and
Technology
Pohang 790-784
South Korea

M. Kambara
Department of Materials Engineering
The University of Tokyo
7-3-1
Hongo
Bunkyo-Ku
Tokyo 113-8656
Japan

Yoshinobu Kawai
Research Institute for Applied
Mechanics
Kyushu University
Kasugakoen 6-1
Kasuga
Fukuoka 816-8580
Japan

W. M. M. Kessels
Department of Applied Physics
Eindhoven University of Technology
P.O. Box 513
5600 MB Eindhoven
The Netherlands

I. E. Kieft
Department of Biomedical Engineering
Eindhoven University of Technology
P.O. Box 513
5600 MB Eindhoven
The Netherlands

Sung Jin Kim
Department of Electronic and Electrical
Engineering
Pohang University of Science and
Technology
Pohang 790-784
South Korea

Kazunori Koga
Graduate School of Information Science
& Electrical Engineering
Kyushu University
Fukuoka 812-8581
Japan

Masuhiro Kogoma
Department of Chemistry
Faculty of Science and Technology
Sophia University
7-1 Kioicho
Chiyoda-ku 102-8554
Japan

Ondřej Kylián
European Commission
Joint Research Centre
Institute for Health and Consumer
Protection
21020 Ispra (VA)
Italy

Sung Hee Lee
Department of Electronic and Electrical
Engineering
Pohang University of Science and
Technology
Pohang 790-784
South Korea

Hae June Lee
Busan National University
Busan 609-735
South Korea

Jae Ko Lee
Department of Electronic and Electrical
Engineering
Pohang University of Science and
Technology
Pohang 790-784
South Korea

Hiroshi Mashima
Nagasaki Research & Development
Center
Mitsubishi Heavy Industries Ltd.
Fukahori
Nagasaki 851-0392
Japan

D. Mataras
Plasma Technology Lab
Department Chemical Engineering
University of Patras
P.O. Box 1407
26504 Patras
Greece

Akihisa Matsuda
Tokyo University of Science
2641 Yamazaki
Noda-shi
Chiba 278-8510
Japan

Andrea Mentrelli
Research Center for Applied
Mathematics (CIRAM)
University of Bologna
Via Saragozza 8
40123 Bologna
Italy

Tarik Meziani
European Commission
Joint Research Centre
Institute for Health and Consumer
Protection
21020 Ispra (VA)
Italy

Antonella Milella
Department of Chemistry
University of Bari
via Orabona 4
70126 Bari
Italy

Marina Nardulli
Department of Chemistry
University of Bari
via Orabona 4
70126 Bari
Italy

F. Palumbo
Institute of Inorganic Methodologies
and Plasmas
IMIP-CNR
via Orabona 4
70126 Bari
Italy

François Rossi
European Commission
Joint Research Centre
Institute for Health and Consumer
Protection
21020 Ispra (VA)
Italy

Eloisa Sardella
Department of Chemistry
University of Bari
via Orabona 4
70126 Bari
Italy

Noriyoshi Sato
Tohoku University (Professor Emeritus)
Kadan 4-17-113
Sendai 980-0815
Japan

D. C. Schram
Department of Applied Physics
Eindhoven University of Technology
P.O. Box 513
5600 MB Eindhoven
The Netherlands

Tsutae Shinoda
Fujitsu Laboratories Ltd.
64 Nishiwaki
Ohkubo-cho
Akashi 674-0054
Japan

Masaharu Shiratani
Graduate School of Information Science
& Electrical Engineering
Kyushu University
Fukuoka 812-8581
Japan

D. W. Slaaf
Department of Biomedical Engineering
Eindhoven University of Technology
P.O. Box 513
5600 MB Eindhoven
The Netherlands

R. E. J. Sladek
Department of Biomedical Engineering
Eindhoven University of Technology
P.O. Box 513
5600 MB Eindhoven
The Netherlands

E. Stoffels
Department of Biomedical Engineering
Eindhoven University of Technology
P.O. Box 513
5600 MB Eindhoven
The Netherlands

Hiromu Takatsuka
Nagasaki Shipyard & Machinery Works
Mitsubishi Heavy Industries Ltd.
Isahaya
Nagasaki 854-0065
Japan

Yoshiaki Takeuchi
Nagasaki Research & Development
Center
Mitsubishi Heavy Industries Ltd.
Fukahori
Nagasaki 851-0392
Japan

Kunihito Tanaka
Department of Chemistry
Faculty of Science and Technology
Sophia University
7-1 Kioicho
Chiyoda-ku 102-8554
Japan

M. Tatoulian
Laboratoire de Génie des Procédés
Plasma et Traitements de Surface
Université Pierre et Marie Curie
ENSCP
11 rue Pierre et Marie Curie
75231 Paris cedex 05
France

Hiroyuki Tobari
Japan Atomic Energy Agency
Naka
Ibaraki 311-0193
Japan

T. Trombetti
Department of Mechanical Engineering
(DIEM)
University of Bologna
Via Saragozza 8
40123 Bologna
Italy

Kuniko Urashima
McMaster University (Professor
Emeritus)
NRB 118 Hamilton
Ontario L8S 4M1
Canada

M. C. M. van de Sanden
Department of Applied Physics
Eindhoven University of Technology
P.O. Box 513
5600 MB Eindhoven
The Netherlands

M. A. M. J. van Zandvoort
Department of Biomedical Engineering
Eindhoven University of Technology
P.O. Box 513
5600 MB Eindhoven
The Netherlands

Yukio Watanabe
Kyushu University (Professor Emeritus)
Iki-Danchi
Fukuoka 819-0042
Japan

Yasuhiro Yamauchi
Nagasaki Shipyard & Machinery Works
Mitsubishi Heavy Industries Ltd.
Isahaya
Nagasaki 854-0065
Japan

T. Yoshida
Department of Materials Engineering
The University of Tokyo
7-3-1
Hongo
Bunkyo-Ku
Tokyo 113-8656
Japan

1
Basic Approaches to Plasma Production and Control
N. Sato

Plasma production and control are of crucial importance for "intelligent" plasma processing in next-stage material and device manufacturing. The author has been concerned with basic experiments on discharge plasmas along this line of research. Here are presented some essential points of basic approaches to plasma production and control. They include works on large-diameter plasma production, electron-temperature and ion-energy controls, and dust particle collection and removal.

At first, two methods of plasma production are presented. They are for high-density electron cyclotron resonance (ECR) and rf plasmas yielding uniform plasma processing in actual manufacturing devices, the diameters of which are larger than several tens of centimeters. These discharge plasmas are produced under low gas pressures. New approaches to medium-pressure and high (atmospheric)-pressure discharge plasmas are also described in some detail.

Electron temperature is continuously controlled in the wide range of one or two orders of magnitude in a region separated from a discharge region. The methods employed might be useful for finding the best conditions for various kinds of plasma processing. In fact, the methods have been proved to be useful for efficient production of negative ions, formation of high-quality diamond particles, and quality increase of a-Si:H film. A good method of ion-energy control should also be established for "intelligent" plasma applications. A new approach is presented for this purpose.

Dust collection and removal are quite important for many kinds of material and device manufacturing. On the basis of fundamental fine-particle behaviors in plasmas, we have proposed a simple method for collection and control of negatively charged fine particles in plasmas. Our collector is often called "NFP-Collector" (negatively charged fine-particle collector). The collector has been proved to be very efficient for collection and removal of dust particles levitating in plasmas, suggesting big effects on plasma processing.

Advanced Plasma Technology. Edited by Riccardo d'Agostino, Pietro Favia, Yoshinobu Kawai, Hideo Ikegami, Noriyoshi Sato, and Farzaneh Arefi-Khonsari
Copyright © 2008 WILEY-VCH Verlag GmbH & Co. KGaA, Weinheim
ISBN: 978-3-527-40591-6

1.1
Plasma Production

1.1.1
Under Low Gas Pressure (<0.1 torr)

Here, two simple methods are presented of plasma production for large-scaled uniform-plasma processing. One of the methods is based on ECR. For the other method, we employ the magnetron-type rf discharge. In both of them, weakly ionized plasmas are produced by low-pressure discharges in a vacuum chamber, the wall of which is separated into two parts. One part is electrically grounded and the other part is used as an antenna or rf electrode. Therefore, in principle, we need no additional electrode for plasma production in the vacuum chamber. Radial plasma profiles are non-uniform in a region of plasma production. But, radial plasma diffusion makes the plasmas uniform at an axial position a little away from the production region. We employ a magnetic field to provide efficient plasma production and to control plasma flow toward the wall (or electrode), which is closely connected with plasma loss and particle sputtering. The magnetic field, which is generated by permanent magnets, is used also to modify electron motions for plasma-profile control, although there is no direct magnetic effect on ions in front of substrates.

A schematic feature of ECR plasma production [1,2] is illustrated in Fig. 1.1(a). The antenna, which is situated at one end of a vacuum chamber, consists of a back plate with permanent magnets behind and a slotted plate separated from the back plate. A microwave of 2.45 GHz is fed through a coaxial waveguide to satisfy the ECR condition (~875 G) in a region near the magnet surfaces in front of the antenna. The slotted plate can be covered with a thin glass plate.

(a)   (b)

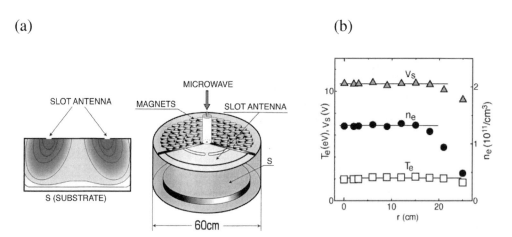

Fig. 1.1 (a) Schematic of ECR plasma production using a plane-slotted antenna with magnets and (b) radial profiles of plasma parameters measured at $z = 10$ cm.

The plasma produced is non-uniform radially in front of the antenna, depending on the positions of the slots and magnets. But, with an increase in z (distance from antenna front), inward plasma diffusion makes the plasma profile flat in the radial direction. Typical results are presented in Fig. 1.1(b), where argon pressure $\approx 1.5 \times 10^{-2}$ torr and microwave power ≈ 1 kW. The plasma of density $n_p \approx 1.3 \times 10^{11}$ cm^{-3} is found to be uniform within 3% in the radial region of 35 cm in diameter at axial distance z of 10 cm. The plasma density is almost proportional to the microwave power. The axial position for the uniform radial plasma profile is controlled by changing the magnetic configuration in front of the antenna.

A reactive plasma produced by this method was confirmed to yield uniform etching of poly-silicon [3]. An antenna system shown in Fig. 1.2 has been proposed for actual plasma processing [4].

A schematic feature of modified magnetron-type (MMT) plasma production [5] is illustrated in Fig. 1.3(a). An rf power of 13.56 MHz is fed to a ring electrode of 55 cm in diameter and 7 cm in length, which is a central part of a cylindrical vacuum chamber of 55 cm in diameter. A discharge is triggered between this powered electrode and the other parts of the vacuum chamber, which are electrically grounded, in the range of argon pressure 5.0×10^{-4}–5.0×10^{-2} torr. Permanent magnets, which are situated just outside the cylinder to construct azimuthal magnet rings, provide magnetic mirrors axially near the inner surface of the ring electrode. This magnetic configuration enhances plasma production because high-energy electrons responsible for ionization move in the azimuthal direction, being well trapped in the magnetic mirrors inside the region near the ring electrode. This motion of electrons reduces a potential drop in front of the electrode, which is closely connected with an interaction of ions with the electrode.

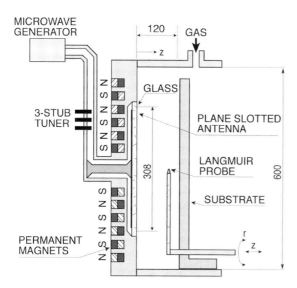

Fig. 1.2 Details of apparatus proposed for ECR plasma production in plasma application.

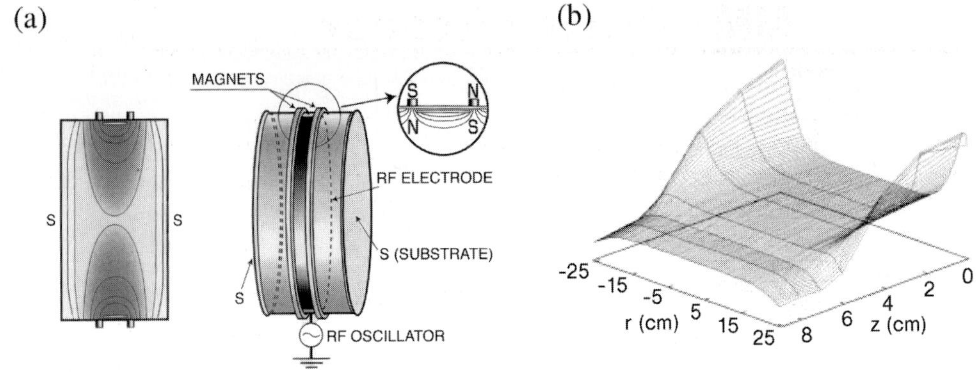

(a)   (b)

Fig. 1.3 (a) Schematic of MMT plasma production and (b) measured variation of radial plasma density profiles in the axial direction.

The plasma density is found to have a peak near the electrode and decreases in the direction toward the radial center. But, with an increase in z (axial distance from machine center), the plasma diffuses toward the radial center, flattening the radial density profile. This MMT rf discharge yields an almost uniform plasma in the radial region of 40 cm in diameter at $z = 6.0$ cm where substrates (S) can be situated, as shown for argon pressure of 1.0×10^{-3} torr and rf power of 200 W in Fig. 1.3(b). Now we can produce a uniform plasma, the diameter of which is larger than 100 cm [6,7]. A feedback control is effective for meter-size uniform processing, where the signal due to the non-uniformity is used as a feedback signal to a small electrode for additional discharge to provide uniform processing.

The potential drop in front of the ring electrode is changed by varying the magnetic strength and configuration. Therefore we can control energies of ions toward substrates [8] and particle sputtering due to high-energy ions accelerated by the potential drop. In the experiment, we could find the condition where there is no appreciable sputtering from the electrode [9]. Figure 1.4 demonstrates the MMT plasma reactor developed by Hitachi Kokusai Electric Inc. for semiconductor manufacturing [10].

1.1.2
Under Medium Gas Pressure (0.1–10 torr)

A parallel-plate rf discharge in this pressure range has been widely used for plasma production in applications. Multi-hollows formed in a cathode (rf powered electrode) are known to be effective for increasing the plasma density. A cathode with isolated hollows (CIH) (see Fig. 1.5(a)) is used in many cases. But, the discharge is often localized in the special hollow(s). There is also a possibility of dust particle trapping in the isolated hollows.

Here, a cathode with connected hollows (CCH) (see Fig. 1.5(b) is employed to eliminate these problems in the CIH [11]. In this case, the hollows are connected

MMT Plasma Reactor

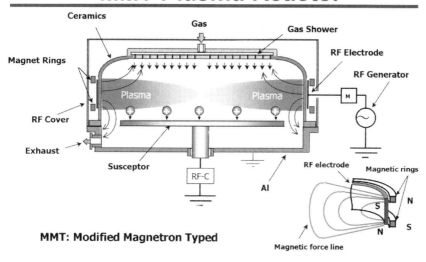

MMT: Modified Magnetron Typed

Fig. 1.4 MMT reactor used in plasma processing for semiconductor manufacturing (Hitachi Kokusai Electric Inc. [10]).

by ditches [3]. The CCH is topologically different from the CIH. Gas-feed holes are made in the bottoms of the hollows and/or between the hollows. An apparatus with the CCH is shown, together with photographs of (a) parallel-plate discharge and (b) CCH discharge, in Fig. 1.6. In the case of the CCH, the discharge brightness is enhanced and the plasma density is twice as high as that in the case of plane parallel-plate discharge at the same input rf power. The density has been confirmed to

(a)   (b)

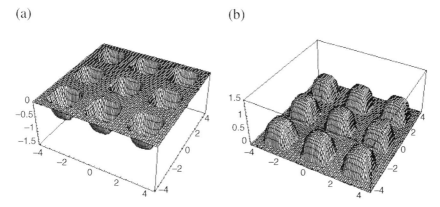

Fig. 1.5 Uneven electrodes: (a) concave-type electrode (CIH) and (b) convex-type electrode (CCH).

(a)   (b)

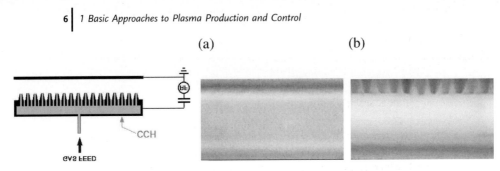

CCH

eV2 LEED

Fig. 1.6 Left: apparatus with CCH. (a) Parallel-plate discharge and (b) CCH discharge.

increase with an increase in the rf power, without localization of the discharge, yielding a uniform plasma for large-scaled processing.

1.1.3
Under High (Atmospheric) Gas Pressure (>10 torr)

Plasma processing using atmospheric plasmas is now quite useful for various kinds of applications. So-called "barrier discharges" are well known as a method of plasma production under high (atmospheric) gas pressure. Electrodes for this discharge are shown in Fig. 1.7(a), where one of the electrodes is covered by dielectric material. An equivalent circuit for this situation of discharge is shown in Fig. 1.7(b).

We have proposed a quite simple method of plasma production under high (atmospheric) gas pressure. Pole-type electrodes, which are coupled with external capacitors, are set near a metal plate. This arrangement is just a direct realization of the circuit in Fig. 1.6(b). This is called capacity-coupled multi-discharge (CCMD) [12]. Under some conditions, the pole length is set to be so short that the electrodes are almost small plates. Being different from the barrier discharges, the discharge power of the CCMD can be externally controlled to increase by increasing the capacity of the capacitors. Measurements have proved that the CCMD provides high-power discharges, suggesting new possibilities for plasma applications in the high (atmospheric) pressure range.

(a)   (b)

ELECTRODE   ELECTRODE

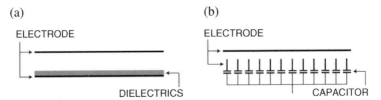

DIELECTRICS   CAPACITOR

Fig. 1.7 (a) Typical barrier-discharge electrodes and (b) electrodes for *capacitor-coupled multidischarge* (CCMD).

1.2
Energy Control

1.2.1
Electron-Temperature Control

Generally speaking, it is quite difficult to change the electron temperature in weakly ionized plasmas. Here are demonstrated two methods adopted for electron-temperature control, which might be useful in next-stage plasma processing. One of them is accomplished by using a pin-hollow cathode while the other employs a mesh grid. Both methods are based on trapping of electrons ionized in discharge-free region, which is provided by varying a local discharge structure. Therefore, in both methods, there appears a region of low electron temperature in addition to the discharge region of high electron temperature. The volume ratio of these regions is important and has to be carefully determined in actual applications for plasma processing. Our methods suggest a general principle of electron-temperature control in low-pressure discharge plasmas.

A schematic feature of the pin-hollow cathode action for low-pressure DC discharges is demonstrated in Fig. 1.8. In a typical setup, the cathode consists of 20 cm diameter stainless steel cylinder with 17 cm diameter hole at the front edge and 0.2 cm diameter pointed stainless steel pins installed inside [13]. The 48 pins, connected electrically with the cylinder, are set with an equal separation on a 16 cm diameter circle. The pin length δ is varied from 0 to 7 cm. A low-pressure gas discharge is triggered between this pin-hollow cathode and a 30 cm diameter anode with 10 cm diameter hole under a weak axial magnetic field of 100–150 G.

As found in Fig. 1.8, for $\delta = 0$ cm, there appears a glowing plasma column, the diameter of which is determined by the cathode front hole in the region up to the anode. With an increase in δ, the glow becomes weak gradually in the radial core part of the plasma. For $\delta = 6$–7 cm, the glow is limited only in the radial edge region. The core

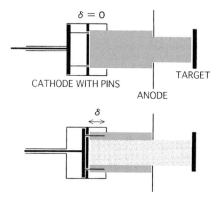

Fig. 1.8 Schematic of discharge using hollow cathode with movable pins inside. Upper: $\delta = 0$ cm; lower: $\delta = 6$ cm.

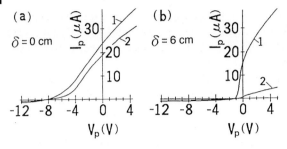

Fig. 1.9 Langmuir probe characteristics at (a) $\delta = 0$ cm and (b) $\delta = 6$ cm for (1) pure Ar gas and (2) for Ar gas + a small amount of CH_4 gas.

plasma passes through the anode hole, being terminated by a target. As δ is increased, the electron temperature T_e is observed to decrease drastically, being accompanied by a slight increase in the plasma density n_p in this core plasma. For argon gas pressure $\leq 1.0 \times 10^{-2}$ torr, T_e (=2–3 eV for $\delta = 0$ cm) decreases by an almost one order of magnitude as δ is increases up to 7 cm. This result is ascribed to trapping of primary electrons in the potential-hill structure formed by the pins in the cathode.

A typical example of electron-temperature effects is presented in Fig. 1.9, where $\delta = 0$ cm (Fig. 1.9(a)) and $\delta = 6$ cm (Fig. 1.9(b)) for discharge of pure Ar with gas flow of 100 sccm ("1") and for discharge of Ar + small amount of CH_4 (Ar with 98 sccm and CH_4 with 2 sccm) ("2") at pressure of 5 mtorr. In the pure Ar discharge, the Langmuir probe shows a clear decrease in T_e. On the other hand, in the presence of a small amount of CH_4, the probe shows a drastic decrease in the negative current, suggesting production of negative ions. Detailed measurements have clarified production of negative hydrogen ions, which is enhanced by the decrease in T_e [14]. There also appears a drastic change in the densities of radical species in this kind of reactive plasma [15].

A grid is employed in the other method, which separates the discharge region (I) from the region (II) for plasma processing [16]. As shown in Fig. 1.10, in the presence of a negative potential applied to the grid, most electrons in the region (I) are reflected

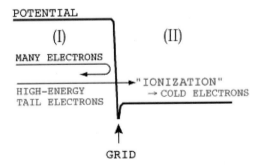

Fig. 1.10 Mechanism of grid control for electron temperature.

by the grid except high-energy tail electrons. Ionization occurs due to the tail electrons in the region (II). Electrons produced there are not responsible for maintaining the discharge, having a low T_e. Therefore, T_e in the region (II) decreases with an increase in the negative grid potential.

In a typical experiment, a plasma of $n_p = 10^9$–10^{10} cm^{-3} and T_e of a few eV is produced by a low-pressure argon discharge in the region (I), diffusing through a coarse mesh grid into the region (II). High-energy electrons passed through the grid are recognized at z (distance from grid toward the region (II)) = 0.2 cm. At $z = 0.4$ cm, however, there appear low-energy electrons. As z is increased, their density increases while high-energy electrons disappear gradually. At $z \geq 2.0$ cm, there are only low-energy electrons with density higher than that in the region (I). There is a quite drastic effect of V_G on T_e. The electron temperature decreases continuously with an increase in V_G, down to $T_e = 0.035$ eV in the region (II), lower by almost two orders of magnitude than that in the region (I). This result is well understood by the mechanism shown in Fig. 1.10. The plasma is produced by a DC discharge in this example. This grid method of electron-temperature control has been confirmed to work also for low-pressure plasmas produced by ECR and rf discharge plasmas [17,18].

The electron temperature can also be controlled by varying a mesh size of the grid at a fixed bias potential [19]. Even if there is no external potential applied to the grid, the electron temperature in the region (II) depends on the mesh size. In general, however, it is difficult to vary the mesh size during machine operation. It is better to make a hole (or slit) in the grid, the size of which is much larger than the mesh size. By varying the hole (or slit) size mechanically, we can control the electron temperature. An example of the grids used is shown in Fig. 1.11(a), where we can vary the slit length L_s. As demonstrated in Fig. 1.11(b), the electron temperature is well controlled by the slit length. This mechanical method might be quite useful for the electron-temperature control in reactive plasmas, where the grid is often covered by thin film of insulator.

(a)   (b)

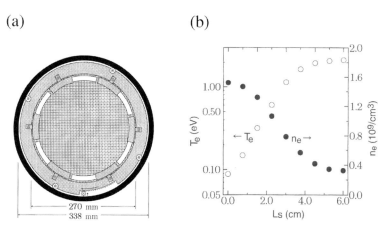

Fig. 1.11 (a) Grid with slits and (b) electron temperature and plasma density against slit length L_s.

(a)   (b)

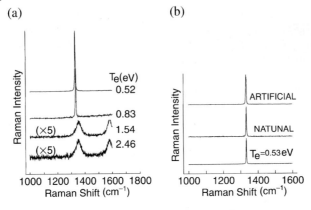

Fig. 1.12 Raman spectroscopic measurements: (a) for diamond particles formed at different electron temperature in $H_2 + CH_4$ gas discharge and (b) comparison with diamonds formed by different methods.

Both of the methods yield continuous T_e controls over one to two orders of magnitude. Drastic effects of T_e on reactive plasmas have been observed in the experiments. A lot of negative hydrogen ions (more than 90%) are produced in the low T_e region in a hydrogen plasma [18], suggesting a possibility of efficient negative ion source, for example, for fusion-oriented plasmas. Figure 1.12(a) presents the effect of T_e on formation of diamond particles in hydrogen–methane plasmas. The diamonds are formed with a decrease in T_e under such a low gas pressure of around 0.1 torr [20]. Their structural quality is much higher than those of diamonds formed under different situations (see Fig. 1.12(b)). T_e control is also useful for producing high-quality a-Si:H films for solar-cell batteries [21].

1.2.2
Ion-Energy Control

The energy of ions flowing toward the substrate often has large effects on plasma processing. The ion energy can be varied by changing the substrate potential with respect to the plasma potential and also depends on the electron temperature. In some applications, however, it is necessary to change the energy without changing sheath structure in front of the substrate.

Here is described a method of ion-energy control at fixed values of the substrate potential and electron temperature. In this method, the so-called double-plasma (DP) technique is adopted. In usual cases, we need two plasma sources for the DP configuration [22]. Here, in contrast to the usual DP technique, one plasma source is used to provide the DP configuration [23]. The plasma produced by the plasma source is supplied into two regions of a vacuum chamber, as illustrated in Fig. 1.13. The each region has the wall with different electric potential. In Fig. 1.13, a short metal cylinder is set in the cylindrical vacuum chamber. The plasma produced by microwave

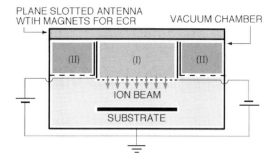

PLANE SLOTTED ANTENNA
WTIH MAGNETS FOR ECR   VACUUM CHAMBER

Fig. 1.13 Schematic for method of ion-energy control.

discharge is divided into two plasmas, i.e., a plasma (I) in the cylinder and a plasma (II) in the region between the cylinder and the vacuum chamber. Under this configuration, an ion beam is supplied from the plasma (I) into the plasma passed through the grid from the region (II). The ion energy is controlled by changing the positive potential applied to the cylindrical wall surrounding the plasma (I) with respect to the vacuum chamber grounded electrically. Then, we can control the energy of ions toward the substrate situated in the diffused plasma. Measurements have well confirmed this control of ion energy. It has to be remarked that, if necessary, the grid method described for electron-temperature control can be employed in addition to the ion-energy control in the plasma supplied from the plasma (II).

The method presented above cannot be used in a situation where the plasma potential is definitely determined by the plasma source, for example, by electrodes connected with the power supply for plasma production. Even in such a situation, a clever method based on the DP technique should be contrived for the plasma processing, on which the ion energy has decisive effects, because the ion energy is easily controlled by this technique.

1.3
Dust Collection and Removal

As an extension of experiments on negative-ion plasmas and fullerene plasmas, the author has been engaged in various basic work on fine-particle plasmas [24,25]. On the basis of the physics clarified, the NFP-Collector has been proposed for collection and removal of fine particles in dusty plasmas [26]. The collector is just a simple electrode with hole(s), which is biased higher than the floating potential in plasmas. Various structures of the collector are adopted, depending on purposes and device configurations. Particles ($<50\,\mu$m), which levitate in plasmas, pass through the hole(s) without impinging the electrode surface into the hole(s). When particles near the collector are collected, particles left away approach the collector in the presence of force balance among particles, being pulled one after another into the collector.

Fine particles levitating in the horizontal plane above a metal plate show a spatial distribution, depending on the potential profile above the plate surface. Under some

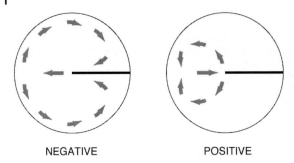

NEGATIVE   POSITIVE

Fig. 1.14 Schematic for vortices of particles, triggered by a small electrode biased negatively and positively.

conditions, there appear vortices of fine particles in plasmas. Examples of the vortices observed are schematically described in Fig. 1.14, where vertical (top) views are shown for the vortices generated in the horizontal plane. These vortices are generated by a cylindrical electrode biased negatively (left) and positively (right), which is situated in a fine-particle cloud above the metal plate for levitation. In usual cases, fine particles are negatively charged in plasmas, and then the shape edge of the electrode biased negatively pushes away fine particles, resulting in such a vortex generation as shown in the left figure in Fig. 1.14. When the electrode is biased positively with respect to the floating potential, fine particles move toward the sharp edge, being also accompanied by vortex generation, without being collected by the electrode. The vortex size is observed to be smaller than in case of the negative potential. Even if the electrode is biased positively, the electrode potential is lower than the plasma potential. There is a sheath potential drop in front of the electrode, which is responsible for reflection of fine particles, resulting in the vortex generation with a direction opposite to that in the case of the negative potential.

Now let us consider a situation of two electrodes separated closely, both of which are biased positively. The situation is shown on the right-hand side of Fig. 1.15. It is not difficult to understand the reason why fine particles flow into the region between the electrodes, where the electrode potential is lower than the plasma potential. This

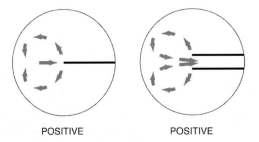

POSITIVE   POSITIVE

Fig. 1.15 Schematic for vortices of particles, triggered by a positively biased small electrode and two positively biased small electrodes separated closely.

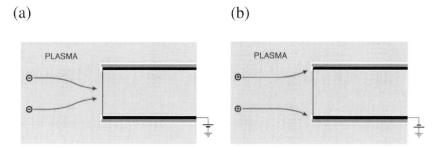

Fig. 1.16 Collection of (a) negatively charged particles and (b) positively charged particles.

behavior of fine particles provides a principle of the NFP-Collector. Under special conditions, fine particles are charged positively in plasmas. A similar collector may be imagined also for positively charged fine particles. Unfortunately, however, we cannot find such a collector in this case, as demonstrated in Fig. 1.16, where a cylinder-type electrode (outside is covered by insulator) is supposed for a collector. In case of negatively charged particles, fine particles flow into the cylinder without hitting the inner wall. In the case of positively charged particles, fine particles are collected by the wall, just as in case of the usual dust collectors. In both cases, the electrode potential is lower than the plasma potential. An essential point of the NFP-Collector is that fine particles are collected without hitting the collector wall. This is possible in the presence of the collector in a plasma.

An example of the NFP-Collector is shown schematically in Fig. 1.17(a). In this configuration, there is a duct surrounding a plasma, which has holes at its inner wall. Behind each hole, a ring-type electrode is situated, which is biased positively to collect externally injected fine particles levitating in a plasma. An experimental observation for collection of 10 µm diameter particles is demonstrated by the photograph in Fig. 1.17(b). Almost all particles are observed to be quickly removed, leaving a dust-free plasma in the vacuum chamber.

In order to enhance the collector action, we use the plate with ditches shaped to yield a potential profile for levitating fine particles to be guided toward the NFP-Collector [27]. Then, the collector can be located at a position far away from the central

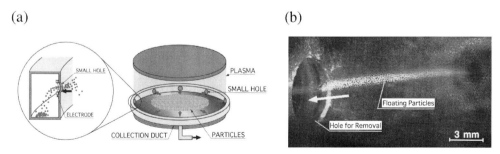

Fig. 1.17 (a) Schematic for example of NFP-Collector and (b) corresponding photograph of dust removal.

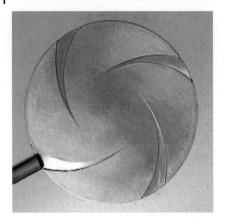

Fig. 1.18 Ditch structure on a 5 cm diam metal plate for particle levitation. Ditch width and depth increase along each ditch for particle guidance in the outward direction.

plasma region, as shown in Fig. 1.18. Even if the ditches are filled with insulator or the plate is covered with a thin insulator film to yield a flat surface, we can obtain almost the same results for the particle removal.

Since the NFP-Collector removes fine particles before they grow large, the collector is useful for suppressing particle growth in reactive plasmas. The quality of a-Si : H film has been proved to increase in the presence of the collector which eliminates dust particles produced during plasma processing [28]. The collector can be adopted for removing dust particles produced in fusion-oriented devices. During the cleaning by a glow discharge, dust particles accumulated on the device wall are sputtered into the glow discharge by applying a potential difference between the plasma and the wall, which is large enough to yield ion sputtering. Those particles are collected and removed by the collector. The collector might also be employed to remove various kinds of fine-particle dusts in the air if there were a good method for plasma production around the collector. There are many works on production of useful fine particles in plasmas. The collector is suggested to be useful for size-separated particle collection because those particles levitate at vertical positions depending on the particle size in plasmas.

Finally, basic plasma experiments have been extended to provide new methods of plasma production and control for plasma applications. They are expected to be useful for "intelligent" plasma processing in the future.

Acknowledgments

The author appreciates the collaborations at Tohoku University, ANELVA Corporation, Hitachi Kokusai Electric Inc., SHARP Corporation, and ADTEC Plasma Technology Co. Ltd.

References

1 Sato, N., Iizuka, S., Nakagawa, Y. and Tsukada, T. (1993) *Appl. Phys. Lett.*, **62**, 1469.

2 Iizuka, S. and Sato, N. (1994) *Jpn. J. Appl. Phys.*, **33** (Part 1), 4221.

3 Ishida, T., Nakagawa, Y., Ono, T., Iizuka, S. and Sato, N. (1994) *Jpn. J. Appl. Phys.*, **33** (Part 1), 4236.

4 Horiuchi, K., Dowaki, S., Iizuka, S. and Sato, N. (1998) *Proc. 15th Symp. on Plasma Processing, Hamamatsu*, 148–151.

5 Li, Y., Iizuka, S. and Sato, N. (1994) *Appl. Phys. Lett.*, **65**, 28.

6 Li, Y., Iizuka, S. and Sato, N. (1997) *Jpn. J. Appl. Phys.*, **36** (Part 1), 4554.

7 Urano, Y., Li, Y., Kanno, K., Iizuka, S. and Sato, N. (1998) *Thin Solid Films*, **316**, 60.

8 Shimizu, T., Li, Y., Iizuka, S. and Sato, N. (1998) *Proc. 15th Symp. on Plasma Processing, Hamamatsu*, 593–596.

9 Li, Y., Iizuka, S. and Sato, N. (1997) *Nucl. Instrum. Methods Phys. Res.*, **B132**, 585.

10 Ogawa, U., Sato, N., Shino, K. and Furukawa, R. (2003) *Hitachi Hyoron*, **85**, 41.

11 Sato, N.Proc. Plasma Sci. Symp. 2005/22th Symp. on Plasma Processing, Nagoya, Japan, 1–4.

12 Mase, H., Fujiwara, T. and Sato, N. (2003) *Appl. Phys. Lett.*, **83**, 5392.

13 Sato, N., Iizuka, S., Koizumi, T. and Takada, T. (1993) *Appl. Phys. Lett.*, **62**, 567.

14 Iizuka, S., Koizumi, T., Takada, T. and Sato, N. (1993) *Appl. Phys. Lett.*, **63**, 1619.

15 Iizuka, S., Takada, T. and Sato, N. (1994) *Appl. Phys. Lett.*, **64**, 1786.

16 Kato, K., Iizuka, S. and Sato, N. (1994) *Appl. Phys. Lett.*, **65**, 816.

17 Kato, K., Iizuka, S., Ganguly, G., Ikeda, T., Matsuda, A. and Sato, N. (1997) *Jpn. J. Appl. Phys.*, **36** (Part 1), 4547.

18 Iizuka, S., Kato, K., Takahashi, A., Nakagomi, K. and Sato, N. (1997) *Jpn. J. Appl. Phys.*, **36** (Part 1), 4551.

19 Kato, K., Shimizu, T., Iizuka, S. and Sato, N. (2000) *Appl. Phys. Lett.*, **76**, 547.

20 Shimizu, T., Iizuka, S. and Sato, N. (2001) *Proc. Plasma Sci. Symp. 2001/18th Symp. on Plasma Processing*, Kyoto, 567–568.

21 Kurimoto, Y., Shimizu, T., Iizuka.S., Suemitsu, M. and Sato, N. (2002) *Thin Solid Film*, **407**, 7.

22 Taylor, R.J., MacKenzie, K.R. and Ikezi, H. (1972) *Rev. Sci. Instrum.*, **43**, 1675.

23 Iizuka, S., Takahashi, A. and Sato, N. (1996) *Proc. 3rd Asia-Pacific Conf. on Plasma Science & Technology, Tokyo*, Vol. 2, 429–433.

24 Sato, N. (1998) *Physics of Dusty Plasmas* (eds M. Horanyi *et al.*), American Institute of Physics, New York, pp. 239–246.

25 Uchida, G., Iizuka, S. and Sato, N. (2000) *Proc. 15th Symp. on Plasma Processing, Nagasaki*, 617–620.

26 Sato, N., Uchida, G. and Iizuka, S. (2000) *IVth European Workshop on Dusty and Colloidal Plasmas, Portugal*.

27 Sato, N. and Koshimizu, T.30th IEEE International Conference on Plasma Science, Korea, 2–5 June 2003, Abstracts p. 195.

28 Kurimoto, Y., Matsuda, N., Uchida, G., Iizuka, S. and Sato, N. (2004) *Thin Solid Films*, **457**, 285.

2
Plasma Sources and Reactor Configurations
P. Colpo, T. Meziani, and F. Rossi

2.1
Introduction

Plasma-assisted material processing now occupies a fundamental place in industry, spanning a wide range of sectors and applications. The concept to modify only the surface of a material to provide a new range of functionality is not new, but the introduction of dry methods like plasma-enhanced chemical vapor deposition (PECVD) or reactive ion etching (RIE) to cite the most common, revolutionized the field. Indeed, these processes allowed for the creation of new materials and structures, high throughput processing, and eco-efficiency.

The creation of gas discharges is commonly done through the application of a high electric field, sufficient to provide the minimum energy for the free electrons to ionize the atoms of the ambient gas. We will not discuss here in detail the mechanisms for igniting and sustaining a plasma as it is very well described in many books [1–3]. But, we will rather try to give an overview of the current trends of plasma sources development, with a focus in particular on inductively coupled plasma (ICP) sources.

The first generation of plasma sources that is still widely used nowadays is the so called "diode" plasma. In this reactor, the plasma discharge is generated by an electric field (from DC to RF) created by applying a high potential difference between two electrodes (anode and cathode) placed under vacuum [4]. Radio frequency discharges are widely employed for industrial applications for their ability to treat semiconductor or insulating materials. Nevertheless, with a plasma density of the order of 10^8 cm^{-3}, the limits of these systems were reached by an ever-more demanding semiconductor industry in terms of performances and throughput. Thus a second generation of plasma sources, the so-called "high density plasma sources" [5,6] was developed to fulfill the current semiconductor industry requirements. It includes mainly the electron cyclotron resonance reactor (ECR), the neutral loop discharge (NLD), the helicon, and the inductively coupled plasma (ICP) [3,7]. Apart from the higher plasma density, a significant advantage brought by these systems is the possibility to apply a

Advanced Plasma Technology. Edited by Riccardo d'Agostino, Pietro Favia, Yoshinobu Kawai, Hideo Ikegami, Noriyoshi Sato, and Farzaneh Arefi-Khonsari
Copyright © 2008 WILEY-VCH Verlag GmbH & Co. KGaA, Weinheim
ISBN: 978-3-527-40591-6

separate biasing of the substrate, and thus to control the ion bombardment energy independently from the ion bombardment flux.

Among the systems cited previously, the ICP is the most widely used for its simplicity of construction, and etching and deposition systems are commercialized by OEMs. Nevertheless, ICP development is still on going for new applications like aeronautic or tool machinery, for which there is still a need for special reactor designs.

In this chapter, taking various examples from our laboratory activity, we will show how the ICP source and reactor configurations could be further developed in order to answer current application needs. The first part contains a brief description of the principle of operation of the ICP source including impedance matching. Then two special reactors designed in our laboratory are described: for metal deposition for aeronautic applications and silica etching for flat panel display applications.

2.2
Characteristics of ICP

2.2.1
Principle

The first work on electrodeless discharges was published in 1884 by Hittorf [8]. It was not clear then whether the discharge was created electrostatically by the high potential difference between the coil ends, or inductively because of the electric field induced by the coil antenna. It was later demonstrated that both modes actually occur, the E-mode and the H-mode respectively. The plasma is always initiated in E-mode. The transition from the E-mode to the H-mode occurring after a particular threshold current (or power) is often visible as a transition from a dim discharge to a bright discharge. The mode shift is characterized by a drastic increase in electron density, typically by an order of magnitude and by a hysteresis. Indeed, the threshold current is not the same when going from an E-mode to an H-mode discharge and from an H-mode to an E-mode discharge. The dynamic of the discharge creation in these sources was explained by Kortshagen *et al.* [9]. The hysteresis and bi-stability phenomena were also described by several authors [9–11].

The oscillating current circulating in the coil creates a RF magnetic field that penetrates the gas chamber through the dielectric. An alternative electric field is induced in that region. The free electrons in the gas are accelerated and gain sufficient energy to ionize the gas and maintain the discharge. The conductivity of the discharge shields the electromagnetic field and limits its penetration to a thin layer named skin depth. The skin depth is the main region of power transfer.

Apart from the ohmic heating process described above where the power is transferred to the discharge by a collisional mechanism, a non-collisional process named "stochastic heating" occurs. In this mechanism related to an anomalous skin effect [12–14], the electrons interact with the oscillating electromagnetic field created by the coil antenna before being thermalized [15–19].

2.2.2
Transformer Model

The configuration of an ICP source is reminiscent of the electrical setup of a transformer with a primary winding being the excitation coil antenna, while the conductive plasma discharge acts as the secondary winding. The analogy of the transformer was developed as a simple model that allows one to describe rather precisely the coupling between the exciting coil and the plasma [20–22]. This schematic circuit includes thus the exciting coil with its inductive and resistive components (R1, L1), and the plasma discharge with its resistive effects described by a resistance value R2, and two inductive elements (L2, Le) that correspond to a geometric component related to the discharge current path and to the electron inertia resulting from the plasma conductivity respectively (Fig. 2.1). The coil to plasma distance generally depending on the dielectric window thickness acts on the mutual inductance between the primary and secondary and has an influence on power transfer efficiency [23].

Common circuit analysis methods allow one to relate the electric parameters of the system to the microscopic characteristics of the discharge such as electron density and collision frequency. A common operation in transformer circuit analysis consists in converting the above circuit shown in Fig. 2.1 to a single RLC circuit by transforming the secondary impedance with the so called coupling coefficient $\omega M^2/Z_2$ (Z_2 representing the complex impedance of the secondary circuit).

Using this model, it is thus easy to characterize the inductive discharge by its electrical components that can be easily measured by electrical probes [21,24–27]. However, our results demonstrate that it is essential to include the parasitic capacitors in the equivalent circuit in order to correctly describe the ICP source (Fig. 2.2) [28,29].

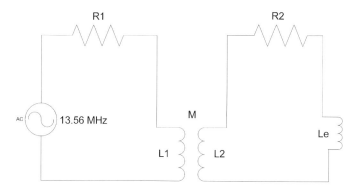

Fig. 2.1 Equivalent circuit of the transformer for an inductively coupled plasma source.

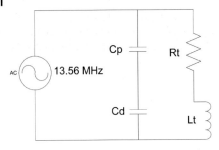

Fig. 2.2 Equivalent circuit of the transformer after impedance transformation to the primary and showing the contribution of the parasitic capacitors (Cp) and the capacitance arising from the dielectric window (Cd).

2.2.3
Technological Aspects

2.2.3.1 Matching

The output impedance of the power supply and the whole system impedance have to be matched in order to maximize the power transfer. The common RF power supply generally presents an output impedance of 50 Ω. The impedance of the load is obviously different and depends on many parameters like the chamber size and geometry, the gas pressure, the electrode and antenna shape and size, etc. Moreover, the load impedance varies with the varying conditions in the plasma when changing the pressure, the gas, or the applied power. A matching circuit is inserted between the power supply and the load in order to adapt the impedance seen by the RF generator to be a total of 50 Ω. The matching network has a specific tuning range depending on its components values (Fig. 2.3).

For inductive sources, the impedance matching can be done with a push–pull transformer [30], or with a classic Γ-type circuit having two variable capacitors but removing the series inductance that is useless in this case. The later solution is the most commonly used.

The electrical layout of an ICP source is presented in Fig. 2.3. In this figure, Z_G is the generator output impedance, Z_N is the matching network impedance, and C_{sh} and C_{se} the shunt and series variable capacitors respectively.

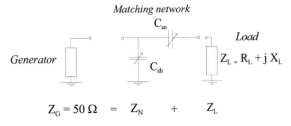

Fig. 2.3 Classic matching network.

For a correct operation, the global impedance Z_T seen by the generator including the load impedance Z_L and the matching impedance Z_N must be equal to the generator impedance Z_G. This results in the following impedance matching condition:

$$Z_T = Z_N + Z_L = Z_G = 50\,\Omega \tag{2.1}$$

After derivation, we find the matching condition function of the load impedance (R_L, X_L):

$$R_L = \frac{Z_G}{(C_{sh} Z_G \omega)^2 + 1} \tag{2.2}$$

$$X_L = \frac{1}{C_{se} \omega} + R_L \sqrt{\frac{Z_G}{R_L} 1} \tag{2.3}$$

One can note that a plasma having a load resistance greater than $50\,\Omega$ cannot be matched without modifying the antenna geometry by decreasing the number of coil turns for instance. Furthermore, relation (2.3) shows that matching a large antenna inductance requires an even lower value of the series capacitance C_{se} that can get down to values lower than 50 pF, i.e. the order of magnitude of parasitic capacitance usually present between ground and the coupling circuit. Impedance matching becomes difficult since the non-controllable parasitic capacitance impedance becomes comparable to the matching network impedance [31].

Matching large area plasma source impedance which could present large plasma resistance and large inductance is then limited without coil geometry modification, which is not suitable since the coil geometry is imposed by plasma density and uniformity as well as process requirements.

One way of tackling this problem is to inductively short the large inductance with a lower inductance to reduce the total impedance seen by the RF power supply. This can be done by connecting a parallel inductance to the matching circuit (Fig. 2.4). By choosing carefully the value of the inductance and its design in order to reduce the ohmic losses, a wide range of impedance can be matched. Indeed, the added inductance X_{sh} being small as compared to X_L, the global impedance seen by the tuner is dramatically decreased and one can verify easily the tuning conditions (2.2) and (2.3). Electrical measurements have been done to illustrate the double inductive device efficiency. Using this setup, the system could be successfully matched in a wide range of discharge conditions.

Several authors carried out measurements of the power losses occurring in the matching network circuitry and cabling for conventional Γ circuits [32], as well as for push–pull transformer matching circuits [30]. In these works, the power losses were carried out with different electrical methods. In our setup, we evaluated the power losses in each of the following elements separately in order to quantify the power losses added by the parallel inductance: the RF power supply, the matching network, the parallel inductance, and the inductive source. This measurement was done by

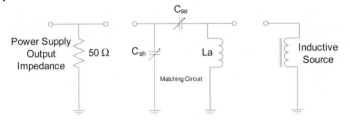

Fig. 2.4 Modified matching network with parallel inductor.

measuring the power dissipated by each of these elements into the cooling water flowing in the whole circuit. In the whole range of power and pressure investigated with an argon discharge (from 3 to 30 mtorr and from 200 to 600 W net power), the average power loss in this parallel inductance is of the order of 4%, whereas the matching network losses account for about 30% of the total power. This should be taken into account to avoid overestimating the total discharge power.

A careful design of the matching system can lead to a significant reduction of the power losses [33]. The solutions taken for the design of the large area inductively plasma source described here are also advantageous for impedance matching. Indeed, the use of multiple inductors connected in parallel presents the advantage of reducing the total inductance value.

2.2.3.2 Capacitive Coupling
Every inductive source is active in capacitive mode and presents some capacitive coupling even when operated in inductive mode. This capacitive coupling is responsible for oscillations in the plasma potential [1,34], which in turn accelerate the ions in the sheath. The energy gained by these ions can cause in particular the heating and sputtering of the dielectric window, and thus be responsible of a contamination of the surface to be treated. To avoid this phenomenon, the capacitive coupling can be removed by electrostatic shielding [23]. Generally, electrostatic shields are constituted by a metallic plate or disc with a certain number of slits allowing the electromagnetic coupling to the plasma [33]. Another possibility to remove totally the capacitive component is to use a metallic surface that does not have any slits at all. The condition for this to work is that the thickness of the metallic sheet is less than the electromagnetic skin depth as defined by Daviet[35] and Godyak et al. [33].

Finally, it was also reported that the addition of an electrostatic shield presents the advantage of decoupling the coil and plasma, thus suppressing the dependence of the power deposition symmetry and plasma uniformity on the discharge conditions [36].

2.2.3.3 Standing Wave Effects
It is quite well known in electromagnetism theory that the length of a conductor plays a capital importance in the power transfer as soon as it becomes of the

order of magnitude of the wavelength at the operating frequency. The conductive element no longer acts as a perfect conductor but rather as a transmission line where a part of the transmitted wave can be reflected. Practically, this means that as soon as the conductor length reaches dimensions close to a quarter wavelength of the operating frequency, the current is no longer constant throughout the conductor. These standing wave effects can cause a significant non-uniformity of the heating profile of the inductive source, and thus of the plasma uniformity [37,38].

2.3
Sources and Reactor Configuration

The source arrangement is generally defined by the application requirements. The main points to be taken into account are the substrate shape and size, and the type and specificity of the operation to perform on it (deposition of conductive or non conductive films, etching, etc).

Subsequently, the coil antenna geometry, the plasma chamber, and the dielectric coupler have to be designed carefully in terms of the application needs.

The antenna is one of the main components of an inductive source. Its shape defines the structure of the electromagnetic field and so, to a certain extent, the main plasma parameters like the electron and ion density, as well as the plasma uniformity [39,40]. Generally, this antenna is made of a copper tube or strap. The tube allows for water cooling when high power density values are used, and a large copper strap is often used in order to minimize the coil resistance.

The antenna can be placed externally to the vacuum chamber, but also internally. When the antenna is placed externally, the dielectric acts generally as a vacuum seal and a certain thickness is needed in order to withstand the pressure difference. Using an internal antenna allows for an increase of the mutual inductance between coil and plasma, and thus to an increase of the power transfer efficiency. This results in a higher plasma density, and thus often to higher treatment rates. When the antenna is in contact with the plasma, it needs to be floating and is thus connected capacitively at both ends to the power supply [41]. This configuration is not very common mainly because of a higher probability of arcing problems.

Other configurations exist where the internal antenna is isolated from the plasma by a dielectric. This is the case for example in the traveling wave-driven ICP designed by Wu and Liebermann [42] where each rod of the antenna is inserted in a dielectric tube, or in the magnetic pole enhanced ICP (MaPE-ICP) source developed by the authors where the antenna is separated from the plasma by a thin dielectric window.

2.3.1
Substrate Shape

The source configuration can be defined mainly by the reactor geometry, and by the shape of the antenna. Often, these two parameters are related in order to optimize

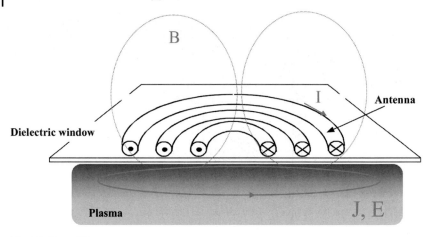

Fig. 2.5 Planar ICP source.

the energy transfer to the system. The arrangement of the source depends mainly on the applications and characteristics of the substrate to be treated.

2.3.1.1 Flat Substrates

The semiconductor industry pushed the development of planar geometry for the treatment of flat substrates like silicon wafers. In this type of reactor first described by Keller [43], the planar antenna is placed above a dielectric window as shown in Fig. 2.5. Various shapes of antenna are possible like for instance a simple loop or more generally a spiral. These sources were extensively studied and their characteristics were reported by many authors [6,7,16,23,37,43–47].

The samples to be treated are generally placed a few skin depths away from the dielectric window in order to avoid the influence of the electromagnetic field.

The aspect ratio of the reactor (length/width) has a fundamental influence on the discharge parameters and in particular on the uniformity of the discharge as will be discussed in Section 2.0. Low aspect ratio configurations are often preferred in order to minimize the charged particles losses to the walls. This aspect has been widely discussed in the literature and will not be developed here [48–50].

2.3.1.2 Complex Three-Dimensional Shapes

The semiconductor industry deals exclusively with flat substrates, i.e. silicon wafers. But surface treatments find many applications in all the different sectors of industry. Often, the pieces to be treated are complex in shape and require a reasonable uniformity over their whole surface. To achieve this, the most useful inductively coupled plasma reactor configuration is the helical ICP.

A typical cylindrical reactor is shown in Fig. 2.6. In this type of reactor, the antenna is wrapped around the vacuum chamber constituted by a dielectric material, generally quartz, to enable electromagnetic coupling.

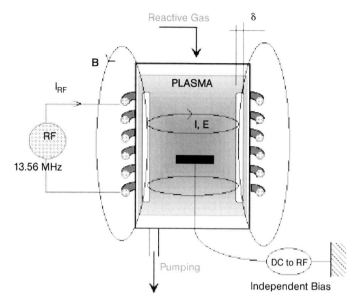

Fig. 2.6 Helical ICP source and cylindrical reactor.

The substrate holder is adapted to avoid shadowing of the sample to be treated. This sample is surrounded by the plasma from all sides and can be treated on its whole area. The uniformity of the treatment depends on many different parameters, mainly the plasma density profile in the volume, the gas injection, and the temperature distribution.

This type of reactor has been successfully employed for the deposition of insulating layers, organic or inorganic [51]. A variation of this kind of source has been developed by Colpo *et al.* in order to heat and control the temperature of the treated sample [52]. Those authors used an inductive heating system by superimposing a second antenna working at a lower frequency, along with specific filters in order to avoid any interference problems due to electromagnetic interaction that could arise between the two inductors.

The major limitation of classic ICP configurations is related mainly to one specific application: the deposition of metal layers by PECVD. The main problem with such a process is that the metal films deposit also on the inside dielectric chamber walls. The deposition of a conductive film on the coupling dielectric of an ICP source causes a drift in the ion current, and in turn modifies some parameters of the discharge, like the plasma potential and sheath voltage [53], modifying the deposition conditions during the process. However, the main problem to face is that once a thick enough conductive layer is covering the inside reactor surface, it shields the penetration of the electromagnetic fields generated by the antenna. The energy absorbed by the electrically conductive film instead of the plasma itself causes the extinction of the plasma that is no longer fed with energy.

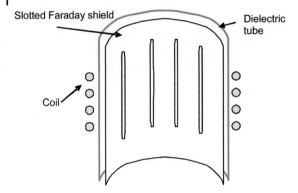

Fig. 2.7 Internal electrostatic shielding for the deposition of conductive layers.

This problem has been tackled in our laboratory where we developed a special reactor for the deposition of metallic films for aeronautic applications (Eureka Project 1229, Inductive Plasma for Anti Erosion Coatings (IPACERC). The solution we have implemented was to arrange a segmented electrostatic shield within the plasma chamber. The electrically conductive film will deposit on this shield and no longer on the dielectric chamber wall. The shield was made of aluminum and is composed of seven slits perpendicular to the coil current to prevent the induced currents from forming a loop in the entire shield periphery (Fig. 2.7). In that manner, the induced current created within each sector of the shield induces itself currents in the plasma. The setup has been successfully tested using WF_6/Ar and $WF_6/Ar/CH_4$ or $WF_6/Ar/C_2H_2$ gas mixtures for the deposition of tungsten and tungsten carbide films (WCx) respectively, for erosion resistance improvement of first-stage compressor blades of helicopter turbines [54].

Electrical measurements using an RF probe have been performed to characterize the effect of the Faraday shield on the electrical efficiency. Results showed that the coil current necessary to sustain the plasma is larger with the shielded ICP source than without. This is due to the increase of the mutual inductance between the coil and the plasma, i.e. the ratio of the magnetic flux effectively acting on the plasma due to the presence of the Faraday shield. In consequence, a larger RF current is needed to sustain the plasma at constant power. This results in an increase of the ohmic losses in the matching network circuitry, decreasing the power transfer efficiency from 95% without to 75% with the shield. An important result was that the presence of the internal Faraday shield does not significantly modify the radial distribution of the ion density, giving the possibility to keep a good process uniformity.

2.3.1.3 Large Area Treatment

The development of flat panel displays, solar cells, and the increasing wafer size in the modern semiconductor industry, together with the need for high-throughput processing drove the development of large area high-density plasma sources.

The helicon and ECR sources showing difficulties inherent to their setup and principle to be scaled up, the ICP appears as one of the most promising candidates for a high-density large area plasma source.

The main difficulty lies in the generation of a plasma with a very good uniformity over the area. For ICP sources, the uniformity of the plasma density is mainly dependent on the reactor geometry and on the inductive antenna shape. These two parameters combined together generally define the plasma density profile within the chamber mainly because:

- the inductor influences directly the electromagnetic field profile and thus the ionization profile
- the vacuum chamber, with in particular its volume, wall distance and aspect ratio, influences the charged particle transport mechanisms [48–50].

The spiral coil is the most commonly used antenna because it radiates the highest intensity flux by inductance value, and has the lowest inductance by surface unit [39,40]. In an attempt to improve the uniformity of the discharge, other shapes of antenna have been investigated by several authors in order to modify the heating profile [39,40,47,55–57].

Alternative coil shapes like the ladder, the shamrock, or the serpentine antenna, provide more uniform heating patterns as shown by Patterson and Wendt [56]. The coupling efficiency of these inductors is, however, limited and this results in a lower plasma density. The coupling efficiency can be further increased by increasing the number of turns. But, for large area sources, this solution can lead to a coil length similar to the excitation RF wavelength and results in standing wave effects. These standing wave effects cause a non-uniform current distribution along the coil, and in turn a non-uniform heating pattern (see below).

The large area plasma source developed in our laboratory makes use of a combination of innovative technological solutions. These solutions are based on:

- the design of a specific antenna providing high heating uniformity
- the amplification of the radiated electromagnetic field intensity by the addition of a magnetic core
- the improvement of the coupling efficiency between coil and plasma by the reduction of the antenna to plasma distance.

In order to provide for a uniform heating pattern of the discharge region, a serpentine-like inductor was chosen. However, the coupling efficiency of such an antenna being quite low, it appears to reach high plasma densities with reasonable power density values. To overcome this problem, the inductive source was modified by embedding the coil antenna in a magnetic core. The magnetic core is made of a particular magnetic material that presents a high magnetic permeability. Using the arrangement shown in Fig. 2.8 then allows one to concentrate the magnetic field only on the load, i.e. the plasma, and reduces the losses in the backpath above the antenna (Fig. 2.9).

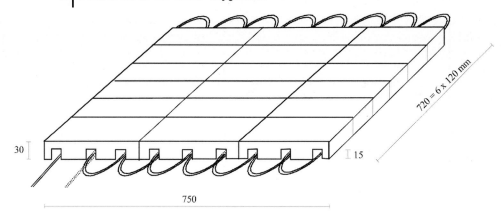

Fig. 2.8 Serpentine antenna inserted in a magnetic core.

This solution was first implemented on a 200 mm diameter ICP reactor and showed that the magnetic core allowed for an increase of the magnetic induction intensity of about 24%, and subsequently to an increase of the plasma density by 50% [58].

Another feature of the so called MaPE-ICP is the reduction of the dielectric window thickness that loses its mechanical function as a vacuum seal. The dielectric window is reduced to a thickness of about 4 mm, which results in a significant increase of the mutual inductance between coil and plasma, and so to a higher coupling efficiency. This integrated solution showed very good performances with a very high plasma density (the ion current density reached 30 mA cm^{-2} for a 700 W argon discharge at 30 mtorr) and a very good plasma uniformity with a 5% deviation within the coil radius [58].

This source was then scaled up to a 72 × 75 cm inductive source in a 1 m^2 reactor. However, in order to avoid standing wave effects that result in a non-constant current

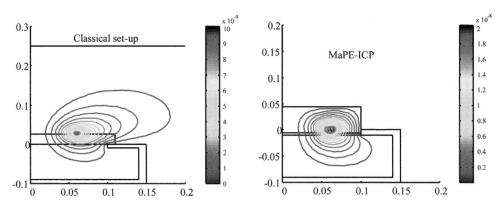

Fig. 2.9 Electromagnetic simulation showing the concentrating effect of the magnetic core (arbitrary units).

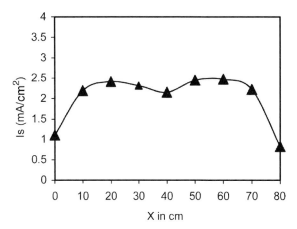

Fig. 2.10 Ion current density profile measured on a large area plasma source. Argon plasma: 1.5 mtorr, $P = 2000\,\text{W}$.

in the coil and so to a non-uniform heating profile, the serpentine-like antenna was actually constituted of three single loop antennas connected in parallel with each other, and the operating frequency was reduced to 2 MHz instead of 13.56 MHz. The choice of this frequency is not fortuitous. It is the frequency for which the magnetic material used in this source presents the highest magnetic permeability value. Indeed, we could observe a significant increase of the plasma density as compared to the results obtained at 13.56 MHz [59].

The relative ion current density profile obtained with three coils in parallel is presented in Fig. 2.10. The uniformity (standard deviation/average) obtained is 7% on 60 cm showing the source capability to generate large area and uniform plasma.

The technical solutions adopted in the MAPE-ICP source present also other advantages. First, the use of multiple coils in parallel results in an overall reduction of the total antenna inductance, which facilitates impedance matching to the RF power supply. Second, the decrease of the operating frequency causes a reduction of the voltage on the coil (proportional to $L\omega$), which in turn causes a decrease of the capacitive coupling.

This is not the case for the 200 mm diameter MaPE-ICP source. The electrical measurements carried out on the MaPE-ICP source show that the magnetic core induced an increase of the root mean square voltage present on the coil [58]. Moreover, the dielectric window is reduced to a low thickness of about 5 mm in the MaPE-ICP technology. Under these circumstances, an increase of the capacitive coupling in the discharge is expected.

An estimation of the level of capacitive coupling in the discharge can be made by measuring the amplitude of the RF component of the floating potential [47]. A more reliable method to assess the level of capacitive coupling consists in measuring the contribution of the harmonic frequencies in the coil current [23]. Indeed, the current harmonics are related to nonlinear effects in the system and the main contribution in the ICP sources is the nonlinear relation between the sheath potential and the displacement current. However, the measurement of the current harmonics gives a good representation of the capacitive coupling only when the capacitive component is the major nonlinearity. The system described here uses a magnetic core that is by nature a nonlinear material, the properties of which (and in particular the magnetic permeability) vary with the applied field. These materials are characterized by the presence of a saturation, and a hysteresis cycle. Another nonlinear effect in the ICP is related to the nonlinear electron transport mechanism caused by the Lorentz force $F = qv \times B$ and the $v\,\mathbf{grad}\,v$ component of the inertial force [33,60]. The resulting forces are proportional to $(\omega^2 + v_{eff}^2)^{-1}\mathrm{grad}\,E^2$ [60,61], and thus this nonlinear effect can be neglected here since it is generally significant only at low pressure (typically around 1 mtorr or lower) and low frequency ($f <$ MHz).

Considering the voltage values present on the coil ($>$kV), the capacitive coupling can contribute to the harmonics in the current. Regarding the nonlinear properties of the magnetic material in use, plotting of the $B = F(I)$ curve showed that this component is operated far from saturation, in a region where the magnetic core appears to be a linear element [58]. Indeed, the core is made of a magnetodielectric material that presents the advantage of having an almost linear behavior, saturating only from field intensities typically around $300\,\mathrm{A\,cm}^{-1}$. Consequently, the harmonics present in the coil current provide a reliable estimation of the capacitive component.

The first, second, and third harmonics were measured and referenced to the fundamental of the coil current and all show a similar behavior. The plot versus net power supplied by the RF generator shows that the capacitive component is reduced with increasing power (Fig. 2.11(a)). This is mainly due to the fact that the discharge character becomes more of the inductive type with increasing power [44]. The use of a magnetic core and the reduction of the dielectric window thickness result in a higher

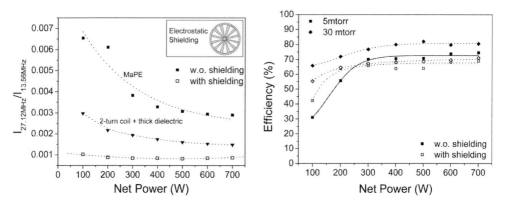

Fig. 2.11 (a) Reduction of the capacitive coupling by electrostatic shielding. (b) Effect of the electrostatic shield on the power transfer efficiency.

capacitive component as compared to a configuration using only a coil above a thick (30 mm) dielectric. This true for all the harmonics measured and Fig. 2.11(a) shows the evolution of the first harmonic referenced to the fundamental. To reduce the capacitive component, an electrostatic shield was constructed.

Instead of using a separate shield, a totally integrated solution was built by depositing a conductive film directly on the dielectric coupling window as shown in the inset of Fig. 2.11(a). The shield seems to be effective since it appears impossible to ignite a plasma without the use of a high-voltage pulse applied on the substrate holder.

The measurement of the current harmonics in this configuration shows that their contribution is strongly reduced and does not vary with the RF power supplied to the inductive source. The power transfer efficiency was measured by the RF impedance probe and appears to be reduced from about 80% to 70% with the use of the electrostatic shielding (Fig. 2.11(b)). This is mainly due to the slight current increase when using the shield, which results in an increase of the ohmic losses in the circuitry.

Alternative solutions were proposed for the production of a large area inductively coupled plasma. To cite just a few, Wu and Lieberman developed a traveling wave-driven ICP [42,62] and achieved good results in terms of plasma density and uniformity. The antenna is constituted of several rods connected in series thus forming a serpentine-like antenna, connected to a special electrical setup that allows one to launch a traveling wave within the antenna. Kawai and co-workers [63,64] described a large area plasma source based on the use of a ladder-shaped antenna. They tackled the problem of low plasma density by increasing the operating frequency, which could cause standing wave effects limiting the plasma uniformity. Finally, Setsuhara *et al.* [65] used a floating internal antenna config-uration that presented reasonable performances in terms of plasma density and uniformity.

2.4
Conclusions

The inductively coupled plasma is a versatile tool that can be used for a wide range of applications. Its flexibility of construction allows the design of a reactor where plasma parameters, such as ion density and energy, can be controlled on very different shapes and dimensions of substrates. The design of the plasma source can be adapted to fulfill specific requirements of each application. In this paper, we described plasma sources developed in our laboratory. These represent only a small selection of plasma sources but underline the ICP flexibility. These sources cover a wide range of processes (from etching to metallic deposition) that can be applied to different fields (aeronautic, flat panel display, etc.).

Many opportunities remain open for the development of plasma sources to meet the demands of new applications such as anticorrosion, solar cells, biotechnology, etc.

References

1 Lieberman, M.A. and Lichtenberg, A.J. (1994) *Principles of Plasma Discharges and Material Processing*, Wiley Interscience.

2 Braithwaite, N.S.J. (2000) *Plasma Sources Sci. Technol.*, **9**, 517–527.

3 Conrads, H. and Schmidt, M. (2000) *Plasma Sources Sci. Technol.*, **9**, 441–454.

4 Francis, G. (1956) in *Handbuch der Physik;* Vol. VVII (ed. S. Flügge), Springer, Berlin.

5 Flamm, D.L. (1991) *Solid State Technol.*, **34**, 47.

6 Popov, O.A. (1995) *High Density Plasma Sources*, Noyes Publications.

7 Hopwood, J. (1992) *Plasma Sources Sci. Technol.*, **1**, 109–116.

8 Hittorf, W. (1884) *Ann. Physik*, **21**, 137.

9 Kortshagen, U., Gibson, N.D. and Lawler, J.E. (1996) *J. Phys. D: Appl. Phys.*, **29**, 1224.

10 Turner, M.M. and Lieberman, M.A. (1999) *Plasma Sources Sci. Technol.*, **8**, 313.

11 Cunge, G., Crowley, B., Vender, D. and Turner, M.M. (1999) *Plasma Sources Sci. Technol.*, **8**, 576–586.

12 Demirkhanov, R.A., Kadysh, I. and Yu, S.K. (1964) *Sov. Phys. JETP*, **19**, 791.

13 Weibel, E.S. (1967) *Phys. Fluids*, **10**, 741.

14 Kolobov, V.I. and Economou, D.J. (1997) *Plasma Sources Sci. Technol.*, **6**, R1.

15 Holstein, T. (1952) *Phys. Rev.*, **88**, 1427.

16 Godyak, V.A., Piejak, R.B. and Alexandrovich, B.M. (1994) *Plasma Sources Sci. Technol.*, **3**, 169.

17 Godyak, V.A. and Piejak, R.B. (1997) *J. Appl. Phys.*, **82**, 5944.

18 Godyak, V.A., Piejak, R.B. and Alexandrovich, B.M. (1998) *Phys. Rev. Lett.*, **80** (15), 3264.

19 Godyak, V.A., Piejak, R.B., Alexandrovich, B.M. and Smolyakov, A. (2001) *Plasma Sources Sci. Technol.*, **10**, 459.

20 Denneman, J.W. (1990) *J. Phys. D: Appl. Phys.*, **23**, 293–298.

21 Piejak, R.B., Godyak, V.A. and Alexandrovich, B.M. (1992) *Plasma Sources Sci. Technol.*, **1**, 179–186.

22 El-Fayoumi, I.M. and Jones, I.R. (1998) *Plasma Sources Sci. Technol.*, **7**, 179–185.

23 Hopwood, J. (1994) *Plasma Sources Sci. Technol.*, **3**, 460.

24 Sobolewski, M.A. (1995) *J. Res. Natl Inst. Stand. Technol.*, **100**, 341.

25 Piejak, R.B., Godyak, V.A. and Alexandrovich, B.M. (1997) *J. Appl. Phys.*, **81**, 3416.

26 El-Fayoumi, I.M. and Jones, I.R. (1997) *Plasma Sources Sci. Technol.*, **6**, 201.

27 El-Fayoumi, I.M. and Jones, I.R. (1998) *Plasma Sources Sci. Technol.*, **7**, 162–178.

28 Colpo, P., Ernst, R. and Rossi, F. (1999) *J. App. Phys.*, **85** (3), 1366.

29 Meziani, T., Colpo, P. and Rossi, F. (2006) *J. Appl. Phys.* **99**, 033303.

30 Godyak, V.A. and Piejak, R.B. (1990) *J. Vac. Sci. Technol. A*, **8**, 3833.

31 Keller, J.H. (1996) *Plasma Sources Sci. Technol.*, **5**, 166–172.

32 Butterbaugh, J.W., Baston, L.D. and Sawin, H.H. (1990) *J. Vac. Sci. Technol. A*, **8**, 916.

33 Godyak, V.A., Piejak, R.B. and Alexandrovich, B.M. (1999) *J. Appl. Phys.*, **85** (2), 703.

34 Sugai, H., Nakamura, K. and Suzuki, K. (1994) *Jpn. J. Appl. Phys.*, **33**, 2189.

35 Daviet, J.F. (2000) US Patent.

36 Khater, M.H. and Overzet, L.J. (2001) *J. Vac. Sci. Technol. A*, **19** (3), 785.

37 Kushner, M.J., Collison, W.Z., Grapperhaus, M.J., Holland, J.P. and Barnes, M.S. (1996) *J. Appl. Phys.*, **80**, 1337.

38 Lamm, A.J. (1997) *J. Vac. Sci. Technol. A*, **15**, 2615.

39 Intrator, T. and Menard, J. (1996) *Plasma Sources Sci. Technol.*, **5**, 371.

40 Menard, J. and Intrator, T. (1996) *Plasma Sources Sci. Technol.*, **5**, 363.

41 Suzuki, K., Konishi, K., Nakamura, K. and Sugai, H. (2000) *Plasma Sources Sci. Technol.*, **9**, 199–204.

42 Wu, Y. and Lieberman, M.A. (1998) *Appl. Phys. Lett.*, **72** (7), 777.

43 Keller, J.H. (1989) *in Proceedings of the 33rd Gaseous Electronics Conference* (unpublished).

44 Hopwood, J., Guarnieri, C.R., Whitehair, S.J. and Cuomo, J.J. (1993) *J. Vac. Sci. Technol. A*, **11** (1), 147.

45 Keller, J.H., Forster, J.C. and Barnes, M.S. (1993) *J. Vac. Sci. Technol. A*, **11** (5), 2487.

46 Lieberman, M.A. and Gottscho, R.A. (1994) in *Physics of Thin Films; Vol. 18* (eds M.H., Francombe and J.L. Vossen), Academic Press, San Diego, CA. pp. 1–119.

47 Forgotson, N., Khemka, V. and Hopwood, J. (1996) *J. Vac. Sci. Technol. B*, **14** (2), 732.

48 Wainman, P.N., Lieberman, M.A., Lichtenberg, A.J., Stewart, R.A. and Lee, C. (1995) *J. Vac. Sci. Technol. A*, **13**, 2464.

49 Stittsworth, J.A. and Wendt, A.E. (1996) *Plasma Sources Sci. Technol.*, **5**, 429.

50 Collison, W.Z., Ni, T.Q. and Barnes, M.S. (1998) *J. Vac. Sci. Technol. A*, **16** (1), 100.

51 Colpo, P., Ceccone, G., Sauvageot, P., Baker, M. and Rossi, F. (2000) *J. Vac. Sci. Technol. A*, **18**, 1096.

52 Colpo, P., Ernst, R. and Keradec, J.P. (1999) *Plasma Sources Sci. Technol.*, **8**, 587–593.

53 Sobolewski, M.A. (2005) *J. Appl. Phys.* **97** (3).

54 Colpo, P., Meziani, T., Sauvageot, P., Ceccone, G., Gibson, P.N., Rossi, F. and Monge-Cadet, P. (2002) *J. Vac. Sci. Technol. A*, **20**, 622.

55 Yu, Z., Gonzales, P. and Collins, G.J. (1995) *J. Vac. Sci. Technol. A*, **13**, 871.

56 Patterson, M.M. and Wendt, A.E. (1999) AVS Fall Conf. Contribution, Seattle.

57 Khater, M.H. and Overzet, L.J. (2000) *Plasma Sources Sci. Technol.*, **9**, 545.

58 Meziani, T., Colpo, P. and Rossi, F. (2001), *Plasma Sources Sci. Technol.*, **10**, 276.

59 Colpo, P., Meziani, T. and Rossi, F. (2005) *J. Vac. Sci. Technol. A*, **23**, 270.

60 Godyak, V.A., Alexandrovich, B.M., Piejak, R.B. and Smolyakov, A. (2000) *Plasma Sources Sci. Technol.*, **9**, 541–544.

61 DiPeso, G., Vahedi, V., Hewett, D.W. and Rognlien, T.D. (1994) *J. Vac. Sci. Technol. A*, **12**, 1387.

62 Wu, Y. and Lieberman, M.A. (2000) *Plasma Sources Sci. Technol.*, **9**, 210.

63 Kawai, Y., Yoshioka, M., Yamane, T., Takeuchi, Y. and Murata, M. (1999) *Surf. Coat. Technol.*, **116–119**, 662.

64 Mashima, H., Murata, M., Takeuchi, Y., Yamakoshi, H., Horioka, T., Yamane, T. and Kawai, Y. (1999) *Jpn. J. Appl. Phys.*, **38**, 4305.

65 Setsuhara, Y., Miyak, S., Sakawa, Y. and Shoji, T. (1999) *Jpn. J. Appl. Phys.*, **38**, 4263.

3
Advanced Simulations for Industrial Plasma Applications

S.J. Kim, F. Iza, N. Babaeva, S.H. Lee, H.J. Lee, and J.K. Lee

Recent advances have significantly increased the credibility and usefulness of time-
and space-resolved fluid and particle simulations of low-temperature plasmas. One
example is the kinetic particle-in-cell Monte Carlo collision (PIC-MCC) model, an
accurate but time-consuming tool that uses particles and a mesh. Advanced PIC-
MCC simulations are frequently used to study capacitively coupled plasmas such as
those used for sub-90 nm dielectric etching. Kinetic simulations elucidate the
electron and ion heating mechanisms as well as the energy and angle distribution
functions of different species. These parameters are closely related to the potential
and density profiles. Recent PIC-MCC simulations [1–4] have produced values for
these kinetic characteristics which compare favorably with available measurements. Particle-in-cell simulations can also be used to investigate plasma-induced
charge-up damage. In contrast, fluid simulations are suitable for modeling high-
pressure plasmas where non-local effects are typically not important. Although the
drift-diffusion approximation is widely used in fluid models, simulation of low-
pressure and/or high-frequency plasmas requires the solution of the full momentum equation in order to account for the inertia of the particles. Since fluid
simulations are faster than PIC simulations, fluid models are suitable for analyzing
large area plasma sources. The two simulation methods (PIC and fluid) comple-
ment each other and, when used appropriately, they provide valuable information
that leads to a better understanding of plasma physics and the development of
improved plasma reactors.

3.1
Introduction

Low-temperature plasmas are widely used in industrial applications such as
semiconductor processing and plasma display panels. Despite the steady progress
in the field during the last decades, plasmas are yet not fully understood and
simulations offer unique capabilities for studying plasma physics and improving

Advanced Plasma Technology. Edited by Riccardo d'Agostino, Pietro Favia, Yoshinobu Kawai, Hideo Ikegami,
Noriyoshi Sato, and Farzaneh Arefi-Khonsari
Copyright © 2008 WILEY-VCH Verlag GmbH & Co. KGaA, Weinheim
ISBN: 978-3-527-40591-6

the performance of low-temperature plasma sources. Physical parameters which are difficult to measure in experiments can be obtained in simulations. Due to computational limitations, however, plasma models require approximations. In spite of these approximations, simulations have advanced our fundamental understanding of plasma physics, plasma processes, and plasma equipment.

Plasma simulations using particles have been studied for half a century [5]. At first, only one-dimensional (1D) simulations with limited numbers of particles were possible. These analyses were suitable for the parallel plate capacitive coupled plasma (CCP) source, the workhorse of the semiconductor industry at that time. The goals of these early 1D simulations were the investigation of basic plasma characteristics and the improvement of modeling techniques. Particle collisions were included in early models although the efficient algorithms used in today's simulations were developed later in the 1990s. More comprehensive two- and three-dimensional fluid models with complex chemistry were introduced in the 1990s to study capacitive coupled plasmas and inductively coupled discharges. Recent advances have significantly increased the credibility and usefulness of time- and space-resolved fluid and particle simulations.

Particle-in-cell and fluid simulations are commonly used as numerical simulation techniques to study low-temperature plasmas. The PIC-MCC model, a self-consistent and fully kinetic approach, is one of the most accurate models for describing the motion of charged particles [5–10]. In this method, the trajectory of particles in phase space is determined by integrating the Newton–Lorentz equation of motion at each time step. The velocity of colliding particles is adjusted by a Monte Carlo collision model [5,6]. PIC-MCC simulations resolve space- and time-dependent velocity distribution functions without any prior assumptions, and reaction rates and transport coefficients are self-consistently modeled. Tracking individual particles, however, is computationally intensive.

In contrast, fluid models describe the plasma using macroscopic properties such as density, mean velocity, and mean energy. These quantities are obtained by solving moments of the Boltzmann equation [11]. The first moments are the particle continuity, the momentum balance, and the energy conservation equations. The solution of this system of equations requires averaged frequencies and transport coefficients such as the ionization frequency, the momentum exchange frequency, the energy exchange frequency, and diffusion constants. Since averaged quantities depend on the electron energy distribution function (EEDF), which is unknown, a distribution function must be assumed. A common approach for estimating the electron energy distribution function is to use the local field approximation (LFA). In the LFA, the energy that the electrons gain from the electric field is locally balanced by collisional losses. As a result, the energy conservation equation does not need to be solved and the various frequencies and transport coefficients can be pre-tabulated as a function of the reduced electric field. Although fluid models are computationally fast, the assumption of an EEDF limits the reliability of fluid models to high-pressure discharges where the electron motion is strongly collisional and the energy distribution at any given point depends only on local conditions.

In this chapter, we present simulation results for various plasma reactors used for semiconductor processing. PIC-MCC and fluid simulations are performed to investigate the underlying plasma physics and to identify possible process and reactor improvements. Applications of PIC-MCC simulations are discussed in Section 3.2. Discharge properties of capacitively coupled Ar/O_2 plasmas are investigated by means of one-dimensional PIC-MCC simulations. Density and temperature profiles as well as ion energy distribution functions (IEDF) are analyzed as a function of the oxygen concentration. A second application of the PIC model is the charge-up simulation of a patterned substrate. Here, a three-dimensional (3D) particle code is used to study the fluxes of charge particles inside a trench. In Section 3.3, two-dimensional (2D) fluid simulations of unmagnetized low-temperature plasmas are presented. A conventional parallel plate CCP and a large area plasma source (LAPS) are analyzed in this section.

3.2
PIC Simulations

PIC simulations resolve self-consistently the energy distribution function of each species in the plasma taking into account local and non-local effects. For example, they can reliably obtain kinetic information such as the electron energy distribution function and the energy and angle distribution functions of ions impinging on a wafer. In this section kinetic characteristics of single- and dual-frequency capacitively coupled discharges and charge-up of narrow trenches are examined by means of PIC-MCC simulations.

3.2.1
Capacitively Coupled O_2/Ar Plasmas

Capacitively coupled radio-frequency (RF) discharges are widely used in material processing for their etching selectivity and process uniformity [12,13]. Conventional CCP reactors with a single RF (13.56 MHz) source have been continuously modified over the last few decades to meet the demands of the semiconductor industry. Controlling etching processes requires control over the flux, energy, and angle of ions bombarding the substrate. To overcome the limitations of conventional single-frequency CCPs, dual-frequency CCPs have been introduced in recent years. The use of two power sources operating at different frequencies offers tools for controlling plasma density (ion flux) and substrate bias (ion energy) independently [14]. Using PIC-MCC simulations, Kim and co-workers [1,3,4,15] have developed an analytic global model for dual-frequency capacitively coupled discharges that elucidates the role of each power source.

In this chapter, we investigate the characteristics of an O_2/Ar plasma as a function of the O_2 concentration and the total gas pressure. Simulation results for single- and dual-frequency CCPs are presented. The PIC-MCC model used tracks electrons and three ion species, namely Ar^+, O_2^+, and O^-. These are the expected dominant ion species in the pressure range of interest (20–100 morr) [16]. Given the low degree of ionization, Ar and O_2 neutrals are assumed to be uniformly distributed in the chamber. Collisions involving metastables are not considered in the model. Since Penning ionization is not possible in Ar/O_2 mixtures, this approximation does not directly affect the generation of modeled ion species. Argon metastables, however, can enhance the population of oxygen metastables leading to an increase of ozone production and O_2^- (byproduct of the dissociative detachment of ozone). These species, however, are typically not dominant in the pressure range of interest and are not included in the model. A description of the 33 collisions considered in the model can be found in [6].

3.2.1.1 Gas Composition

Plasma density profiles in an O_2/Ar discharge are shown in Fig. 3.1(a)–(d) for three different gas compositions. The discharge is sustained in a single-frequency parallel plate reactor with a gap between the electrodes of 1.6 cm. The reactor is asymmetric and the electrode area ratio is ~1.7. The large electrode is grounded while the small one is powered with a 500 V/27 MHz source. The total gas pressure inside the chamber is 40 morr and the concentration of O_2 in the feed gas is varied from 0% (pure argon discharge) to 100% (pure oxygen discharge). In the bulk, quasineutrality holds as the total numbers of positive and negative charges are the same. Figure 3.1(a) and (c) correspond to pure oxygen and pure argon discharges, respectively, and Fig. 3.1(b) to a 50%–50% mixture. Due to electron attachment, the electron density decreases and the sheaths widen with increasing O_2 concentration (Fig. 3.1(a)–(c)). Charged particle densities (peak values) as a function of the gas composition are shown in Fig. 3.1(d). Ar^+ and electron densities decrease while O_2^+ and O^- densities increase with increasing oxygen content. Due to the close ionization potentials of argon and oxygen (15.7 and 12.6 eV respectively), the plasma transition from a pure argon discharge to a pure oxygen discharge is quite smooth. At 50% O_2, argon and O_2^+ densities are comparable. The maximum effective electron temperature and the plasma potential are shown as a function of the gas composition in Fig. 3.1(e) and (f). In contrast with the electron and ion density profiles, the electron temperature and the plasma potential are not significantly affected by the gas composition. A relatively constant electron temperature has also been observed in experiments, and has been attributed to the close ionization potentials of argon and oxygen [17]. The plasma potential is insensitive to the gas composition because the discharge is driven by a constant voltage source (500 V). The relatively low average plasma potential (~120 V) is due to the negative self-bias of the powered electrode.

The ion energy distribution function (IEDF) on the powered electrode as a function of the O_2 concentration in the plasma is shown in Fig. 3.2(a) and (b). The IEDF depends on the sheath length, the sheath voltage, the applied RF frequency, the ion mean free path, and the ion velocity at the sheath edge. The shape of the IEDF is determined by the ratio of the ion transit time to the RF period. When the ions require several RF cycles to transit the sheath, they respond to the average sheath potential, and, in a collisionless case, the IEDF displays a single energy peak. When the ion

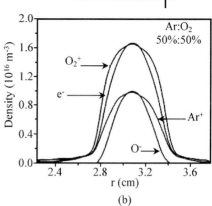

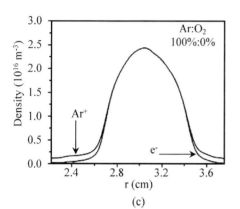

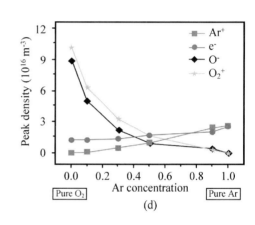

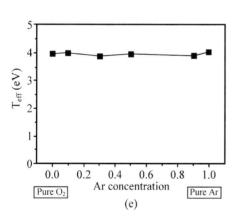

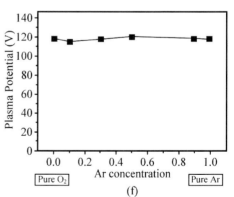

Fig. 3.1 Plasma density as a function of the O_2 concentration in an Ar/O_2 plasma: (a) pure O_2, (b) Ar : O_2 = 1 : 1, and (c) pure Ar. (d) Peak density, (e) effective electron temperature in the bulk plasma, and (f) plasma potential as a function of the gas composition. Total pressure and electrode gap are 40 morr and 1.6 cm, respectively. The discharge is sustained at 500 V/27 MHz.

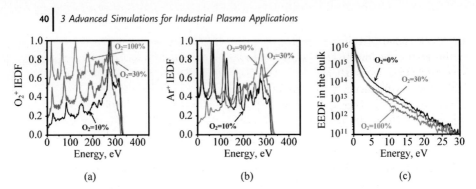

Fig. 3.2 Ion energy distribution functions on the powered electrode: (a) O_2^+ and (b) Ar^+ ions. (c) Electron energy distribution functions in the bulk. Data shown for various O_2/Ar plasmas at 40 morr in a single-frequency CCP reactor driven at 500 V/27 MHz.

transit time is comparable to the RF frequency, ions start responding to the instantaneous sheath voltage and the previous single energy peak widens and two maxima appear. These maxima correspond to ions entering the sheath when the sheath voltage is at its maximum and minimum values [18]. Collisions in the sheath, however, significantly change the shape of the IEDF. Collisions, especially resonant charge exchange collisions, limit the maximum energy of the ions bombarding the electrodes, extend the IEDF towards low energy values, and introduce additional peaks [19,20]. Since the IEDF depends on the ion transit time and on the collisions in the sheath, ions with different masses and cross sections display different distributions. (Compare Fig. 3.2(a) and (b) for O_2^+ and Ar^+ respectively.) For the discharge conditions of our study, collisions take place in the sheath as evidenced by the extension of the IEDF towards low-energy values and the presence of multiple peaks. O_2^+ and Ar^+ ions, however, display different trends. As the ratio of O_2 concentration increases, the number of low-energy O_2^+ ions increases while the number of low-energy Ar^+ ions decreases. At low oxygen concentrations, resonant charge exchange collisions of O_2^+ ions with O_2 molecules are rare due to the low oxygen partial pressure. On the other hand, Ar^+ ions undergo multiple resonant charge exchange collisions due to the high argon partial pressure. As the O_2 concentration increases and the Ar concentration decreases, the scenario reverses. At high O_2 concentrations, resonant charge exchange collisions of O_2^+ ions are frequent while resonant argon charge exchange collisions are rare.

The electron energy distribution function (EEDF) in the bulk plasma as a function of the O_2 concentration is shown in Fig. 3.2(c). The EEDF displays a bi-Maxwellian distribution with two electron temperatures. This distribution is typical of low-pressure capacitively coupled argon discharges and results from the selective heating of high-energy electrons at the sheaths [1,2]. As the O_2 concentration increases, the temperature of high-energy electrons remains nearly constant while that of low-energy electrons decreases.

3.2.1.2 Pressure Effect in Ar/O₂ Plasmas

In this section we investigate the effect of total gas pressure in Ar/O$_2$ discharges. The discharges are sustained in a dual-frequency parallel plate reactor. The high- and low-frequency power sources supply 500 V at 27 MHz and 400 V at 2 MHz, respectively. The gap size between the electrodes is 2 cm and the pressure is varied from 20 to 100 morr. An argon plasma with a small admixture of O$_2$ (Ar/O$_2$ = 97.3/2.7) and an oxygen plasma with a small admixture of argon (Ar/O$_2$ = 5/95) are considered.

Electron density and electron temperature in the bulk plasma as a function of the total pressure are shown in Fig. 3.3 for the two discharges. The plasma density increases with increasing pressure. The electron density in the argon plasma with a small O$_2$ admixture is larger than that in the oxygen plasma with a small argon admixture due to the electronegativity of oxygen. As pressure increases, particle diffusion to the chamber walls decreases and a lower electron temperature is required to sustain the discharge (Figs. 3.3(a) and (b)). Simulation results show a remarkably low effective electron temperature ($T_{eff} < 1$ eV) for the discharge with high oxygen content in the pressure range 60–100 morr. Although, according to a global model analysis, an electron temperature of ~2 eV is required for an oxygen discharge in which the main loss mechanism of O$_2{}^+$ ions is volume recombination with O$^-$, it should be noted that such analysis assumes a Maxwellian electron distribution. As shown in Fig. 3.2(c), however, the EEDF in the pressure range of interest is bi-Maxwellian. In this case, the effective electron temperature (shown in Fig. 3.3) is determined mainly by the abundant low energy electrons while the ionization rate is set by the high energy tail. As a result, a bi-Maxwellian distribution with a given effective temperature T_{eff}, has an ionization rate equivalent to a Maxwellian distribution with a much larger electron temperature. Although this argument holds true both for argon and oxygen discharges, vibrational excitation processes in an

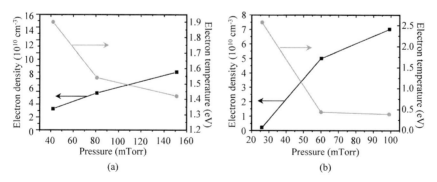

Fig. 3.3 Density (squares) and electron temperature (circles) as functions of pressure: (a) argon discharge with a small admixture of oxygen, Ar : O$_2$ = 97.3 : 2.7 and (b) oxygen discharge with a small admixture of argon, Ar : O$_2$ = 5 : 95. The discharge is sustained in a dual-frequency CCP reactor powered by a 500 V/ 27 MHz and 400 V/2 MHz sources.

oxygen discharge accentuate the bi-Maxwellian EEDF (Fig. 3.2(c)) and further reduce the effective electron temperature.

3.2.2
Three-Dimensional (3D) Charge-up Simulation

As devices shrink down to the nanoscale, plasma-induced damage (PID) such as bowing, trenching, reactive ion etching lag, and notching during plasma processing becomes an important issue that affects the reliability and reproducibility of the device [21,22]. Charging of the material being processed is one of the factors contributing to notching in reactive ion etching. Charge-up of narrow trenches results from the different motions of ions and electrons. While ion motion is predominantly perpendicular to the substrate enabling anisotropic etching, electron motion is isotropic. Thus, the electron flux to the bottom of a high-aspect-ratio trench is much smaller than the ion flux and the bottom of the trench charges positively. The potential induced inside the trench by these uneven fluxes changes the trajectory of subsequent incoming ions and causes undesirable etching profiles [23–25].

3.2.2.1 Description of 3D Charge-up Simulations
The uneven charge accumulation inside a trench due to the different electron and ion motions induces a local potential gradient. Using a 3D charge-up simulation [26,27], we have studied the potential profiles generated in high-aspect-ratio trenches. The simulation procedure is divided in two parts to account for the different scales of the plasma reactor (>1 cm) and the trench (<1 μm). The electron and ion fluxes to the substrate are determined from a PIC-MCC simulation of a reactor following the same procedure as in Section 3.2.1. The resulting fluxes are then passed to a 3D PIC simulator to study the charge-up of microtrenches. The simulation domain of this second PIC module is shown in Fig. 3.4(a). The boundary conditions for the field solver are the following. The Neumann boundary condition ($\nabla V = 0$) is adopted at the left ($x = 0$) and right ($x = x_{max}$) boundaries above the trench. A periodic boundary condition is adopted at the front ($z = 0$) and back ($z = z_{max}$) surfaces. Finally, Dirichlet boundary conditions ($V = 0$) are assumed at the top ($y = y_{max}$) and bottom ($y = 0$) boundaries. Ions and electrons are injected from the top plane reproducing the fluxes obtained in the simulation of the reactor. The dielectric material is assumed to be a perfect dielectric, i.e., impinging charged particles are absorbed and do not redistribute. Considering the symmetry of the problem, the simulation domain can be reduced to one quarter of that shown in Fig. 3.4(a).

One computational cycle consists of injecting particles, tracking their trajectories, and updating the potential inside the trench. The energy and angle of injected ions is determined by randomly sampling the ion energy distribution obtained in the large-scale simulation of the reactor. The electrons are assumed to be Maxwellian although they could also be sampled from another distribution. Since a plasma in steady state has equal average electron and ion fluxes, the same number of electrons and ions are injected in each cycle. The electric field due to a single particle is very strong in the nanometer range, so the potential inside a trench induced by a limited number of

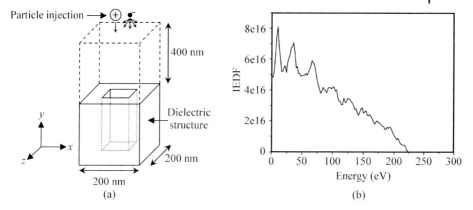

Fig. 3.4 (a) Schematic of the simulation domain for the 3D charge-up simulations. (b) Energy distribution function of injected ions. The IEDF is obtained from a PIC-MCC simulation of a single-frequency CCP source.

particles can be very noisy. To mitigate the noise problem, the 3D charge-up simulation uses computational particles (also referred as superparticles) instead of real ions and electrons. In contrast with typical PIC simulations where super-particles represent multiple real particles, in this simulation superparticles represent a fraction of a real particle. For example, if each superparticle represent 0.1 real particles, then the contribution of a real particle can be considered to be distributed over ten different simulations. Using statistically averaged particles reduces the noise level in the simulation and allows the trajectory of subsequent incoming particles to be resolved.

Since we are interested in the potential gradient induced by accumulated surface charges, space charge in the simulation domain is neglected. In fact the effect of the space charge is already accounted for in the distribution of the injected particles because their velocities are sampled from the results of a self-consistent PIC simulation. Thus, the effects of the bias potential and the space charge in the sheath region are included in the incident ion energy, and, since the length of the simulation domain is much smaller than the sheath length, the potential difference between the top and bottom boundaries can be assumed to be zero. Therefore, since the potential profiles evolve as charges accumulate on the dielectric in each cycle, the Laplace equation ($\nabla^2 V = 0$) must be solved.

Collisions in the simulation domain are neglected because the electron and ion mean free paths (λ_e, λ_i) are much longer than the simulation domain. The typical charge-up time in high-aspect-ratio trenches is of the order of a tenth of a second while the etching takes place in a longer time scale (hundreds of nanometers per second). Therefore the etching during the charge-up time is negligible, and the trench structure does not change during this time.

The IEDF used in the charge-up simulation is shown in Fig. 3.4(b). The distribution was obtained from a 1D PIC-MCC simulation of a CCP driven by a

400 V/27 MHz voltage source. The discharge was sustained in argon at 50 morr and the gap between the electrodes was 2 cm. Under these conditions the sheath is collisional and ions hit the substrate with a broad energy range (0–225 V).

3.2.2.2 Effects of Secondary Electron Emission

The different motion of electrons and ions results in an uneven charge distribution that induces a potential gradient within the trench. The electron shading effect, i.e. the appearance of a negative potential at the top of a trench, further prevents electrons from entering the trench. When ions bombard the substrate, however, secondary electrons can be emitted. The ratio of emitted electrons to bombarding ions is referred as the secondary electron emission coefficient (SEEC), γ_i. In general, the SEEC depends on the substrate material, the gas species, and the energy and angle of bombarding ions [28]. In the simulation, we assume a constant averaged SEEC. Secondary electrons generated inside the trench drift towards the bottom due to the existing potential gradient and can reduce the charge-up effect.

The effect of secondary electrons is shown in Fig. 3.5, where a simulation without secondary electrons and one with a SEEC of 0.4 are compared in terms of their

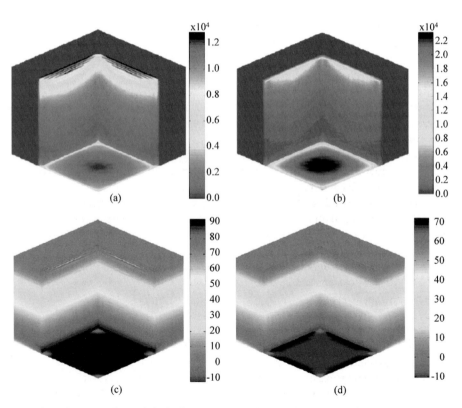

Fig. 3.5 (a, c) Ion flux and (b, d) charge-up potential distributions. (a, b) SEEC = 0 and (c, d) SEEC = 0.4.

incident ion fluxes and resulting charge-up potentials. In both cases, electrons and ions are injected according to a Maxwellian distribution of 1 eV and the IEDF shown in Fig. 3.4(b), respectively. The saturation potential and charge-up time depends on the IEDF and the trench geometry. Secondary electron emission, however, leads to longer charge-up times and a reduction of the potential difference along the trench. Most secondary electrons generated in the trench are trapped by the large potential gradient and end up being collected at the bottom of the trench. This electron flux to the bottom of the trench reduces the final potential difference between the top and bottom and extends the charge-up time. Since the potential gradient is changed, the trajectory of subsequent incoming ions is also affected. As shown in Fig. 3.5, secondary electron emission leads to a reduction of the ion bombardment at the top of the trench and an increase of the ion flux to the bottom. Since the ion flux determines the rate and anisotropy of the etching process, secondary electrons can reduce sidewall etching and counteract the decrease of etching rate in high aspect ratio features. Thus, a high SEEC contributes to a reduction of the charge-up damage. Although the SEEC is an intrinsic value of the material, it can be controlled by selecting the composition, energy, and angle of bombarding ions.

3.2.2.3 Negative Ion Extraction

The reduction of charge-up damage using anisotropic negative ions instead of isotropic electrons has been studied in the past. It is difficult, however, to extract negative ions from a plasma because the difference in mass and temperature between positive ions and electrons leads to a positive plasma potential. This positive plasma potential prevents negative ions from entering the sheath, so they cannot be extracted from the bulk plasma. Ion–ion plasmas depleted of electrons, however, can be generated using a time-modulated power source. In this case, a sheath inversion during the off phase makes it possible to extract negative ions [29–32].

We have studied negative ion extraction in a dual-frequency CCP using a 1D PIC-MCC code (as in Section 3.2.1). An oxygen discharge is sustained at 50 morr in a reactor with an electrode of 3 cm. The high-frequency source (300 V/100 MHz) is modulated with on and off times of 20 and 150 µs, respectively. The flux and energy of negative ions is controlled by a low-frequency source (300 V/500 kHz). This source applies a square wave with rising and falling times of 50 ns. The resulting energy and angle distribution functions of positive ions (O_2^+), negative ions (O^-), and electrons are used as input parameters in the 3D charge-up simulation.

Time integrated ion fluxes and IEDFs on the biased electrode are shown in Fig. 3.6. Negative ions can be extracted during the afterglow in the off phase of the modulated high-frequency source. The extraction of negative ions after the high frequency is turned off is delayed due to the initial presence of electrons. Only after most electrons have been lost either to the wall or attached to oxygen atoms it is possible to extract negative ions from the plasma. Fluorine containing gases such as CF_4 and SF_6, with their strong electron attachment cross-sections, have shorter delays. During the off phase, the positive ion flux decreases due to the continuous loss of positive ions. The IEDF of O_2^+ during the off phase is different from that of the on phase (compare

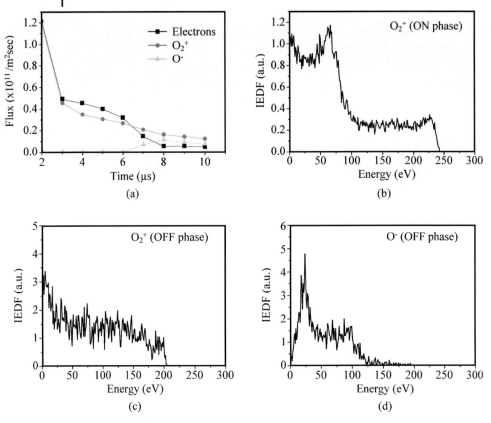

Fig. 3.6 (a) Particle fluxes as a function of time in a time
modulated CCP. Time integrated energy distributions of (b) O_2^+
during the on phase, (c) O_2^+ during the off phase and (d) O^-
during the off phase.

Fig. 3.6(b) and (c)) and has a lower maximum energy. The IEDF of negative O^- ions
during the off phase is shown in Figure 3.6(d). The average incidence angle of O_2^+
ions is ~1.5°. Negative O^- ions also display anisotropic motion with an average
incidence angle of ~17°. Electrons, on the other hand, arrive at the substrate with an
average angle of ~40°. Figure 3.7 shows the 3D potential profiles obtained when the
high frequency source is operated in a continuous-wave (CW) mode and when it is
time-modulated. The potential at the bottom of the trench reduces from 120 V to
95 V when the source is modulated. (Compare Fig. 3.7(a) and (b)) This reduction
due to the anisotropic flux of negative ions indicates that a time-modulated power
source can be used to reduce the charge-up damage. Although the etching rate is
affected by the reduction of ion flux, the extraction of negative ions in a time-
modulated plasma is an attractive means for reducing charge-up damages such as
notching, trenching, and bowing.

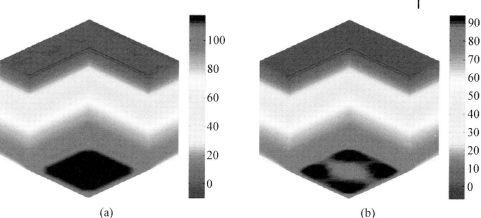

Fig. 3.7 Three-dimensional potential profiles (in V).
(a) Continuous-wave power source. (b) Time-modulated
power source.

3.3
Fluid Simulations

Although PIC-MCC simulations capture local and non-local particle kinetics in detail, they are very intensive computationally. For this reason, fluid simulations are often used because they are much faster. In this work we use a 2D full-momentum drift-diffusion (FMDC) fluid code. In this code, either the momentum conservation equation or the drift-diffusion approximation can be selected for determining the ion motion. For low-pressure high-frequency discharges like the ones presented here, the full momentum equation (Eq. 3.1) is used to account for the inertia of the ions:

$$\frac{\partial}{\partial t}(n_i \vec{v}_i) + \nabla \bullet (n_i \vec{v}_i \vec{v}_i) = \frac{Z_i}{M_i} n_i \vec{E} - \frac{\nabla(n_i k T)}{M_i} - n_i v_{in} \vec{v}_i \tag{3.1}$$

Here, n_i and v_i are the density and the average velocity of the ions, v_{in} is the ion–neutral momentum transfer collision frequency, and Z_i, k, M_i, and E are the ion charge, the Boltzmann constant, the ion mass, and the electric field, respectively. For the less massive electrons, the drift-diffusion approximation is used. Ions are assumed to be at room temperature ($T_i = 0.026$ eV) and therefore the energy equation is solved only for electrons. The code has options for changing the electrode structure and the power supply configuration (bottom–bottom or top–bottom).

In this study, we focus on low-pressure low-temperature plasmas for material processing. Plasma sources for large area processing, where plasma uniformity is an important issue, are simulated using the 2D fluid model. A comparison between the simulation results of a capacitively coupled plasma obtained with the 2D FMDC code and a 1D PIC-MCC code is also presented.

3.3.1
Capacitively Coupled Discharges

An asymmetric dual-frequency CCP reactor with an electrode gap of 2 cm is analyzed with the described 2D axisymmetric FMDC fluid code. A schematic of the reactor is shown in Fig. 3.8(a). Only the high-frequency power source (300 V/27 MHz) is used in this simulation. The source is connected to the top electrode which has a diameter of 16.8 cm. The low-frequency source connected to the bottom electrode (9.4 cm in diameter) is not powered. The bottom electrode self-biases negatively due to the different area of the electrodes. The reactor is assumed to have dielectric walls. 2D Ar plasma density contours are shown in Fig. 3.8(b). The maximum plasma density develops at the edge of the electrodes and not at the center of the discharge as one may have anticipated. This is attributed to the relatively high pressure of the discharge (100 morr) and the presence of an intense electric field between the electrodes and the walls.

A comparison between the 2D fluid simulation and a 1D PIC-MCC simulation is presented in Fig. 3.9. To account for the different electrode area, the 1D PIC-MCC simulation is performed in cylindrical coordinates (along r). Density profiles and power per unit volume absorbed by electrons and ions are shown in Fig. 3.9. Although the absorbed power profiles obtained with the 2D fluid simulation differ from those of the 1D PIC-MCC simulation, the axial density profiles are similar. The discrepancy in electron absorbed power is due to non-local electron kinetics that is captured in the PIC model but not in the fluid model.

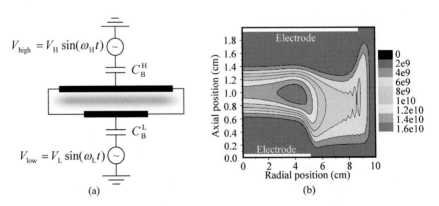

(a)   (b)

Fig. 3.8 (a) Schematic of the asymmetric single-frequency CCP reactor used in the 2D fluid simulation. The gap between the electrodes is 2 cm and the electrodes are 16.8 cm and 9.4 cm in diameter. (b) 2D plasma density contours (in cm^{-3}) for a 100 morr argon discharge at 300 V/27 MHz.

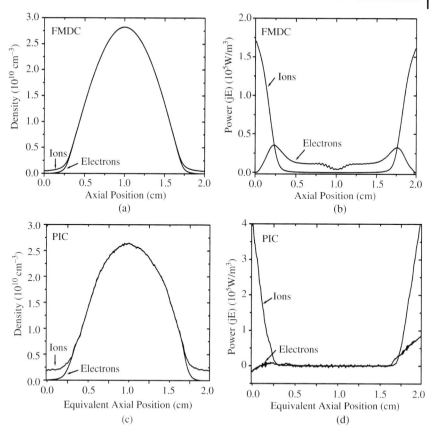

Fig. 3.9 (a) Average electron and ion density profiles and (b) power per unit volume absorbed by electrons and ions in the 2D fluid simulation. (c) Average density profiles and (d) power per unit volume in the 1D PIC-MCC simulation. Discharge conditions: argon 80 mtorr, 100 V/27 MHz.

3.3.2
Large Area Plasma Source

As the semiconductor and display industries move to larger wafers and glass substrates, larger plasma reactors are required. Inductively coupled sources where RF or microwave power is coupled to the plasma across a dielectric window have been used as high-density low-pressure plasma sources during the last decade [33,34]. External planar inductively coupled plasma (ICP) sources, however, present scale-up limitations due to the high impedance of a large antenna and the thickness of the required dielectric window [35].

These limitations have triggered the search for alternative plasma sources capable of generating large-area high-density plasmas without the scaling and uniformity problems associated with conventional ICP sources. A schematic of one such source

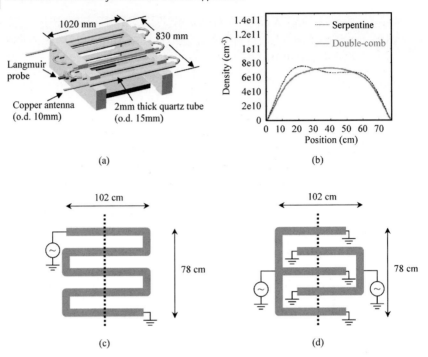

(a)

(b)

(c)

(d)

Fig. 3.10 (a) Schematic of a large area plasma source.
(b) Electron densities 8 cm below antenna for two antenna
configurations. Schematics of (c) the serpentine and (d) the
double-comb antennas. Hatched squares represent connections
to ground and filled squares connections to the power supply. The
dotted lines indicate the planes of simulation.

is shown in Fig. 3.10(a). A linear antenna is immersed in the plasma eliminating the
need for a large dielectric window. The antenna is made out of copper and has a
diameter of 10 mm. A 2 mm thick quartz tube (outer diameter 15 mm) isolates and
protects the antenna from the plasma. The antenna couples power to the plasma
inductively and a high-density large-area plasma can be generated. We simulated
two antenna configurations and studied the uniformity of this large area plasma
source (LAPS). The reactor, with dimensions of 1020 mm × 830 mm × 300 mm, is
intended for flat panel display processing [36,37]. The first antenna is serpentine
with five linear segments connected in series (Fig. 3.10(c)). The distance between
neighboring segments is 10.2 cm. The second antenna is a double-comb antenna
(Fig. 3.10(d)). In this configuration, the five linear segments are connected in parallel,
resulting in a shorter antenna length that reduces standing wave effects. In both cases
the discharge is sustained by a 13.56 MHz power source that delivers 600 W. The two
combs of the double-comb antenna are powered with the same in-phase signal.

The 2D fluid simulations are carried out on a plane perpendicular to the linear
segments of the antennas (Fig. 3.10(c) and (d)). The power coupled to the plasma and
the induced electric field are determined by solving a wave equation. Figure 3.10(b)
shows the density profiles across the reactor 8 cm below the antennas. The plasma

uniformity improves when a double-comb antenna is used. This antenna has a lower inductance and its shorter length reduces standing wave effects. The simulation results are in good agreement with experimental measurements [31].

3.4
Summary

The increasing credibility and usefulness of time- and space-resolved fluid and particle simulations of low-temperature plasmas are being recognized in industry and academia. Simulations provide a valuable means to study plasmas and reactors without the cost associated with building prototypes. Plasma modeling, however, is complex. Plasma has many physical aspects that need to be considered in a self-consistent analysis with significantly different spatial and temporal scales. Despite the difficulties, simulation is a very useful tool for understanding plasma physics and improving the performance of plasma equipment. PIC-MCC and fluid simulations are typically employed to simulate low-temperature plasmas. Although not discussed in this chapter, adequate numerical parameters such as cell size (Δx) and time step (Δt) should be considered for stability and accuracy of the simulations.

PIC-MCC simulations capture accurately the kinetic behavior of charged particles. Local and non-local kinetics can be studied by this simulation technique. PIC-MCC simulations, however, are computationally very intensive and thus are slow. For this reason, PIC-MCC codes are mainly employed to simulate low-pressure high-frequency plasmas where non-local effects must be considered. We have performed 1D PIC-MCC simulations of O_2/Ar plasmas in single- and dual-frequency CCPs and 3D simulations of the charge-up of high-aspect-ratio features in continuous and time modulated plasmas. In O_2/Ar plasmas the oxygen concentration affects the electron density and the EEDF. In the pressure range investigated (20–100 morr) the EEDF is bi-Maxwellian and effective temperatures below 1 eV can be generated in high oxygen content discharges. The three dimensional charge-up simulations show that secondary electron emissions and anisotropic negative ion bombardment enabled by a time modulated power source can reduce the charge-up damage.

Fluid simulations describe the plasma in terms of macroscopic quantities such as density, mean velocity, and mean energy. They are suitable for modeling high-pressure discharges where the EEDF can be estimated from local conditions and non-local effects can be neglected. Fluid simulations are faster than PIC-MCC simulations. This faster performance allows 2D and 3D analyses with complex chemistry that are prohibitively expensive for PIC-MCC simulations. To account for ion inertia in low-pressure and/or high-frequency plasmas, the full ion momentum equation must be considered. Two-dimensional simulations of an asymmetric CCP show that the density profile peaks at the edge of the electrodes instead of at the center of the chamber. This profile is attributed to the relative high pressure of the discharge and the intense electric field generated between the electrodes and the dielectric walls. The axial density profile in a 2D fluid simulation is in good agreement with the density profile obtained in a 1D PIC-MCC simulation. Due to their faster computational speed, fluid codes are suitable for simulating large area plasma sources (LAPS), where plasma uniformity is an important issue. LAPS simulations show that a double-comb antenna improves the plasma uniformity due to its lower impedance and the reduction of standing wave effects.

Acknowledgments

The authors would like to thank Dr. O. Manuilenko, Dr. S.S. Yang, Prof. M.J. Kushner, Prof. F.F. Chen, Prof. M.A. Lieberman and Prof. J.P. Verboncoeur for helpful comments and discussions. This work was supported in part by Samsung Electronics, Jusung Engineering and the Korean Science and Engineering Foundation.

References

1 Kim, H.C., Iza, F., Yang, S.S., Radmilovic-Radjenovic, M. and Lee, J.K. (2005) *J. Phys. D: Appl. Phys.*, **38**, R283.

2 Kim, H.C. and Lee, J.K. (2004) *Phys. Rev. Lett.*, **93**, 085003.

3 Kim, H.C., Lee, J.K. and Shon, J.W. (2004) *Appl. Phys. Lett.*, **84**, 864.

4 Lee, J.K., Babaeva, N.Yu, Kim, H.C., Manuilenko, O. and Shon, J.W. (2004) *IEEE Trans. Plasma Sci.*, **32**, 47.

5 Birdsall, C.K. (1991) *IEEE Trans. Plasma Sci.*, **19**, 65.

6 Vahedi, V. and Surendra, M. (1995) *Comput. Phys. Commun.*, **87**, 179.

7 Birdsall, C.K. and Langdon, A.B. (1991) *Plasma Physics via Computer Simulation*, IOP Publishing, Bristol.

8 Hockney, R.W. and Eastwood, J.W. (1988) *Computer Simulation Using Particle*, Adam Hilger, Bristol.

9 Tajima, T. (1988) *Computational Plasma Physics*, Addison-Wesley, Redwood City.

10 Verboncoeur, J.P., Alves, M.V., Vahedi, V. and Birdsall, C.K. (1993) *J. Comput. Phys.*, **104**, 321.

11 Makabe, T. (2002) *Advances in Low Temperature RF Plasmas: Basis for Process Design*, Elsevier.

12 Lieberman, M.A. and Lichtenberg, A.J. (2005) *Principles of Plasma Discharges and Material Processing*, Wiley, New York.

13 Chen, F.F. and Chang, J.P. (2003) *Lecture Notes on Principles of Plasma Processing*, Kluwer/Plenum, New York.

14 Goto, H.H., Lowe, H.-D. and Ohmi, T. (1992) *J. Vac. Sci. Technol. A*, **10**, 3048.

15 Kim, H.C., Lee, J.K. and Shon, J.W. (2003) *Phys. Plasmas*, **10**, 4545.

16 Gudmundsson, J.T., Kouznetsov, I.G., Patel, K.K. and Lieberman, M.A. (2001) *J. Phys. D: Appl. Phys.*, **34**, 1100.

17 Taylor, K.J. and Tynan, G.R. (2005) *J. Vac. Sci. Technol. A*, **23**, 643.

18 Kawamura, E., Vahedi, V., Lieberman, M.A. and Birdsall, C.K. (1999) *Plasma Sources Sci. Technol.* **8**, R45.

19 Babaeva, N.Y., Shon, J.W., Hudson, E.A. and Lee, J.K. (2005) *J. Vac. Sci. Technol. A*, **24**, 699.

20 Lee, J.K., Manuilenko, O.V., Babaeva, N.Y., Kim, H.C. and Shon, J.W. (2005) *Plasma Sources Sci. Technol.*, **14**, 89.

21 Gottscho, R.A., Jurgensen, C.W. and Vitkavage, D.J. (1992) *J. Vac. Sci. Technol. B*, **10**, 2133.

22 Hashimoto, K. (1994) *Jpn. J. Appl. Phys.*, **33**, 6013.

23 Kinoshita, T., Hane, M. and Mcvittie, J.P. (1996) *J. Vac. Sci. Technol. B*, **14**, 560.

24 Matsui, J., Nakano, N., Petrovic, Z.L. and Makabe, T. (2001) *Appl. Phys. Lett.*, **78**, 883.

25 Hwang, G.S. and Giapis, K.P. (1999) *Appl. Phys. Lett.*, **74**, 932.

26 Park, H.S., Kim, S.J., Wu, Y.Q. and Lee, J.K. (2003) *IEEE Trans. Plasma Sci.*, **31**, 703.

27 Kim, S.J., Lee, H.J., Yeom, G.Y. and Lee, J.K. (2004) *Jpn J. Appl. Phys.*, **43**, 7261.

28 Phelps, A.V. and Petrović, Z.L. (1999) *Plasma Sources Sci. Technol.*, **8**, R21.

29 Shibayama, T., Shindo, H. and Horiike, Y. (1996) *Plasma Sources Sci. Technol.*, **5**, 254.

30 Midha, V. and Economou, D.J. (2000) *Plasma Sources Sci. Technol.*, **9**, 256.

31 Kanakasabapathy, S.K., Khater, M.H. and Overzet, L.J. (2001) *Appl. Phys. Lett.*, **79**, 1769.

32 Dai, Z.L. and Wang, Y.N. (2002) *J. Appl. Phys.*, **92**, 6428.

33 Kushner, M.J. (2003) *J. Appl. Phys.*, **94**, 1436.

34 Hoekstra, R.J. and Kushner, M.J. (1996) *J. Appl. Phys.*, **79**, 2275.

35 Wu, Y. and Lieberman, M.A. (2000) *Plasma Source Sci. Technol.*, **9**, 210.

36 Park, S.E., Cho, B.U., Lee, J.K., Lee, Y.J. and Yeom, G.Y. (2003) *IEEE Trans. Plasma Sci.*, **31**, 628.

37 Kim, K.N., Lee, Y.J., Jung, S.J. and Yeom, G.Y. (2004) *Jpn J. Appl. Phys.*, **43**, 4373.

4
Modeling and Diagnostics of He Discharges for Treatment of Polymers

E. Amanatides and D. Mataras

An exhaustive study of the effect of the substrate bias voltage on polyethylene terephthalate (PET) film treatment from He discharges was performed by applying a combination of different experimental techniques together with a two-dimensional (2D) self-consistent modeling of He plasmas. The excellent agreement between model and experiment allowed the investigation of the main mechanisms that lead to PET surface modification. Ion bombardment was favored at conditions were the substrate was negatively biased and resulted in much more intense surface treatment. In these conditions the enhanced production of species coming from the decomposition of the surface was followed by quenching of the He metastables. Ion bombardment was identified as the main mechanism of PET surface modification and under certain conditions can fully account for the observed etch rate.

4.1
Introduction

In the last few decades, the industrial applications of polymeric materials have increased rapidly due to their unique physical, chemical and mechanical properties [1–3]. Automotive, aerospace, microelectronics, biomedical and food industries are a few examples of industries where polymers have found technological applications [4–6]. However, certain modern applications of polymeric materials and especially of low-cost polymers like PET require an improvement of their surface properties like wettability, printability and biocompatibility [7,8]. Among several techniques that have been proposed for the improvement of these properties, plasma processing presents some major advantages: i.e. it is a dry, clean and very fast process with a very low consumption of chemicals and energy, while it modifies only the surface properties without affecting the bulk material [9,10].

Several gas discharges, such as He, Ar, O_2, N_2, NH_3 and H_2O and mixtures of them, have been applied for the surface modification of different polymers [11,14].

Advanced Plasma Technology. Edited by Riccardo d'Agostino, Pietro Favia, Yoshinobu Kawai, Hideo Ikegami, Noriyoshi Sato, and Farzaneh Arefi-Khonsari
Copyright © 2008 WILEY-VCH Verlag GmbH & Co. KGaA, Weinheim
ISBN: 978-3-527-40591-6

The application of reactive gases (O_2, N_2, NH_3 and H_2O) results mainly in grafting of new functional groups on the treated surface, a process known as functionalization [15].

The use of inert gas discharges (He, Ar) results in chain scission of the outermost layer of the polymer followed by cross-linking, a process known as CASING (cross-linking via inert gas activated species) [16]. This process has the advantage of simplicity while in some cases it can be as effective as reactive gas discharges in terms of wettability improvement. As a consequence, there are an increasing number of studies dealing with the interaction of RF capacitively coupled He plasmas with various polymeric materials (PET, polycarbonate, poly(methyl methacrylate)) [17–19]. However, these investigations are mostly focused on the output of the process and only a small part of the existing literature refers to the gas-phase mechanisms and the nature of species that play a major role in the surface modification [20,21].

Thus, the present chapter aims at the investigation of the gas-phase mechanisms for producing species that finally lead to the modification of the surface. For this purpose, two different approaches were followed. In the first one, the process was investigated by applying a series of plasma diagnostics as RF power and impedance measurements, spatially resolved emission spectroscopy, mass spectrometry and *in situ* etch rate measurements via laser reflectance interferometry. This work was complemented by the development of a 2D self-consistent simulation of He discharges that modeled the variation of the microscopic plasma parameters due to the presence of these polymeric surfaces, during their treatment.

In both approaches the effect of the substrate bias voltage (-30 to $+30$ V) on the power actually dissipated in the discharge, the production of excited He^* species, the formation of byproducts of the PET treatment and the etch rate was recorded systematically. A comparison between the model and the experimental results was then performed and an excellent agreement was found for the power dissipation, the discharge current and the spatial distribution of excited species in the discharge. This agreement confirms the model applicability at least in this range of conditions that permitted the use of the simulation for the prediction of parameters that are not easily accessible experimentally and to identify the mechanism governing the PET treatment.

4.2
Experimental

The experimental setup is built around a fully characterized cell from the electrical point of view. It consists of a 160 mm stainless steel high-vacuum chamber having two parallel round stainless steel electrodes with a diameter of 55 mm, equipped with four 50 mm diameter quartz observation windows [22]. The interelectrode distance is kept constant at 25 mm throughout these experiments. A layer of PET film with a thickness of 23 μm is mounted on the surface of the grounded electrode. A 13.56 MHz generator is used to supply power to the system through an SWR meter and a proper impedance matching network.

A base vacuum of 10^{-6} torr is achieved by a diffusion pump before the introduction of pure helium. The gas flow is controlled using mass flow controllers while the desired pressure is independently adjusted by a downstream throttle valve using the feedback of a capacitive manometer.

The amount of RF power actually fed into the discharge chamber is determined using an accurate method employing Fourier transform voltage and current wave-form measurements (FTVCWM). Namely, the RF voltage and the discharge current waveforms are measured on the powered electrode lead, using a high-impedance 1 : 100 attenuation voltage probe and a 0.1 Ω transfer impedance RF current probe, and then processed as described in Ref. [23].

Spatially resolved optical emission profiles (OES) were obtained by moving the chamber and then recording the intensity value at a certain position of the inter-electrode space [24], with a resolution of 0.5 mm. The light emitted from the discharge was collected using a 10 cm focal length lens to the entrance slit of a monochromator equipped with a photomultiplier tube.

In addition to the measurements, a quadrupole mass spectrometer was used to analyze the gas, after 70 eV electron impact ionization [25], in a molecular mass range from 1 to 50 Da.

Finally, laser reflectance interferometry (LRI) was employed for the *in situ* etch rate measurement. For this, a green, solid-state diode laser emitting at 532 nm was directed upon the polymer film and the reflected beam was collected with a suitable lens and recorded with a photodiode. The method is ideal for monitoring the variation of surface thickness since it is practiced during the treatment, without disturbing the discharge without exposing the sample to the atmosphere.

4.3
Model Description

The 2D self-consistent model has been described in detail in Refs. [26,27]. Briefly, the model uses the particle, momentum and energy balances obtained from moments of the Boltzmann transport equation, coupled with Poisson's equations for a self-consistent calculation of the electric field. The particle balance for electrons, ions and neutrals is described by the continuity equation

$$\frac{\partial n_j}{\partial t} + \nabla \cdot \vec{\Gamma}_j = S \tag{4.1}$$

where n_j is the density, $\vec{\Gamma}_j$ is the flux of particle j and S is the source of particles consumed or produced in chemical reactions.

The drift diffusion approximation replaces the momentum balance for the charged species:

$$\vec{\Gamma}_j = -D_j \nabla n_j + \mu_j n_j \nabla V \tag{4.2}$$

where μ_j is the charged species mobility, D_j is the diffusion coefficient and V is the electrostatic potential.

The energy balance is solved only for electrons, assuming that ions have equal energy with neutrals. The electron temperature T_e is derived from the electron energy balance

$$\frac{3}{2}\frac{\partial}{\partial t}(n_e T_e) + \nabla \cdot \left(\frac{5}{2}T_e\vec{\Gamma}_e - \frac{5}{2}n_e D_e \cdot \nabla T_e\right) = P - n_e \sum_i N_i K_i \qquad (4.3)$$

where the energy transfer is due to convective flux and thermodiffusion (e, i, p and o subscripts are used for electron, ion, neutral particle and for the normal direction respectively), P is the pressure gradient and the last term on the right-hand side is the rate of electron production/consumption by electron impact collisions.

For a self-consistent calculation of the electric field/potential, Poisson's equation is solved simultaneously with the fluid equations:

$$\nabla^2 V = -\frac{e}{\varepsilon_0}\left(\sum_{i=\text{ions}} q_i n_i - n_e\right); \quad E = -\nabla V \qquad (4.4)$$

where ε_0 is the permittivity of free space, q_i is the sign of the charge of ion i and E is the electric field.

The system of equations (4.1)–(4.4) is complemented by a set of boundary conditions for the densities of the species and the electric potential. The voltage applied at the driven and the grounded electrodes is defined respectively as

$$V_{\text{RF}} = V_0 \sin 2\pi F; \quad V_{\text{sub}} = V_{\text{bias}}, \quad V_{\text{sub}} = V_{\text{bias}} \qquad (4.5)$$

The electron flux normal to the walls (assuming no reflection or secondary emission) is given by

$$\vec{\Gamma}_{e,n} = \frac{1}{4}n_e v_{e,\text{th}}; \quad v_{e,\text{th}} = \left[\frac{8 K_B T_e}{\pi m_e}\right]^{1/2} \qquad (4.6)$$

where $v_{e,\text{th}}$ is the electron thermal velocity, k_B is the Boltzmann constant and m_e is electron mass.

The velocity of ions is determined by the local value of the electric field and their flux to the walls is supposed to be pure drift when the velocity is directed to the walls,

$$\vec{\Gamma}_{i,n} = q_i \mu_i E_n n_i \qquad (4.7)$$

and zero otherwise. At dielectric surfaces, the net surface charge σ_s is obtained from the particle fluxes:

$$\frac{\partial \sigma_s}{\partial t} = e\left(\sum_i q_i \vec{\Gamma}_{i,n} - \vec{\Gamma}_{e,n}\right); \quad \varepsilon_0 E_n = \sigma_s \qquad (4.8)$$

The fluxes of gas-phase species at the surface are modified through surface reactions and their flux balance at the surface, yielding

$$n \cdot \vec{\Gamma}_p = m_p S_p \tag{4.9}$$

where n is a unit vector normal to the surface and S_p is the surface production (consumption rate per unit area).

The PET etch rate in Å s^{-1} was finally calculated using the relation

$$R = 10^8 \frac{m}{N_A \rho} \left(\sum_n s_n D_n \frac{\partial n_n}{\partial x} + \sum_i s_i n_i u_i \right) \tag{4.10}$$

where m is the molar mass and ρ the mass density of PET and N_A is the Avogadro number. The term in parentheses is the sum of the neutral species and ions flux that cause PET etching, while s_n and s_i are the etching probabilities of a neutral or an ion respectively.

The gas-phase chemistry module of the simulation takes into account 16 reactions (electron–molecule, de-excitation, quenching) that are summarized in Table 4.1, while the plasma–PET interaction is described by 14 reactions that are given in Table 4.2. Besides He chemistry, reactions (R4–R10) between electrons and species that are produced from the decomposition of PET (H, CO, CO_2, C_2H_4) were also included. Collision cross-sections were used for the calculation of electron–molecule collision rates. In addition, quenching of He metastables from the products of the PET treatment was taken into account (R11–R14) and the quench rate was estimated by

Tab. 4.1 Electron Impact Ionization, Momentum Transfer and Quenching Included in the Gas-Phase Model.

R	Process	Rate constant[a]	Reference
1	$e^- + He \rightarrow e^- + He$	Cross-section	[28]
2	$e^- + He \rightarrow e^- + He^* \, (^3S - {}^3P^o)$	Cross-section	[28]
3	$e^- + He \rightarrow 2e^- + He^+$	Cross-section	[28]
4	$e^- + CO_2 \rightarrow 2e^- + CO_2^+$	Cross-section	[28]
5	$e^- + CO_2 \rightarrow e^- + CO + O$	Cross-section	[28]
6	$e^- + CO_2 \rightarrow e^- + CO_2$	Cross-section	[28]
7	$e^- + CO \rightarrow e^- + CO^+$	Cross-section	[28]
8	$e^- + C_2H_4 \rightarrow e^- + C_2H_4$	Cross-section	[28]
9	$e^- + C_2H_4O \rightarrow e^- + C_2H_4O$	Cross-section	[28]
10	$e^- + H \rightarrow e^- + H^* \, (n=4 \rightarrow 2)$	Cross-section	[28]
11	$He^* + CO_2 \rightarrow He + CO_2$	1.8×10^{-11}	[29]
12	$He^* + CO \rightarrow He + CO$	1.14×10^{-11}	[29]
13	$He^* + C_2H_4 \rightarrow He + C_2H_4$	1.14×10^{-11}	[29]
14	$He^* + C_2H_4O \rightarrow He + C_2H_4$	1.8×10^{-11}	[29]
15	$He^* \rightarrow He$	1.5×10^7	[29]
16	$H^* \rightarrow H$	2×10^8	[30]

[a]Rate constants of reactions 15 and 16 have units of s^{-1}.

Tab. 4.2 Frequencies ν and Activation Energies E_a for Reactions Between Surface Sites and Probabilities of Radical–Surface Reactions Included in the Surface Model.

R	Surface reactions	ν (s^{-1}); E_a (J)	Ref.
1	$2C_{10}H_7O_4 \rightarrow C_{20}H_{14}O_8$	1.4×10^{13}; 0.11	[31]
2	$2C_7H_4O_2 \rightarrow C_{14}H_8O_4$	1.3×10^{14}; 0.06	[31]
3	$2C_7H_4O \rightarrow C_{14}H_8O_2$	1.6×0^{13}; 0.18	[31]
4	$C_{10}H_7O_4 + C_7H_4O_2 \rightarrow C_{17}H_{11}O_6$	3.0×10^{13}; 0.89	[31]
5	$C_7H_4O_2 + C_7H_4O \rightarrow C_{14}H_8O_3$	1.2×10^{12}; 0.94	[31]
6	$C_{10}H_7O_4 + C_7H_4O \rightarrow C_{17}H_{11}O_5$	8.5×10^{11}; 0.30	[31]

R	Surface-species reactions	Probabilities	
7	$He^+ + C_{10}H_8O_4 \rightarrow HE + C_2H_4 + CO_2 + C_7H_4O_2$	$P_1 = 0.06$	Est.
8	$He^* + C_{10}H_8O_4 \rightarrow HE + C_2H_4 + CO_2 + C_7H_4O_2$	$P_2 = 0.06$	Est.
9	$He^+ + C_{10}H_8O_4 \rightarrow HE + H + C_{10}H_7O_4$	$P_3 = 0.04$	Est.
10	$He^* + C_{10}H_8O_4 \rightarrow HE + H + C_{10}H_7O_4$	$P_4 = 0.04$	Est.
11	$He^+ + C_{10}H_8O_4 \rightarrow HE + C_2H_4O + CO_2 + C_7H_4O$	$P_5 = 0.05$	Est.
12	$He^* + C_{10}H_8O_4 \rightarrow HE + C_2H_4O + CO_2 + C_7H_4O$	$P_6 = 0.05$	Est.
13	$He^+ \rightarrow He$	$P_7 = 0.6$	Est.
14	$He^* \rightarrow He$	$P_8 = 0.6$	Est.

assuming that 10% of He*–molecule collisions are effective. A parametric study on the validity of this assumption was performed and the results are presented in Section 4.4.2.

Moreover, on the PET surface, sputtering as well as interaction of the He metastables with the PET monomer was taken in account. The result of these interactions was considered to be a bond break in the PET chain (C–C, C–H, C–O) that in turn produces a surface site ($C_7H_4O_2$, $C_{10}H_7O_4$, C_7H_4O). Furthermore, these sites react to produce a surface with a chemical composition ($-C_{20}H_{14}O_8-$, $-C_{17}H_{11}O_6-$, $-C_{14}H_8O_3-$, $C_{17}H_{11}O_5$) that is different from that of PET. The frequency and the activation energy of most of these surface reactions were taken from Ref. [31], while in some cases where data were not available (R2 and R5) frequency and activation energy of similar reactions was used instead. Concerning the probability of an ion or an excited He atom to break a bond, it was initially estimated that 15% of these species can contribute to the surface treatment. The probability of these species to break a C–C, C–H or C–O bond was then adjusted according to the bond energy. Finally, for the interaction of He discharges with the stainless steel parts of the reactor the last two reactions (R13, R14) were considered.

4.4

Results and Discussion

The above mentioned plasma diagnostic techniques and the discharge simulator were applied for the investigation of the effect of substrate bias voltage on the PET treatment in pure He discharges. The conditions were: pressure 500 mtorr, applied peak to peak voltage 300 V, He gas flow 20 sccm and interelectrode space 25 mm. The substrate DC bias was varied from −30 to +30 V with a step of 10 V. The effect of the variation of the bias potential on the electrical properties of the discharge, the gas-phase chemistry and the PET etch rate and treatment were monitored/simulated and the results are presented and discussed in the following sections.

4.4.1
Electrical Properties

The first step of the investigation was to study the effect of the bias potential on the power dissipation in the discharge while keeping a constant RF voltage. Figure 4.1(a) presents a map of the average power dissipation in the reactor for the case of −30 V bias, as predicted by the model. The simulated area represents a part of the reactor, where the experiments were performed and includes the powered electrode, a dielectric (Teflon) part that is used for RF shielding, the PET thin film and a stainless steel ring (SS ring) that is used to hold the polymer on the electrode. The map of the power consumption clearly depicts the well known "edge effects" [32] due to the shield of the RF electrode, resulting in much higher power dissipation in the area

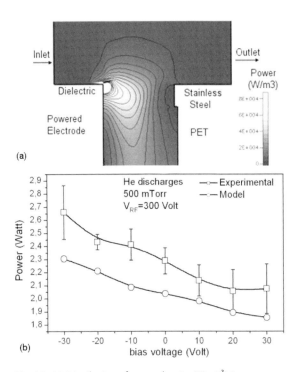

Fig. 4.1 (a) Distribution of power density (W m^{-3}) in 0.5 torr He discharges and −30 V substrate bias potential. (b) Measured and calculated total power consumed in the discharge as a function of the substrate bias voltage.

close to the powered electrode and the dielectric material interface. In addition, a significant amount of power is consumed in the discharge closer to the powered electrode, indicating that sheath ohmic heating is the main mechanism through which the electrons gain energy. In contrast, the power dissipated close to the surfaces (PET, SS ring) is very low, revealing that the power spent for the acceleration of ions is almost negligible compared to the electron power despite the application of the substrate bias. It is also worth noting that the variation of the bias potential does not affect the distribution of the power dissipation in the discharge.

However, the total power consumed in the discharge as measured from FTVCWM (Fig. 4.2(b)) drops as we go from negative to positive bias potential. Actually, the model underestimates slightly the power consumption but reproduces very well the power decrease with the variation of the substrate bias.

In order to interpret the change of the RF power with the bias voltage, the variation of the discharge current at the same conditions was studied. Figure 4.2(a) shows (left axis) model results of the variation of the displacement, electron, ion and total discharge current during the RF cycle, in the RF electrode and for the case of $-30\,V$ substrate bias. The right-hand axis includes the temporal variation of the applied RF voltage, which is of the form $V = -V_{dc} + V_{RF} \sin\omega t$. One can observe that the

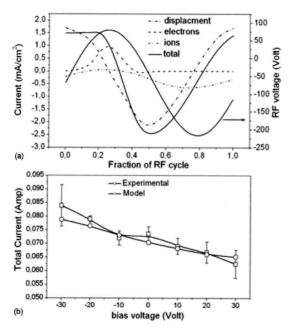

Fig. 4.2 (a) Displacement, conduction and total current (left axis) and applied RF voltage (right axis) during the RF cycle in 0.5 torr He discharges, $-30\,V$ substrate bias voltage. (b) Measured and calculated total power consumed in the discharge as a function of the substrate bias voltage.

displacement current dominates by far conduction currents due to the capacitive nature of the sheath and appears at phase shift of $-\pi/2$ relative to the applied voltage. The electron current presents a sharp peak at $\pi/2$ of the RF cycle where the applied voltage has the most positive value and remains almost zero for the larger part of the RF period. The ion current does not have a pure DC behavior due to the high mobility of the light He ions and varies with the RF cycle. In fact, ions are attracted by the RF electrode during the larger part of the RF cycle and the ion current is maximized at $3\pi/2$ where the applied voltage has the most negative value. Ions are pushed away from the electrode for only a small portion of the RF period around $\pi/2$. This behavior is rather interesting as it directly affects the ion bombardment of the PET surface and will be discussed later in detail.

In addition, the sum of the displacement and the conduction currents results in a rather anharmonic total discharge current with a phase shift of $-82°$ relative to the applied voltage. This value is very close to the experimentally determined discharge phase impedance, which is $-78°$ at the specific conditions, while the anharmonicity of the discharge current was also confirmed by the current measurements on the RF electrode.

Furthermore, besides the good agreement between model and experiment concerning the phase impedance and the current anharmonicity, there is also an excellent agreement in the measured and the calculated values of the total current amplitude (Fig. 4.2(b)). In both cases, the amplitude was calculated by applying Fast Fourier Transform on the current waveforms. The effect of bias on the discharge current is similar to the effect on the power and the application of positive substrate bias result in a drop of the discharge current. The excellent agreement between model and experiment concerning the discharge current and at the same time the underestimation of the total power dissipation by the model implies that the discrepancy may come from either the small difference of the discharge phase impedance or from the electric field calculation and consequently from the voltage distribution in the discharge.

Thus, Fig. 4.3(a) presents the spatial variation of the potential for the different values of substrate bias, resulting by averaging the calculated values of voltage in the y-direction in time. It can be clearly observed that by changing the substrate bias from negative to positive the voltage become less negative at the powered electrode and more positive in the bulk. This means that the application of a bias voltage from -30 to 30 V lifts the potential at each and every point of the discharge. However, this increase is not in one-to-one relation with the change of the substrate bias. This is more clearly illustrated in Fig. 4.3(b) where both the calculated and the experimental values of the self-bias potential (V_{dc}, left axis) and the peak value of the plasma potential (V_p, right axis) are plotted. The calculated values of V_{dc} become less negative by 45 V while the plasma potential increases by 45 V as the substrate bias changes by 60 V. In addition if one considers the experimental measurements of V_{dc} and V_p the change of the potential is even slower (25 V) compared to the bias potential. This deviation from a one to one relation is the result of the change in the relation between electron and ion densities and consequently the electron and ion loss rates towards the powered electrode, the substrate and the reactor walls. Electron losses are mainly determined by their high thermal velocities and their diffusion towards surfaces.

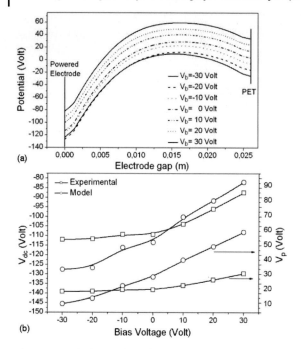

Fig. 4.3 (a) Calculated values of the distribution of the voltage in the space between the two electrodes at different values of the substrate bias voltage. (b) Measured and calculated values of the self-bias potential (left axis) and plasma potential (right axis) as a function of the substrate bias potential.

Thus, the loss rate of electrons will not be seriously affected by the changes in the substrate bias potential. On the other hand, the ion loss rate, which is exclusively determined by the drift in the sheath field will be reduced due to the application of the positive bias. In consequence, ion density will be enhanced relative to electron density at such conditions, thus altering the one-to-one relation between the change in V_{dc} and V_p with the variation of the substrate bias.

The slower response of V_{dc} and V_p to the changes of the substrate bias denotes an increase of the electric field in both the powered electrode and the substrate holder in the case of negative biasing. This enhancement is also sketched in Fig. 4.4(a) and (b) that present the spatial distribution of the average electric field in the y-direction at two different moments of the RF cycle, $\pi/2$ and $3\pi/2$, respectively. In both cases, the electric field is higher for the -30 V bias, while the lowest values are observed for the 30 V biasing. It is also remarkable that for $\pi/2$ the electric field in the powered electrode is positive, which means that electrons are attracted from the electrode during this portion of the RF cycle. The same is true for the substrate holder at the $3\pi/2$ case and this is attributed to the high mobility of He ions, leading to rather high loss rates to the surfaces and consequently imposes an increase of the electron flux in

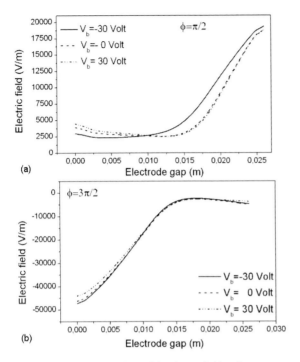

Fig. 4.4 (a) Calculated values of the electric field in the space between the two electrodes at $\varphi = \pi/2$ at different values of the substrate bias voltage. (b) Calculated values of the electric field in the space between the two electrodes at $\varphi = 3\pi/2$ at different values of the substrate bias voltage.

order to compensate for the positive ion losses [33]. This necessity is higher for the case of positive bias, which is in agreement with the above discussion concerning the relative change of V_{dc} and V_p with bias voltage.

The change of the electric field intensity with bias voltage is expected to affect also the electron energy acquired through sheath ohmic heating. Electron–molecule collision processes as ionization and consequently the plasma density will be affected by these changes. Thus, Fig. 4.5(a) presents the distribution of electron density in the discharge for the case of $-30\,$V substrate bias. The electron density due to the "edge effects" is maximized outside the plasma volume and in the middle of the two electrodes. In addition, Fig. 4.5(b) shows that the change of the bias potential from -30 to $30\,$V result in a drop of the space- and time-averaged electron density, which is in agreement with the decrease of the discharge current that was presented above. The drop of the electron density with positive bias is related to the drop of the electric field and the ionization rate; however, the production of charged species through the R4 and R5 reactions (Table 4.1) may also play a role.

The decrease of plasma density and electric field intensity as the bias voltage changes from -30 to $30\,$V implies a drop of the charged species flux towards the surfaces. This is shown in Fig. 4.6(a) (left axis), where the ion current towards the PET

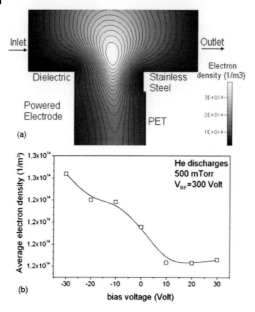

Fig. 4.5 (a) Distribution of electron density (m^{-3}) in 0.5 torr He discharges and $-30\,V$ substrate bias voltage. (b) Calculated values of the average electron density as a function of the substrate bias voltage.

surface is plotted. The right axis of the same figure presents the variation of the ion bombardment on the PET surface as calculated from the product of the electric field with the ion current. Ion bombardment drops with the change from negative to positive bias, while it also has an important variation during the RF cycle. Thus, Fig. 4.6(b) presents the variation of ion bombardment per surface site during the RF cycle, for -30, 0 and $+30\,V$ bias voltage. Ion bombardment is maximized at about $\pi/2$ of the RF cycle and is almost zero for about a half of the RF cycle between π and 2π. The energy that is transferred to a surface site from ions is higher at $-30\,V$ while the fraction of the RF cycle when the surface is bombarded is also larger.

One must note that for the case of the positive bias voltage, the energy that is transferred to the surface is rather low, slightly exceeding $5\,eV\,s^{-1}$ at $\pi/2$. If one considers that the lowest energy required for a bond break of PET is 3.5 eV, it is quite clear that ion bombardment will be less effective in the case of positive bias.

4.4.2
Gas-Phase Chemistry

The changes in power, electric field and plasma density that were predicted from both experimental measurements and modeling will have a definitive effect on the rate of all electron–molecule reactions. In fact, it has already been shown in the previous section that the ionization rate drops as the bias voltage changes from negative to

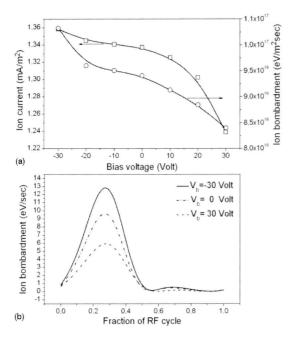

Fig. 4.6 (a) Calculated value of ion current towards PET surface (left axis) and ion bombardment as a function of the substrate bias voltage. (b) Calculated values of the ion bombardment per surface site during the RF cycle for different values of the substrate bias voltage.

positive values and this is followed by a drop of the ion flux and bombardment of the PET surface.

Another important process during PET treatment in He plasmas is the production of He^* metastables, which was suggested as being the main reason of PET surface modification [34]. In our case, the variation of the He^* metastables density with the substrate bias was followed by applying spatially resolved optical emission spectroscopy. The line at 388.9 nm, corresponding to the $^3S - {}^3P^o$ transition having a lifetime of the order of microseconds, was recorded. In the simulation, the total production of He metastables was considered by using the cross-section for total excitation (R2) while the de-excitation rate and quenching of these molecules were also taken into account (R11–R15). Figure 4.7(a) presents a typical 2D map of the distribution of He metastables that was predicted by the model for the case of -30 V bias voltage. As in the case of electron density (Fig. 4.5(a)), the He metastable density is maximized outside the plasma volume and closer to the RF electrode. The concentration of these species is much lower close to all surfaces as both stainless steel and PET surfaces consume these species (Table 4.2; R8, R10, R12, R14). Figure 4.7(b) shows the change of the total He^* density as calculated by the model (left axis) and as measured with optical emission spectroscopy with varying substrate bias (right axis). Unlike the plasma density, He metastable density increases as the substrate voltage is changed

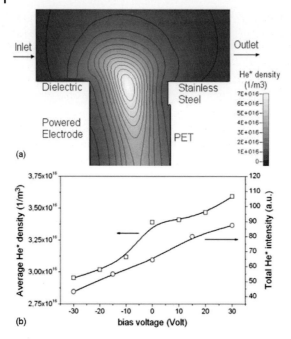

Fig. 4.7 (a) Distribution of He metastable density (m^{-3}) in 0.5 torr He discharges and -30 V substrate bias voltage. (b) Calculated values of the average He metastable density (left axis) and measured values of the total He* emission intensity (right axis) as a function of the substrate bias voltage.

from negative to positive values. This increase is clearly due to the decrease of the quench rate (R11–R14) in the case of positive bias, as the excitation rate (R2) decreases and the de-excitation rate is not affected by the substrate bias. It has to be mentioned that for the calculation of the quench rate, it was considered that 10% of the collisions of He metastables with the byproducts of PET treatment finally lead to quenching.

In order to have a better estimation of the way that this assumption affects the results, a parametric study of the rate constant of R11–R14 was performed. The rate constant of these reactions was changed from zero (no quenching) up to $\sim 9 \times 10^{-11}$ m^3 s^{-1} (80% of collisions lead to quenching) and the model outputs concerning the spatial distribution of He metastable de-excitation were compared to the experimental measurements. Figure 4.8(a) summarizes these results for conditions of -30 V bias. The measured spatial profile of He metastables presents a maximum that is located closer to the RF electrode (0.7 cm), while there is a sharp drop of the emission intensity close to both the powered electrode and the PET surface. On the other hand, the spatial profile that is predicted from the simulation if no quenching is included is of triangular shape with a maximum at 1.5 cm and a rather smooth drop of He* density close to the surface. The assumption that 10% of collisions of He* molecules with byproducts of the PET treatment lead to quenching of the metastables is followed by a displacement of the profile towards the RF electrode (maximum at 1.2 cm); however, the shape of the profile is not seriously affected. Finally, further increase of the quenching rate (80% of collisions) results in a spatial profile that reproduces fairly well the experimental measurements. The maximum is located at 0.7 cm and there is a good agreement of the drop of the de-excitation rate from the maximum to the PET surface. However, even at these conditions the model fails to reproduce the sharp drop of the de-excitation rate close to the surface and this is attributed to the moderate values of the probability of He metastables to interact with either stainless steel or PET surfaces adopted (0.6 and 0.15, respectively). Whatever the case, it is clearly proved that quenching of He* metastables governs their spatial distribution in the discharge and consequently their density. This is consistent with previous results of this group, where the variation of He metastable with the presence or not of PET thin films in the discharge was presented [35].

Moreover, quenching will also affect the metastable flux, and the energy transfer towards the PET surface as presented in Fig. 4.8(b) as a function of the substrate bias

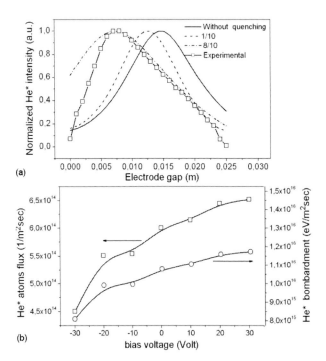

(a)

(b)

Fig. 4.8 (a) Spatial distribution of normalized measured and calculated values of He* density in 0.5 torr He discharges and -30 V substrate bias voltage and different values of quench rate. (b) He metastable flux (left axis) and He metastable bombardment to the PET surface as a function of the substrate bias voltage.

voltage. Thus, although the excitation rate drops as we go from −30 to +30 V, the flux of metastables increases (left axis) with positive biasing. In turn, the energy that is transferred to the PET surface is higher for the +30 V substrate bias, which is opposite to what is observed for the energy transfer to the surface from ions (Fig. 4.6(a)). In addition, the power supplied by the metastables to the PET surface is about an order of magnitude lower compared to ions and this automatically limits their role in the PET treatment. For the calculations of the power transferred from He metastables to the surface a rather optimistic value of 20 eV was used.

The production of molecule fragments of the treated PET can seriously affect the composition of the flux of precursors of PET etching as proven above. Thus, it is useful to monitor the distribution of these species in the discharge and the variation of their density with the substrate bias potential. Figure 4.9(a) presents a map of the CO_2 density in the discharge for the −30 V substrate bias potential. CO_2 is a product of most of the surface reactions (R7, R8, R11, R12) and its density is higher compared to all other molecules that are produced [36]. Concerning the distribution of this molecule, we can observe that CO_2 density is not uniformly spread over the entire plasma volume but it is higher close to the PET surface where it is produced. This explains the fact that the quenching of He metastables is stronger close to the PET surface [35,36] and also explains the shape of the emission profiles. Similar space distributions were obtained for all the PET treatment byproducts (C_2H_4O, C_2H_4, H)

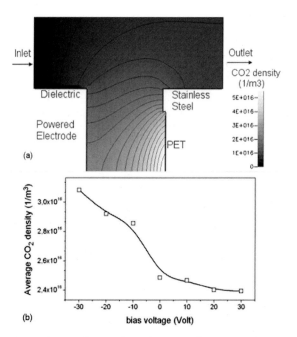

Fig. 4.9 (a) Distribution of CO_2 density (m^{-3}) in 0.5 torr He discharges and −30 V substrate bias voltage.
(b) Calculated values of the average CO_2 density as a function of the substrate bias voltage.

with the non-uniform distribution being affected by the mass and consequently the diffusion coefficient of these species. Thus lighter species (H atoms) are spread more uniformly in the plasma volume than heavier species.

Furthermore, the variation of the average production of CO_2 molecules with the substrate bias voltage was calculated and plotted in Fig. 4.9(b). The change of the substrate bias from negative to positive values results in a continuous drop of the CO_2 density which can be attributed to the decrease of the ion flux and the ion bombardment presented in Section 4.4.1 and the consequent limitation of the surface reactions R7 and R11. The increase of He metastable flux which was presented above cannot lead to an increase of the CO_2 density through the surface reactions R8 and R12, because their flux is much lower compared to ions. Moreover, the highest CO_2 density in the case of negative substrate bias corresponds to an enhanced quenching rate leading to the lowest He metastable density.

4.4.3
Plasma–Surface Interactions

Experimental and model results concerning the electrical properties and the gas-phase chemistry of He discharges have predicted a more intense PET treatment under negatively biased substrate conditions. This was also confirmed from the experimental measurements of the etch rate. The etch rate was measured *in situ* by applying laser reflectance interferometry and *ex situ* from the mass difference of the PET thin film before and after the plasma treatment. The variation of the etch rate as a function of the substrate voltage is presented in Fig. 4.10(a) together with the model prediction of the etch rate. Both experimental and simulation results show a drop of the etch rate as the bias voltage changes from negative to positive values. This drop can be attributed to the reduction of the ion flux and the ion bombardment with increasingly positive bias. However, the model predicts an etch rate of about 30 to 40% lower compared to the experimental measurements over the entire range of substrate bias voltage. This difference may well be a result of the ion etching probability which was adopted in the simulation (15%). This probability is *a priori* not known and is an interesting subject requiring further study. However, it is important to note that according to these results ions could account for the total observed etch rate, if one adopts an etching probability higher than ~25%. On the other hand, the role of the UV radiation that has been reported by other groups as the main path of PET treatment [37] cannot be estimated so far by the simulation, as the model does not include the radiation problem. The certain result of this investigation is that ions play a very important role in the PET treatment.

Finally, the increase of the etch rate and the more intense PET treatment is followed by an enhancement of the cross-linked surface sites under negative bias conditions. This is presented in Fig. 4.10(b), where the fraction of the surface sites ($-C_{20}H_{14}O_8-$, $-C_{14}H_8O_4-$, $-C_{14}H_8O_2-$) as a function of the substrate bias voltage is plotted. The change of the bias potential from -30 to $+30$ V results in a drop of the fraction of all cross-linked surface sites. The fraction of the $-C_{14}H_8O_4-$ site is higher compared to all other sites as the frequency of the surface reaction that leads to this site is much

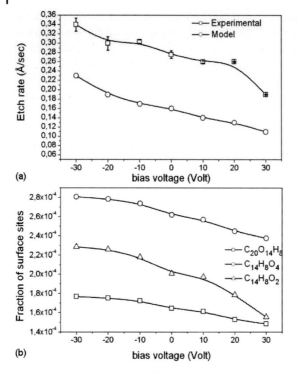

Fig. 4.10 (a) PET etch rate (Å s^{-1}) and (b) fraction of surface sites as a function of the substrate bias voltage.

higher. It is also obvious that the fraction of the cross-linked sites is rather low (<0.028%) while the sum of all cross-linked sites (Table 4.1, R1–R6) with the sum of all intermediate sites (Table 4.1, R7–R12) slightly exceeds 0.1% of the total surface. This observation indicates that most of PET surface remains unmodified and this is basically due to the small treatment times that were simulated (1 s). However, the most important feature that results from the calculation of the fraction of the surface sites as a function of the substrate bias is the enhancement of the fraction of the cross-linked surface sites at negative bias potential. This in turn can explain the higher stability of PET films treated under these conditions, against the well-known aging effect [38,39].

4.5
Conclusions

A thorough study of the effect of the substrate bias voltage on the PET treatment from He discharges was performed by applying a series of plasma diagnostics (electrical, spectroscopic and etch rate measurements) together with self-consistent modeling of He plasmas.

The change of the substrate bias potential from negative to positive values for constant applied RF voltage was followed by a slight drop of the total power dissipated in the discharge and the total discharge current. The model reproduces this trend fairly well and reveals that the drop of the power consumption and the discharge current are the result of the simultaneous drop of both the electron density and the electric field. As a consequence of this decrease, the ion flux and the ion bombardment of the PET surface is not favored by positive biasing.

In contrast, the He metastable density was favored by the change of bias voltage from -30 to $+30$ V although the excitation rate decreases. This was attributed to the quenching of He metastable molecules from the byproducts of PET surface modification which is much more intense in the case of negative bias. A parametric study of the quench rate of He metastables has shown that the model and the experimentally recorded spatial profiles of He^* are in good agreement only if 8 to 10 collisions of He metastable with byproducts of the surface treatment lead to quenching of the former. The rather high quench rates in combination with the decrease of the byproduct density as the bias potential changes from -30 to $+30$ V lead to an increase of the flux of He metastables towards the PET surface and of the energy that is transferred to the surface by these species. However, both flux and energy are much lower compared to ions and this severely limits the role of He^* in the PET surface modification.

In addition, PET etch rates decrease with the change from negative to positive PET substrate bias, indicating the significant role of ion bombardment. Model results have also shown that if the ion etching probability was about 25% then the ions could be exclusively responsible for the observed etch rate. Finally, the much more intense PET treatment under conditions of negative substrate bias results in an increase of the cross-linked surface sites, which in turn can be responsible for the higher resistance of PET thin films treated under such conditions, against the aging effect.

References

1 Carlsson, C.M. and Johansson, K.S. (1993) *Surf. Interface Anal.*, **20**, 441.

2 Dai, L., Griesser, H.J. and Mau, A.W.H. (1997) *J. Phys. Chem. B*, **101**, 9548.

3 Arefi, F., Andre, V., Montazer-Rahmati, P. and Amourouz, J. (1992) *Pure Appl. Chem.*, **64**, 715.

4 Xie, Y., Sproule, T., Li, Y., Powell, H., Lannuti, J. and Kniss, D.A. (2002) *J Biomed. Mater. Res.*, **61**, 234.

5 Friedrich, J.F. *et al.* (1995) *J. Adhesion Sci. Technol.*, **9**, 1165.

6 Inagaki, N., Narushim, K., Tuchida, N. and Miyazaki, K. (2004) *J. Polym. Sci. B: Polym. Phys.*, **42**, 3727.

7 Jie-Rong, C., Xue-Yan, W. and Tomiji, W. (1999) *J. Appl. Polym. Sci.*, **72**, 1327.

8 Koen, M.C., Lehmann, R., Groening, P. and Schlapbach, L. (2003) *Appl. Surf. Sci.*, **207**, 276.

9 Laurens, P., Petit, S. and Arefi-Khonsari, F. (2003) *Plasmas Polym.*, **8**, 281.

10 Arefi-Khonsari, F., Kurdi, J. and Tatoulian, M. (2001) *J. Amouroux, Surf. Coat. Technol.*, **142–144**, 437.

11 Goldblatt, R.D. *et al.* (1992) *J. Appl. Polym. Sci.*, **46**, 2189.

12 Tahara, M., Cuong, N.K. and Nakashima, Y. (2003) *Surf. Coat. Technol.*, **173–174**, 826.

13 Drachev, A.I., Gil'man, A.B., Pak, V.M. and Kuznetsov, A.A. (2002) *High Energ. Chem.*, **36**, 116.

14 Guruvenket, S., Mohan Rao, G., Komath, M. and Raichur, A.M. (2004) *Appl. Surf. Sci.*, **236**, 278.

15 Liston, E.M. (1989) *Adhes. Age*, **30**, 199.

16 Carlotti, S. and Mas, A. (1998) *J. Appl. Polym. Sci.*, **69**, 2321.

17 Hegemann, D., Brunner, H. and Oehr, C. (2003) *Nucl. Instrum. Methods Phys. Res. B*, **208**, 281.

18 Cioffi, M.O.H., Voorwald, H.J.C. and Mota, R.P. (2003) *Mater. Charact.*, **50**, 209.

19 Kim, B.K., Kim, K.S., Park, C.E. and Ryu, C.M. (2002) *J. Adhes. Sci. Technol.*, **16**, 509.

20 Shi, M.K., Graff, G.L., Gross, M.E. and Martin, P.M. (1999) *Plasmas Polym.*, **4**, 247.

21 Wilken, R. and Holländer, A. (1999) J. Behnisch, *Surf. Coat. Technol.*, **116–119**, 991.

22 Mataras, D., Cavadias, S. and Rapakoulias, D.E. (1989) *J. Appl. Phys.*, **66**, 119.

23 Spiliopoulos, N., Mataras, D. and Rapakoulias, D.E. (1996) *J. Vac. Sci. Technol. A*, **14**, 2757.

24 Mataras, D., Cavadias, S. and Rapakoulias, D.E. (1993) *J. Vac. Sci. Technol.*, **11**, 664.

25 Spiliopoulos, N., Mataras, D. and Rapakoulias, D. (1997) *J. Electrochem. Soc.*, **144**, 634.

26 Lyka, B., Amanatides, E. and Mataras, D. (2004) *Proc. 19th European Photovoltaic Solar Cell Energy Conf. and Exhibition, Paris.*

27 Amanatides, E., Lykas, B. and Mataras, D. (2005) *IEEE Trans. Plasma Sci.*, **33**, 372.

28 ftp://jila.colorado.edu/collision_ data/.

29 Dubreuil, B. and Prigent, P. (1985) *J. Phys. B-At. Mol. Opt.*, **18**, 4597.

30 Tochikubo, F., Makabe, T., Kakuta, S. and Suzuki, A. (1992) *J. Appl. Phys.*, **71**, 2143.

31 http://kinetics.nist.gov/index.php.

32 Boeuf, J.P. and Pitchford, L.C. (1995) *Phys. Rev. E*, **51**, 1376.

33 Czarnetzki, U., Luggenholscher, D. and Dobele, H.F. (2001) *Appl. Phys. A*, **10**, 1007.

34 Arefi-Khonsari, F., Placinta, G., Amouroux, J. and Popa, G. (1998) *Eur. Phys. J. Appl. Phys.*, **4**, 193.

35 Papakonstantinou, D., Mataras, D. and Arefi-Khonsari, F. (2001) *J. Physique IV*, **11** (Pr3), 357.

36 Papakonstantinou, D. and Mataras, D. (2001) *15th International Symposium on Plasma Chemistry, Orleans*, 2421.

37 Holländer, A., Wilken, R. and Behnisch, J. (1999) *Surf. Coat. Technol.*, **116–119**, 788.

38 Kaminska, A., Kaczmarek, H. and Kowalonek, J. (2002) *Eur. Polym. J.*, **38**, 1915.

39 Chatelier, R.C., Xie, X., Gengenbach, T. and Griesser, H.J. (1995) *Langmuir*, **11**, 2576.

5

Three-Dimensional Modeling of Thermal Plasmas (RF and Transferred Arc) for the Design of Sources and Industrial Processes

V. Colombo, E. Ghedini, A. Mentrelli, and T. Trombetti

A three-dimensional (3-D) model for the simulation of inductively coupled plasma torches (ICPTs) working at atmospheric pressure has been developed at the University of Bologna, using the customized CFD commercial code FLUENT®. The helicoidal coil is taken into account in its actual 3-D shape, showing its effects on the plasma discharge for various geometric, electric and operating conditions without axisymmetric hypotheses of simplification. Simulations have been performed for Ar plasmas. The gas injection section of an industrial TEKNA PL-35 plasma torch is included in the model without geometry simplifications, refining the mesh at the injection points, in order to perform a more realistic simulation of the inlet region of the discharge, taking into account also turbulence effects. Metallic and ceramic particle axial injection in the discharge through a carrier gas by means of a probe is simulated as well, taking into account the energy and momentum transfer between the continuous and the discrete phases and the effect of particle turbulent dispersion. The behavior of transferred arc thermal plasma sources operating at atmospheric pressure for the treatment of a substrate material (for waste treatment purposes and for metallic substrate cutting or hardening) has also been investigated by means of a 3-D time-dependent numerical model, using a customized version of the CFD commercial code FLUENT®. Unsteady flow and heat transfer equations are solved with coupled electromagnetic ones, for an Ar optically thin plasma under conditions of laminar flow and local thermodynamic equilibrium (LTE). The transient effects of an imposed external magnetic field on the shape of the single torch arc are investigated. The importance of fully investigating plasma velocity and temperature fields in high-power twin torch transferred arc systems designed for waste treatment purposes is outlined with reference to a plasma source designed and operated by Centro Sviluppo Materiali (CSM SpA) in Castel Romano, Rome. All calculations have been performed using PlasMac, a cluster of workstations available at CIRAM & DIEM, University of Bologna, so allowing for a large reduction in computational time as well as for the treatment of complex computational domains otherwise not manageable with traditional personal computers.

Advanced Plasma Technology. Edited by Riccardo d'Agostino, Pietro Favia, Yoshinobu Kawai, Hideo Ikegami, Noriyoshi Sato, and Farzaneh Arefi-Khonsari
Copyright © 2008 WILEY-VCH Verlag GmbH & Co. KGaA, Weinheim
ISBN: 978-3-527-40591-6

5.1
Introduction

In recent years, ICPTs have played a role of increasing importance in many technological processes, as a clean and effective means to produce plasma jets with high enthalpy content that can be usefully employed in a wide range of applications, such as plasma spray deposition of materials, densification and spheroidization of powders, chemical synthesis of nanoparticles, waste treatment and others [1–11]. Since the success of a given process depends directly on the plasma temperature and velocity fields in the discharge and/or in the jet, which in turn depend on the geometric and operating parameters of the system, the characterization of the torch and the knowledge of the influence of such parameters on the plasma properties are of primary importance. However, the detailed diagnostics of ICPTs is very hard to perform, due to the high temperatures involved and to the difficulty of reaching the internal zones of the device without perturbing the discharge. As an alternative, mathematical modeling represents a valid and powerful tool to predict the characteristics of these kinds of systems, also due to the relevant progress recently made in computer technology, which allows for the implementation of more and more sophisticated modeling approaches. In this frame, various 2-D models have been proposed in the past by different authors [12–19] to simulate the physical behavior of the plasma in ICPTs, also using an extended grid approach [15–19] for the description of the electromagnetic field. However, as all these models assume the torch to be axisymmetric, important 3-D effects due to the actual shape of the induction coil or to the non-axisymmetric distribution of the inlet gases could not be revealed. Moreover, 2-D models do not permit one to study torches with non-circular cross-section. A first step in trying to highlight some of the 3-D effects caused by the typical helicoidal shape of the coil was the work of Xue *et al.* [19] which also took into account within a 2-D modeling the axial component of the induction current flowing with a given inclination angle through an idealized axysymmetric cylindrical coil. 3-D simulation aimed at the study of gas mixing in the downstream region of an ICPT was also performed by Mostaghimi and co-workers [20], within a computational domain that did not include the induction coil and its electromagnetic effects. In order to obtain a more realistic description of ICPTs, a fully 3-D model [21–27] which completely removes the assumption of axisymmetry has been recently developed. The model has been implemented in the framework of the CFD commercial software FLUENT®, using a grid extending also outside the plasma region for the treatment of the electromagnetic field. A user-defined scalar (UDS) technique [18] has been adopted to suitably customize the basic FLUENT® code in order to add the Maxwell equations to its built-in fluid dynamic module. In this chapter, simulation results obtained by means of the 3-D model will be presented for the Tekna PL-35 Plasma Torch working under typical industrial operating conditions, taking also into account the injection of powders to be treated in the plasma and collected in a reaction chamber downstream of the torch exit.

Thermal plasma devices include also DC transferred arc plasma torches, which are widely used in industrial processes [28]. For example, transferred arc torches may find useful applications in the surface treatment of metallic materials. In this regard, the deflection of the electric arc by means of an external applied magnetic field generated by a current flowing in a wire may find interesting applications for the treatment of large area anodic surfaces [29–31]. Another interesting area in which DC transferred arc systems find application is the processing of waste materials. The *twin torch* device presented in this chapter, for example, has been the object of recent research to model its behavior inside a plasma furnace for hazardous waste incineration and asbestos inertization [32]. In the past research on the subject of DC transferred arc *twin torch* systems has been done [33] concerning 3-D simulation of flows inside a plasma furnace devoted to combustion/vitrification of radioactive wastes, based on a molten glass bath under an electric arc transferred between two metallic parallel electrodes (cathode and anode) under the hypothesis that the arc shape closes itself through the molten bath. The arc domain was modeled on a 3-D structured mesh integrating two parallel torches, providing results on the plasma side for electrical potential, arc shape, and temperature of the plasma and comparing them with some experimental results such as emission spectroscopy or arc observation with a high-speed camera. Another thermal plasma device whose numerical simulation is presented in this chapter is the cutting torch, widely used in industrial cutting processes on a wide range of metallic materials due to the high productivity achievable [34–37]. The plasma arc cutting process is characterized by a transferred electric arc that is established between an electrode that is part of the cutting torch (the cathode) and another electrode that is the metallic workpiece to be cut (the anode). In order to obtain a high-quality cut and a high productivity (high cutting velocity) the plasma jet must be, among other things, as collimated as possible and must have a high achievable power density. In this regard, modeling and numerical simulation may be very useful tools for the investigation of the characteristics of the plasma discharge generated in these kinds of devices, as well as for optimization of industrial cutting torches. In this chapter, a brief review of the recent advances in transferred arc plasma torch modeling is presented, including also a selection of the results of numerical simulations of magnetically deflected arc, twin torch and cutting plasma torches.

5.2
Inductively Coupled Plasma Torches

5.2.1
Modeling Approach

5.2.1.1 Modeling Assumptions
The physical behavior of the plasma, treated as a fluid continuum, has recently been modeled removing the axisymmetric assumption that has been extensively used in most of the previous studies concerning ICPTs. This leads to a full 3-D model, which has been implemented by the authors in the FLUENT® environment [21–27].

The following basic assumptions have been employed throughout the numerical study:

- the flow is steady;
- the plasma is optically thin and in LTE;
- the viscous dissipation term in the energy equation is neglected;
- the displacement currents are neglected; and
- the thermodynamic and transport properties are functions only of the temperature [38,39].

An additional assumption related to the plasma flow regime is necessary. Two different assumptions have been considered, depending on the particular simulation to be performed:

- the plasma flow is laminar, or
- the plasma flow is turbulent and it is describable by the Reynolds stress model (RSM).

In numerical simulations the injection of powders in the torch may also be taken into account, with the aim of evaluating the trajectories and thermal histories of these powders during their residence time in the torch region and in the reaction chamber region (downstream of the torch exit) [40–44]. In these simulations, the injected powders are treated as a discrete phase, dispersed in the continuum phase representing the plasma.

When injection of powders is considered, the basic assumptions made to model the continuum phase are unchanged with respect to those explained earlier. Concerning the discrete phase, the basic following assumptions are made:

- the powders are axially injected with velocity equal to that of the carrier gas;
- the powders are assumed to be spherical; and
- the internal resistance to heat transfer of the powders is neglected.

5.2.1.2 Governing Equations of the Continuum Phase

The equations for the transport of mass, momentum and energy for the continuum phase representing the plasma are written as follows.

Conservation equation of mass:

$$\nabla \cdot (\rho u) = 0 \tag{5.1}$$

Conservation equation of momentum:

$$\nabla \cdot (\rho uu) = -\nabla + \nabla \cdot \left[\mu(\nabla u + \nabla u^T) - \frac{2}{3}\mu\nabla \cdot (uI)\right] + \rho g + J \times B \tag{5.2}$$

Conservation equation of energy:

$$\nabla \cdot (\rho uh) = \nabla \cdot \left(\frac{k}{c_p}\nabla h\right) + J \times E - R \tag{5.3}$$

where ρ is the plasma density, p is the pressure, h is the enthalpy, T is the temperature, $\mathbf{u}$ is the velocity, k is the thermal conductivity, c_p is the specific heat at constant pressure, μ is the molecular viscosity (when the laminar model is adopted) or the sum of the molecular and turbulent viscosity (when the turbulent model is adopted), $\mathbf{g}$ is the gravitational force, $\mathbf{E}$ is the electric field, $\mathbf{B}$ is the magnetic induction, $\mathbf{J}$ is the current density induced in the plasma and R is the volumetric radiative losses.

The electromagnetic field generated by the current flowing in the coil ($\mathbf{J}^{coil}$) and by the induced currents in the plasma ($\mathbf{J}$) can be described by means of the Maxwell's equations written in their vector potential formulation:

$$\nabla^2 A - i\omega\mu_0\sigma A + \mu_0 J^{coil} = 0 \tag{5.4}$$

where μ_0 is the magnetic permeability of free space ($4\pi \times 10^{-7}\,\mathrm{H\,m^{-1}}$), σ is the plasma electrical conductivity and $\omega = 2\pi f$ (f being the frequency of the electromagnetic field). In the presented model the simplified Ohm's law $\mathbf{J} = \mathbf{E}$ has been employed, as suggested in [16]. The electric field $\mathbf{E}$ and the magnetic field $\mathbf{B}$ which appear in Eqs. (5.2) and (5.3) are obtained from the vector potential $\mathbf{A}$ with the following expressions: $\mathbf{E} = -i\omega\mathbf{A}$ and $\mathbf{B} = \lambda \times \mathbf{A}$.

The Reynolds stress transport equations, coupled with the previous governing equations and to be solved when a turbulent description of the plasma flow is desired, may be found in [45].

5.2.1.3 Governing Equations of the Discrete Phase

When numerical simulations of the trajectory and heating history of powders injected with a carrier gas into ICPTs working at atmospheric pressure are to be performed, the fully 3-D FLUENT®-based model previously described must be suitably modified in order to take into account the appropriate source–sink terms due to the interaction between the continuum phase (i.e. the plasma) and the discrete phase (i.e. the powders) in the continuity, momentum and energy equations.

In this frame, the effects of the particle injection on the plasma are suitably taken into account by adding proper exchange terms in the fluid dynamics equations, as explained in [46–50].

The particles trajectory is obtained by solving the following equation of motion:

$$\rho_p \frac{d\mathbf{v}_p}{dt} = \left(\frac{3\rho_\infty C_D}{4d_p}\right)|\mathbf{v}_\infty - \mathbf{v}_p|(\mathbf{v}_\infty - \mathbf{v}_p) + \mathbf{g}(\rho_p - \rho_\infty) \tag{5.5}$$

where $\mathbf{v}_\infty$ and ρ_∞ are the velocity and density of the plasma; $\mathbf{v}_p$, d_p, ρ_p are the velocity, diameter and density of the particle, respectively; $\mathbf{g}$ is the gravitational acceleration; and C_D is the drag coefficient which is calculated as in [51] but neglecting the Knudsen (rarefaction) effect:

$$C_D = \gamma f(Re_\infty)\left(\frac{Re_\infty}{Re_w}\right)^{0.1} = \gamma f(Re_\infty)\left(\frac{\nu_\infty}{\nu_w}\right)^{0.1} \tag{5.6}$$

In Eq. (5.6), ν_∞ and ν_w are the values of the gas kinematic viscosity calculated at the plasma and particle temperatures, T_∞ and T_p, respectively; Re_∞ is the following function of the Reynolds number $Re_\infty = d_p|\mathbf{v}_\infty|/\nu_\infty$:

$$f(Re_\infty) = \begin{cases} \dfrac{24}{Re_\infty} & Re_\infty < 0.2 \\[2ex] \left(\dfrac{24}{Re_\infty}\right)(1+0.1875\,Re_\infty) & 0.2 \le Re_\infty < 2 \\[2ex] \left(\dfrac{24}{Re_\infty}\right)(1+0.11Re_\infty^{0.81}) & 2 \le Re_\infty < 21 \\[2ex] \left(\dfrac{24}{Re_\infty}\right)(1+0.189Re_\infty^{0.632}) & 21 \le Re_\infty < 200 \end{cases} \tag{5.7}$$

and γ is a correction factor accounting for the effect due to the particle evaporation (if it occurs):

$$\gamma = \frac{\lambda_v}{S_\infty - S_w} \int_{T_w}^{T_\infty} \frac{k_\infty}{h_\infty - h_w + \lambda_v} dT \tag{5.8}$$

where λ_v is the latent heat of evaporation of the particle material, γ, h_∞, h_w are the values of the gas specific enthalpy evaluated at the plasma and particle temperatures, respectively, k_∞ is the plasma thermal conductivity and S_∞, S_w are the heat conduction potentials calculated at S_∞ and T_p following the definition of $S(T)$:

$$S(T) = \int_{T_0}^{T} k(T)dT \tag{5.9}$$

T_0 being an arbitrary reference temperature. The heating history of the particle in the solid phase is obtained by solving the energy balance equation:

$$m_p c_p \frac{dT_p}{dt} = A_p h_c(T_\infty - T_p) - A_p \varepsilon \sigma(T_p^4 - T_a^4) \tag{5.10}$$

where m_p and A_p are the mass and the surface area of the particle; c_p and ε are the specific heat and the emissivity of the particle material; σ is the Stefan–Boltzmann constant $(5.67 \times 10^{-8}\,\mathrm{Wm^{-2}K^{-4}})$; T_a is room temperature (300 K); and h_c is the convective coefficient given by

$$h_c = \gamma \frac{Nu(S_\infty - S_w)}{d_p(T_\infty - T_p)} \tag{5.11}$$

where Nu is the Nusselt number calculated as in [52]:

$$Nu = 2\left[1 + 0.63Re_\infty Pr_\infty^{0.8}\left(\frac{Pr_w}{Pr_\infty}\right)^{0.42}\left(\frac{\rho_\infty\mu_\infty}{\rho_w\mu_w}\right)^{0.52}C^2\right]^{0.5} \tag{5.12}$$

where ρ_∞, μ_∞, Pr_∞ are the density, the dynamic viscosity and the Prandtl number ($Pr = \mu c_p / k$) of the gas, respectively, evaluated at the plasma temperature; ρ_w, μ_w, Pr_w are the same quantities calculated at the particle temperature; while C is a factor whose expression is

$$C = \frac{1 - (h_w/h_\infty)^{1.14}}{1 - (h_w/h_\infty)^2} \tag{5.13}$$

As soon as the particle temperature reaches the melting point, it is assumed to remain constant and the liquid-phase fraction, x, is calculated by integrating the equation

$$\frac{dx}{dt} = \frac{6q}{\rho_p d_p \lambda_m} \tag{5.14}$$

where q is the net specific heat flux transferred to the particle, given by the right-hand side of Eq. (5.10) and λ_m is the melting latent heat of the particle material. Once the particle is completely melted ($x = 1$), its temperature is allowed to follow again Eq. (5.10). As soon as the evaporation point is achieved, the temperature of the particle is kept constant while its diameter reduces itself according to the following law:

$$\frac{dd_p}{dt} = -\frac{6q}{\rho_p \lambda_v} \tag{5.15}$$

It is worth noting that, in the present model, no mass transfer between particles and plasma is accounted for; i.e. the eventual evaporated fraction of the particles does not alter the composition of the surrounding plasma gas, which is considered to be always pure argon.

5.2.1.4 Computational Domain and Boundary Conditions

The geometry of the Tekna PL-35 plasma torch, the behavior of which has been numerically simulated, is represented in Fig. 5.1.

The actual torch geometry has been accurately modeled in order to take into account the real geometry of the gas inlet region, of the non-axisymmetric coil and of the exit region of the torch. In Fig. 5.1, the injection ports of the axial sheath gas and of the tangential plasma gas may be appreciated. These geometrical details are commonly not included in computational models, where the hypothesis of uniformly injected gas is usually performed in order to simplify the model. The performance of such hyper-realistic simulations has only recently become possible because of the availability of parallel computing resources. The details of the geometrical configuration, including the dimensions of the various components, may be found in [42].

Boundary conditions for the conservation equations of mass (Eq. (5.1)), momentum (Eq. (5.2)) and energy (Eq. (5.3)) are the following. On the inner wall of the confinement tube, the no-slip condition is imposed, while on the outer wall of the confinement tube a fixed temperature value of 300 K is imposed. At the torch inlet, uniform velocity profiles of gas (calculated on the basis of the given flow rates) are

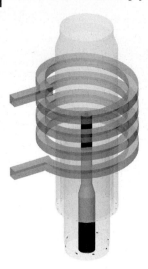

Fig. 5.1 Three-dimensional schematic of the inductively coupled plasma torch with detailed gas injection section.

assumed. At the torch exit the FLUENT® outflow condition (which corresponds to a fully developed flow condition [45]) was adopted. Equation (5.2) is solved in a domain extending also outside the torch region, using vanishing boundary conditions for the vector potential [18].

When the numerical simulation of the trajectory and heating history of powders injected in the inductively coupled plasma torch is performed, the computational domain is extended downstream of the torch region, in order to take into account the expansion of the plasma jet in a reaction chamber where to powders are collected. In this kind of simulation, the powders are assumed to be injected with velocity and temperature equal to that of carrier gas. When a particle hits the inner wall of the confinement tube of the torch it is assumed to be trapped by the wall and the particle is not included further in the computations.

The computational grid is a hybrid one composed by tetrahedrons, hexahedrons and wedges, and it is built by means of the GAMBIT© software package and then imported into the FLUENT® environment. The approximate number of cells composing the grid is 4.5×10^5 (slightly varying depending on the torch and coil configuration).

Simulations have mainly been carried on a cluster of workstations, i.e. splitting up the grids and data in various partitions and then assigning each partition to a different compute process.

5.2.2
Selected Simulation Results

5.2.2.1 **High-Definition Numerical Simulation of Industrial ICPTs**
In Fig. 5.2(a) the temperature field is shown in a 3-D view in order to underline the effects on the discharge of the non-axisymmetric coil shape. The views on two perpendicular planes passing through the axis of the torch are presented in Fig. 5.2(b)

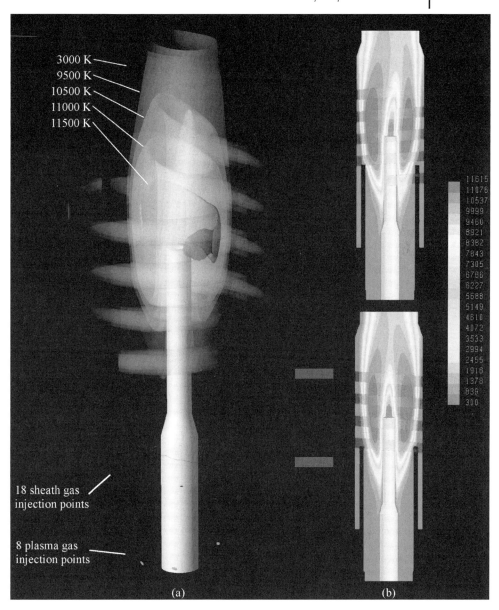

Fig. 5.2 Temperature fields [K] in a fully 3-D view and on two perpendicular planes passing through the axis of the torch. Operating conditions for the ICPT Tekna PL-35 are: discharge power of 15 kW, RF induction frequency of 3 MHz and inlet mass flow rates of 60, 15 and 2.5 slpm for the sheath, plasma and carrier argon gases, respectively.

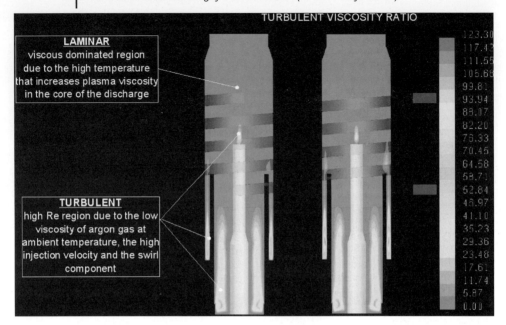

Fig. 5.3 Turbulent viscosity ratio field on two perpendicular planes passing through the axis of the torch. Operating conditions as in Fig. 5.2.

for a more detailed representation of the temperature field. The turbulence-affected region is shown in Fig. 5.3 where the turbulence viscosity ratio field in shown on the same two perpendicular planes for which the plasma temperature field is given in Fig. 5.2(b).

From Fig. 5.3 it may be appreciated that the inlet region of the plasma and sheath gas inlet sections is characterized by non-negligible turbulence effects, while the discharge is quite laminar due to the high viscosity of the plasma [52].

5.2.2.2 Numerical Simulation of the Trajectories and Thermal Histories of Powders Injected in Industrial ICPTs

In this section, simulation results are presented concerning the injection of $20\,\mathrm{g\,min}^{-1}$ of Al_2O_3 particles (with diameter $d = 25\,\mu m$) in a Tekna PL-35 torch working under the operating conditions described, with reference to the torch drawing reported in [42], by the following parameters: $Q_1 = 0.5$ slpm; $Q_2 = 15$ slpm; $Q_3 = 60$ slpm; $f = 3$ MHz; $P_0 = 10$ kW.

In Fig. 5.4, a 3-D isocontour of plasma temperature together with the trajectories of 10 particles injected in the discharge from the same injection point are shown.

The trajectories are shaded by the particle temperature, showing also that, for the present configuration and operating conditions, most of the particles reach the bottom of the reaction chamber in a molten state (the melting and boiling temperatures of alumina, respectively, being $T_m = 2323$ K and $T_b = 3800$ K).

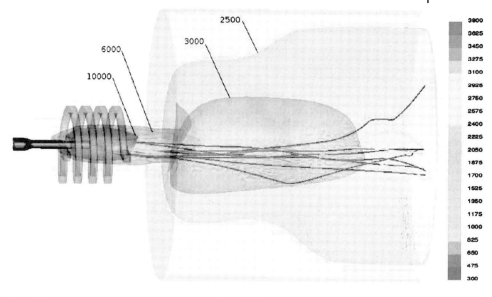

Fig. 5.4 Isocontours of the plasma temperature [K] and 3-D schematic of the trajectories of 10 particles injected from the same injection point shaded by particle temperature [K]. Gravity vector is along the axis of the torch pointing downstream.

Fig. 5.5(a) shows plasma temperature isocontours on a plane passing through the axis of the torch, highlighting the substantial local cooling effect of the particle injection on the plasma (the so-called *loading effect*) and the non-axisymmetric nature of the discharge due to both the non-axisymmetric coil configuration and the intrinsically 3-D nature of the turbulent dispersion phenomenon.

In Figs. 5.5(b) and (c) visualizations of particle concentration and of plasma velocity magnitude, respectively, in the torch and reaction chamber regions, under the same operating conditions of Fig. 5.4, are shown.

5.3
DC Transferred Arc Plasma Torches

Thermal plasma devices of DC transferred arc type have been much modeled and numerically simulated in the recent past [53–61].

5.3.1
Modeling Approach

5.3.1.1 Modeling Assumptions
The 3-D computational model useful for the simulation of a DC transferred arc plasma torch involves the simultaneous solution of the coupled set of nonlinear fluid

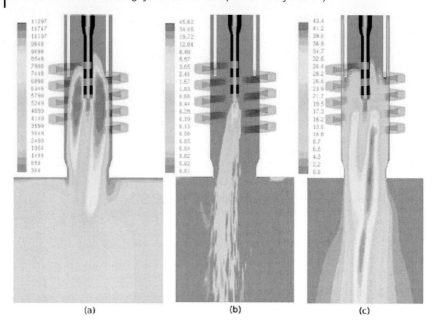

Fig. 5.5 (a) Plasma temperature field [K], (b) particle concentration [kg m^{-3}] and (c) plasma velocity magnitude [m s^{-1}] in a plane passing through the axis of the torch whose position is evidenced by coil view.

dynamic, electromagnetic and energy transfer equations. The main simplifying assumptions on which the developed model relies are the following:

- the flow is steady;
- the plasma is described as a continuum mono-component phase;
- the plasma is optically thin and in LTE;
- the viscous dissipation term in the energy equation is neglected;
- the thermodynamic and transport properties are functions only of the temperature; and
- the turbulent plasma flow is described by the renormalization group (RNG) turbulence model.

5.3.1.2 Governing Equations
Taking into account the previous assumptions, the equations of the conservation of mass and momentum take the following form:

$$\nabla \cdot (\rho \mathbf{v}) = 0 \tag{5.16}$$

$$\nabla \cdot \rho \mathbf{v}\mathbf{v} = -\nabla p + \nabla \cdot \tau + F_{\mathrm{L}} \tag{5.17}$$

where ρ is the density of the fluid, $\mathbf{v}$ the velocity of the fluid, p the pressure, τ the stress tensor and F_{L} the Lorentz force due to the interaction of the conductive fluid and the electromagnetic field.

The selected turbulence model is the FLUENT®-implemented RNG-based k- one, which is a two-equation model with a form similar to the standard k- model:

$$\frac{\partial}{\partial t}(\rho k) + \frac{\partial}{\partial x_i}(\rho k u_i) = \frac{\partial}{\partial x_j}\left(\alpha_k \mu_{\text{eff}} \frac{\partial k}{\partial x_j}\right) + G_k + G_b - \rho\varepsilon - Y_M \tag{5.18}$$

$$\frac{\partial}{\partial t}(\rho\varepsilon) + \frac{\partial}{\partial x_i}(\rho\varepsilon u_i) = \frac{\partial}{\partial x_j}\left(\alpha_\varepsilon \mu_{\text{eff}} \frac{\partial \varepsilon}{\partial x_j}\right) + C_{1\varepsilon}\frac{\varepsilon}{k}(G_k + C_{3\varepsilon} G_b) - C_{2\varepsilon}\rho\frac{\varepsilon^2}{k} - R_\varepsilon \tag{5.19}$$

In these equations, G_k is the generation of turbulence kinetic energy due to the mean velocity gradients; G_b is the generation of turbulence kinetic energy due to buoyancy; and Y_M represents the contribution of the fluctuating dilatation in compressible turbulence to the overall dissipation rate. Finally, α_k and α_ε are the inverse effective Prandtl numbers for k and ε, respectively [45].

The energy equation is solved without taking into account the viscous dissipation, which can safely been neglected when studying thermal plasmas:

$$\nabla \cdot \rho\mathbf{v}h - \nabla \cdot k\nabla T - \frac{5}{2}\frac{k_B}{e}\left(\mathbf{j} \cdot \frac{1}{C_p}\nabla h\right) = QJ - Q_R \tag{5.20}$$

where h is the enthalpy of the fluid, k is the thermal conductivity of the fluid, k_B is the Boltzmann constant, e is the electron charge, $\mathbf{j}$ is the current density and C_p is the specific heat at constant pressure of the fluid. The last term on the lefthand side of the equation represents the enthalpy transport due to stream of the conductive electrons. Finally, Q_J is the energy dissipated in the discharge by the Joule effect and Q_R represents the radiative losses.

The equation for the calculation of the electrostatic potential is

$$\nabla \cdot \sigma\nabla V = 0 \tag{5.21}$$

where σ is the electrical conductivity of the plasma and V is the electrostatic potential. The vector potential $\mathbf{A}$, useful for the evaluation of the electromagnetic field, is calculated by means of the following vectorial equation:

$$\nabla^2\mathbf{A} + \mu_0\mathbf{j} = 0 \tag{5.22}$$

where $\mathbf{j} = \mathbf{E} = -\lambda V$ represents the current density and μ_0 is the magnetic permeability of a vacuum.

The electromagnetic equations are solved, by the FLUENT® solver, by means of a user-defined scalar (UDS) approach, in a way similar to that described in [18].

5.3.1.3 Computational Domain and Boundary Conditions

Magnetically Deflected Transferred Arc The current density is imposed on the flat surface of the cathode tip as a boundary condition:

$$\mathbf{j}(r) = -\sigma\nabla V = j_{\max}e^{-br}\mathbf{n} \tag{5.23}$$

where $j_{max} = 1.4 \times 10^8 \, A \, m^{-2}$ is the maximum current density on the cathode tip, r is the distance from the axis of symmetry, b is a parameter whose value allows one to have the desired total electric current I on the surface of the cathode and $\mathbf{n}$ is the unit vector perpendicular to the surface. The value of j_{max} is a realistic value typical of these kinds of torch configurations.

On the anodic bottom surface, the electric potential V is given as a boundary condition ($V = 0$), while on the remaining part of the boundary of the computational domain the given condition is on the value of the current ($\mathbf{j} = 0$).

The boundary conditions for the temperature are as follows: a temperature of 300 K is imposed on the bottom anodic surface and on the lateral surface of the computational domain, while a temperature of 3500 K is imposed the cathodic surface.

The boundary condition applied on the solid walls for the momentum equation is the traditional *no-slip* condition. The FLUENT[®]-implemented pressure-outlet condition is applied on the lateral boundaries.

Regarding the vector potential equation, the component of the derivative of $\mathbf{A}$ along the direction perpendicular to all the surfaces is supposed to be zero.

The deflecting current, whose intensity is I_C, is supposed to flow in a leading wire parallel to the axis of the cathode at a distance of 1 cm from this axis.

The computational mesh used for the calculations, for the case with a distance L between the tip of the cathode and the anodic surface equal to 11 mm, is made up of 6.5×10^5 tetrahedral cells.

The Twin Torch   In the case of the *twin torch* setup, the computational domain does not extend to the whole reactor, since the simulation of such a complete 3-D configuration would be computationally too expensive. The domain is thus limited to the region between the cathode and anode where the discharge takes place, as shown in Figs. 5.8 and 5.9.

As in the magnetically deflected arc configuration, the profile of the current density on the cathodic surface is given as a boundary condition. In this case the profile is set as parabolic:

$$\mathbf{j}(r) = -\sigma \nabla V = -j_{max} \left[1 - \left(\frac{r}{R_0} \right)^2 \right] \tag{5.24}$$

where $j_{max} = 0.8 \times 10^7 \, A \, m^{-2}$ is the maximum value of the current density on the cathode, r is the distance from the axis of the cathode and R_0 is a reference value calculated imposing a given value of the total current, I, on the cathodic surface.

On the anodic surface the electrical potential is assumed to be zero, while on all the other surfaces of the domain the current flux is assumed to be zero.

The temperature value on the anodic and cathodic surfaces is an input coming from previous calculations developed by CSM SpA [33]; on all the other surfaces the temperature is fixed at 500 K. The FLUENT[®]-implemented pressure-outlet condition is selected as a boundary condition for the momentum equation on the external boundaries of the computational domain, while the traditional *no-slip* condition is imposed on all the solid walls.

In the performed *twin torch* simulations the working gas is either pure Ar or a mixture of Ar–H_2 with an H_2 content equal to 5% by volume. For such a H_2 content, the diffusion or demixing effects are negligible [62,63], allowing one to treat the plasma as a one-component fluid.

The Cutting Torch   The 3-D model presented in the previous section has been applied to the CP-200 CEBORA plasma cutting torch operating at atmospheric pressure, which has been modeled taking suitably into account all the geometrical details of the inlet gas section.

Fluid flow and heat transfer equations are solved together with the coupled electromagnetic ones for an optically thin air plasma assumed to be in LTE. The details of the gas injection section and of the cathode chamber are included in the computational domain in order to determine the effects of the geometry on the flow field characteristics of the discharge and to optimize it in the design phase. The metallic substrate with the keyhole is included in the computational domain only for electrical and fluid dynamics purposes, in order to take into account its effect on the flow field of the discharge at the nozzle exit. In the model developed for the plasma cutting torch presented here, turbulence phenomena are taken into account by means of a k-e RNG model in order to better describe the flow field inside the device.

5.3.2
Selected Simulation Results

5.3.2.1   Magnetically Deflected Transferred Arc
In Figs. 5.6(a) and (b), the plasma temperature fields on two planes perpendicular to each other and passing through the axis of the cathode are shown for the case of an undeflected arc configuration (the current flowing in the external conductor is $I_C = 0$). In Figs. 5.6(c) and (d) the temperature fields inside the metallic substrate (anode) are shown, on the same two planes of Figs. 5.6(a) and (b), respectively, and under the same operating conditions, while in Fig. 5.6(e), the temperature field on the upper surface of the anode (plasma–anode interface) is shown. It may be noticed that the temperature field in both the discharge region and the anodic region is axisymmetric because of full axisymmetric configuration of the torch.

In Fig. 5.7 the same temperature fields of Fig. 5.6 are shown for the case in which the electric arc is deflected by the presence of a magnetic field induced by a current flowing in a wire parallel to the axis of the cathode and with intensity $I_C = 50$ A. In this case, the symmetry of the discharge and of the temperature distribution in the substrate is lost as a consequence of the presence of the deflecting external magnetic field. In Fig. 5.7(e) the position of the leading wire in which the current flows is evidenced by the black dot.

5.3.2.2   The Twin Torch
Results in this section are presented in order to characterize the fluid flow and the temperature field of this kind of device. The electrical current of the arc I (imposed on the surface of the cathode as a boundary condition) is equal to 1500 A and the anodic

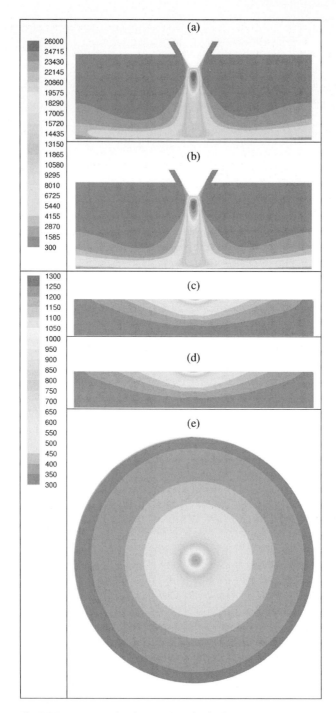

Fig. 5.6 Temperature distribution [K] in the discharge region (a, b) and inside the anode (c, d) on two planes perpendicular to each other and passing through the axis of the cathode, and on the upper surface of the anode (e) in the absence of deflecting current.

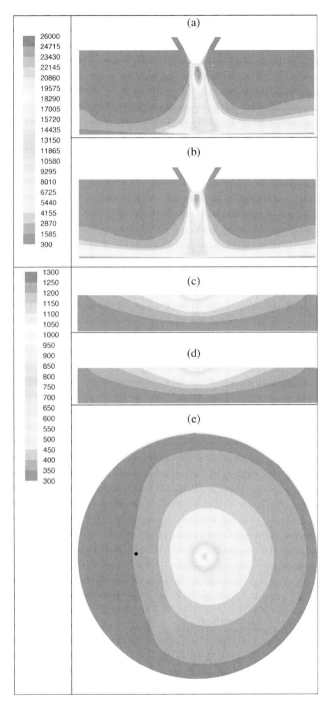

Fig. 5.7 Temperature distribution [K] in the discharge region (a, b) and inside the anode (c, d) on two planes perpendicular to each other and passing through the axis of the cathode, and on the upper surface of the anode (e). The position of the wire is evidenced by the black dot in (e).

(a)   (b)

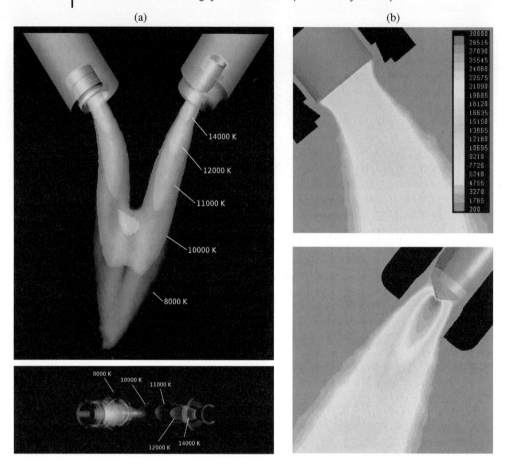

Fig. 5.8 Isotemperature surfaces in the discharge (a) and details of the plasma temperature [K] in the region close to the anodic (top) and cathodic (bottom) surfaces (b). Operating conditions: $I = 1500\,A$; pure Ar as working gas.

and cathodic gas flow rates are, respectively, 5.34 and $4.74\,m^3\,h^{-1}$, the working gas being either pure Ar or an Ar/H_2 mixture.

A 3-D visualization of the isotemperature surfaces in the pure Ar discharge is reported in Fig. 5.8(a), in which the deflection of the anodic and cathodic plasma jets due to the repulsive effects induced by the Lorentz force is appreciable. A more detailed visualization of the plasma temperature fields in the regions close to the two electrodes is reported in Fig. 5.8(b).

The maximum plasma temperature, located close to the cathodic surface, is nearly 30 000 K for the case presented in Fig. 5.8, while in the proximity of the anodic surface the plasma temperature reaches a value close to 12 000 K. It is interesting to note that these values of the plasma temperature close to the electrodes are mainly influenced by the maximum value of the current density on the surfaces of the

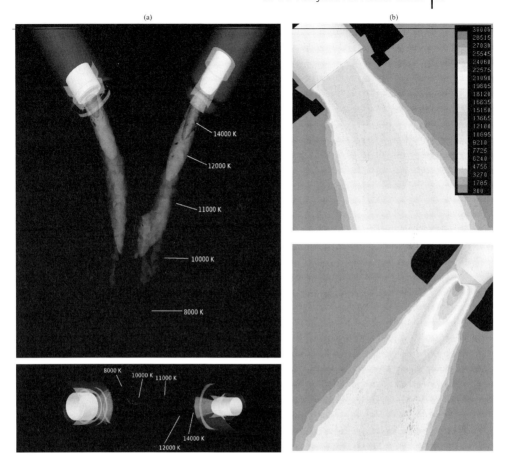

Fig. 5.9 Isotemperature surfaces in the discharge (a) and details of the plasma temperature [K] in the region close to the anodic (top) and cathodic (bottom) surfaces (b). Operating conditions: $I = 1500\,\text{A}$; Ar/H$_2$ mixture (5% H$_2$ content) as working gas.

electrodes, and are not noticeably influenced by the total electric current of the arc. The *twin torch* configuration has also been simulated under operating conditions in which a homogeneous mixture of Ar and H$_2$ is injected as working gas. In this case, the diffusion of the two species has been neglected due to the low content of H$_2$ considered in the present case (equal to 5% by volume).

In Fig. 5.9(a) a 3-D visualization of the plasma temperature isosurfaces for the *twin torch* operating with the Ar/H$_2$ mixture is shown. The main difference between the plasma temperature field presented in Fig. 5.9(a) and the one presented in Fig. 5.8(a) consists in higher values close to the anodic region. In fact, in the simulation with Ar/H$_2$ mixture a value close to 14 000 K is found, while for the case with pure Ar in the same region a temperature close to 12 000 K is found. From the comparison of these two figures, it may also be noticed that the macroscopic behavior of the discharge is not considerably affected by the injection of H_2. A detailed visualization of the plasma temperature fields in the anodic and cathodic regions is shown in Fig. 5.9(b).

For some particular geometries and operating conditions, an oscillation of the computed fields during iterations has been observed. Since the code used in this work is a steady-state one, the observed oscillating behavior does not follow the physics of the device in its time progress. Nevertheless, it can suggest the existence of unsteady phenomena in the discharge. Simulations can also give important information on non-axisymmetric anode attachment under particular operating conditions, allowing the use of this modeling tool to predict the plasma discharge behavior when anode disruption occurs under critical operating conditions as an effect of gas entrainment in the anode region.

5.3.2.3 The Cutting Torch
A detail of the temperature field resulting from 3-D simulation of the CP-200 CEBORA cutting plasma torch is shown in Fig. 5.10. The inclusion of the inlet region allows a better understanding of the characteristics of the flow in the upstream region of the plasma chamber and to perform a better design of the torch components.

Fig. 5.10 Detail of the temperature field [K] in the nozzle region of the CP200 plasma cutting device for a 1.8 mm nozzle diameter, 160 A of cathode current and 400 kPa (absolute) at the inlet.

References

1 Boulos, M.I. (1997) *High Temp. Mater. Process.*, **1**, 17.

2 Dignard, N.M. and Boulos, M.I., (1997) Ceramic powders spheroidization under induction plasma conditions, *Proceedings of 13th International Symposium on Plasma Chemistry (ISPC-13), Bejing, China, 18–22 August 1997, Vol. III,* 1031–1036.

3 Dignard, N.M. and Boulos, M.I. (1998) Metallic and ceramic powder spheroidization by induction plasmas, *Int. Thermal Spray Conf. (ITSC-98), Nice, France.*

4 Fauchais, P. and Vardelle, A. (1997) *IEEE Trans. Plasma Sci.*, **25**, 6.

5 Nutsch, G. (2003) in *Progress in Plasma Processing of Materials 2003* (ed P. Fauchais), Begell House, New York, pp. 401–408.

6 McKelliget, J.W. and El-Kaddah, N. (1988) *J. Appl. Phys.*, **64**, 2948.

7 Salvati, F., Tolve, P., Masala, M., Peisino, E. and Broglio, D. (1991) Application of plasma system for tundish heating, *Proceedings of the 1st European Conference on Continuous Casting, Firenze, Italy.*

8 Salvati, F. (2002) Development of thermal technologies for pyrolysis and combustion of waste with vitrification of ash, *Proceedings of the 22nd International Conference on Incineration and Thermal Treatment Technologies, Orlando, FL.*

9 Panciatichi, C., Cocito, P. and DeLeo, M.C.N. (1999) in *Progress in Plasma Processing of Materials 1999* (eds P. Fauchais and J. Amoroux), Begell House, New York, pp. 885–890.

10 Venkatramani, N. (2002) *Curr. Sci.*, **83**, 3.

11 Fauchais, P. (2004) *J. Phys. D: Appl. Phys.*, **37**, R86.

12 Mostaghimi, J. and Boulos, M.I. (1989) *Plasma Chem. Plasma Process.*, **9**, 25.

13 Chen, X. and Pfender, E. (1991) *Plasma Chem. Plasma Process.*, **11**, 103.

14 Proulx, P., Mostaghimi, J. and Boulos, M.I. (1991) *Int. J. Heat Mass Transfer*, **34**, 2571.

15 Colombo, V., Panciatichi, C., Zazo, A., Cocito, G. and Cognolato, L. (1997) *IEEE Trans. Plasma Sci.*, **25**, 1073.

16 Xue, S., Proulx, P. and Boulos, M.I. (2001) *J. Phys. D: Appl. Phys.*, **34**, 1897.

17 Boulos, M.I. (2001) *J. Visualization*, **4**, 19.

18 Bernardi, D., Colombo, V., Ghedini, E. and Mentrelli, A. (2003) *Eur. Phys. J. D*, **27**, 55.

19 Xue, S., Proulx, P. and Boulos, M.I. (2003) *Plasma Chem. Plasma Process.*, **23**, 245.

20 Njah, Z., Mostaghimi, J. and Boulos, M. (1993) *Int. J. Heat Mass Transfer*, **36**, 3909.

21 Bernardi, D., Colombo, V., Ghedini, E., Mentrelli, A., Tolve, P., Masala, M., Pcisino, E. and Broglio, D. (2003) *Eur. Phys. J. D*, **22**, 119.

22 Bernardi, D., Colombo, V., Ghedini, E. and Mentrelli, A. (2003) *Eur. Phys. J. D*, **25**, 271.

23 Bernardi, D., Colombo, V., Ghedini, E. and Mentrelli, A. (2003) *Eur. Phys. J. D*, **25**, 279.

24 Bernardi, D., Colombo, V., Ghedini, E. and Mentrelli, A. (2003) Three-dimensional effects in the design of inductively coupled plasma torches, Atti del XVI Congresso dell' Associazione Italiana del Vuoto, ed. Compositori, Bologna, 267–272.

25 Bernardi, D., Colombo, V., Ghedini, E. and Mentrelli, A. (2004) Time dependent 3-D modelling of inductively coupled plasma torches, *16th International Vacuum Congress (IVC-16), Venezia, Italy.*

26 Bernardi, D., Colombo, V., Ghedini, E. and Mentrelli, A. (2005) *IEEE Trans. Plasma Sci.*, **33**, 426.

27 Bernardi, D., Colombo, V., Ghedini, E. and Mentrelli, A. (2005) *Pure Appl. Chem.*, **77**, 359.

28 Ushio, M., Tanaka, M. and Lowke, J.J. (2004) *IEEE Trans. Plasma Sci.*, **32**, 1.

29 Blais, A., Proulx, P. and Boulos, M.I. (2003) *J. Phys. D: Appl. Phys.*, **36**, 488.

30 Franceries, X., Lago, F., Gonzalez, J.J., Freton, P. and Masquere, M. (2005) *IEEE Trans. Plasma Sci.*, **33**, 432.

31 Bernardi, D., Colombo, V., Ghedini, E., Melini, S. and Mentrelli, A. (2005) *IEEE Trans. Plasma Sci.*, **33**, 428.

32 Colombo, V., Ghedini, E., Mentrelli, A. and Malfa, E. (2005) 3-D Modelling of DC transferred arc twin torch for asbestos inertization, in *Nuclear Reactor Physics. A Collection of Papers Dedicated to Silvio Edoardo Corno*, ed. CLUT, Torino, Italy, 167–192.

33 Barthelemy, B., Girold, C., Delalondre, C., Paya, B. and Baronnet, J.M. (2003) Modeling a pilot-scale combustion/vitrification furnace under oxygen plasma arc transferred between twin torches, *Proceedings of the 16th International Symposium on Plasma Chemistry (ISPC-16), Taormina, Italy, 22–27 June.*

34 Nemchinsky, V.A. (1998) *J. Phys. D: Appl. Phys.*, **31**, 3102.

35 Gonzalez-Aguilar, J., Pardo Sanjurjo, C., Rodriguez-Yunta, A. and Angel Garcia Calderon, M. (1999) *IEEE Trans. Plasma Sci.*, **27**, 1.

36 Freton, P. (2002) Etude d'un arc de découpe par plasma d'oxygène. Modélisation – expérience, PhD thesis, Universite Paul Sabatier, Toulouse III, France. [In French].

37 Freton, P., Gonzalez, J.J., Gleizes, A., Camy Peyret, F., Caillibotte, G. and Delzenne, M. (2002) *J. Phys. D: Appl. Phys.*, **35**, 5131.

38 Murphy, A.B. and Arundell, C.J. (1994) *Plasma Chem. Plasma Process.*, **14**, 451.

39 Murphy, A.B. (2000) *Plasma Chem. Plasma Process*, **20**, 279.

40 Bernardi, D., Colombo, V., Ghedini, E., Mentrelli, A. and Trombetti, T. (2003) Powders trajectory and thermal history within 3-d modelling of inductively coupled plasma torches, *Proceedings of the IV Int. Conf. Plasma Physics Plasma Technology (PPPT-4), Minsk, Belarus, 15–19 September 2003, vol. 2, 463–464.*

41 Bernardi, D., Colombo, V., Ghedini, E., Mentrelli, A. and Trombetti, T. (2003) Powders trajectory and thermal history within 3-D ICPTs modelling for spheroidization and purification purposes, *Proceedings of the 48th Internationales Wissenschaftliches Kolloquium (48.IWK), Ilmenau, Germany, 22–25 September 2003, 285–286.*

42 Bernardi, D., Colombo, V., Ghedini, E., Mentrelli, A. and Trombetti, T. (2004) *Eur. Phys. J. D*, **28**, 423.

43 Bernardi, D., Colombo, V., Ghedini, E. and Mentrelli, A. (2004) 3-D numerical analysis of powder injection in various ICPT configurations, *16th International Vacuum Congress (IVC-16), Venezia, Italy.*

44 Bernardi, D., Colombo, V., Ghedini, E., Mentrelli, A. and Trombetti, T. (2005) *IEEE Trans. Plasma Sci.*, **33**, 424.

45 FLUENT 6.1 User's Guide, Fluent Inc., Lebanon, NH (2003).

46 Proulx, P., Mostaghimi, J. and Boulos, M.I. (1985) *Int. J. Heat Mass Transfer*, **28**, 1327.

47 Proulx, P., Mostaghimi, J. and Boulos, M.I. (1987) *Plasma Chem. Plasma Process.*, **7**, 29.

48 Huang, P.C., Heberlein, J. and Pfender, E. (1995) *Surf. Coat. Technol.*, **73**, 142.

49 Ye, R., Proulx, P. and Boulos, M.I. (2000) *J. Phys. D: Appl. Phys.*, **33**, 2154.

50 Xu, D.-Y., Chen, X. and Cheng, K. (2003) *J. Appl. Phys.*, **36**, 1583.

51 Li, H. and Chen, X. (2002) *Plasma Chem. Plasma Process*, **22**, 27.

52 Chen, K. and Boulos, M.I. (1994) *J. Phys. D: Appl. Phys.*, **27**, 946.

53 Hsu, K.C., Etemadi, K. and Pfender, E. (1983) *J. Appl. Phys.*, **54**, 1293.

54 Zhu, P., Lowke, J.J. and Morrow, R. (1992) *J. Phys. D: Appl. Phys.*, **25**, 1221.

55 Speckhofer, G. and Schmidt, H.-P. (1996) *IEEE Trans. Plasma Sci.*, **24**, 1239.

56 Lowke, J.J., Morrow, R. and Haidar, J. (1997) *J. Phys. D: Appl. Phys.*, **30**, (2033).

57 Freton, P., Gonzalez, J.J. and Gleizes, A. (2000) *J. Phys. D: Appl. Phys.*, **33**, 2442.

58 Chen, X. and He-Ping, L. (2001) *Int. J. Heat Mass Transfer*, **44**, 2541.

59 Freton, P., Gonzalez, J.J. and Gleizes, A. (2002) *J. Phys. D: Appl. Phys.*, **35**, 3181.

60 Gleizes, A., Gonzalez, J.J. and Freton, P. (2005) *J. Phys. D: Appl. Phys.*, **38**, R153.

61 Lago, F., Freton, P. and Gonzalez, J.J. (2005) *IEEE Trans. Plasma Sci.*, **33**, 434.

62 Murphy, A.B. (1997) *Phys. Rev. E*, **55**, 7473.

63 Murphy, A.B. (2001) *J. Phys. D: Appl. Phys.*, **34**, R151.

6
Radiofrequency Plasma Sources for Semiconductor Processing
F. F. Chen

6.1
Introduction

In the etching and deposition steps in the production of semiconductor chips, plasma processing is required for three main reasons. First, electrons are used to dissociate the input gas into atoms. Second, the etch rate is greatly enhanced by ion bombardment, which breaks the bonds in the first few monolayers of the surface, allowing the etchant atoms, usually Cl or F, to combine with substrate atoms to form volatile molecules. And third, most importantly, the electric field of the plasma sheath straightens the orbits of the bombarding ions so that the etching is anisotropic, allowing the creation of features approaching nanometer dimensions.

The plasma sources used in the semiconductor industry were originally developed by trial and error, with little basic understanding of how they work. To achieve this understanding, many challenging physics problems had to be solved. This chapter is an introduction to the science of radiofrequency (rf) plasma sources, which are by far the most common. Sources operating at zero or other frequencies, such as 2.45 GHz microwaves, lie outside our scope. Most rf sources use the 13.56 MHz industrial standard frequency. Among these, there are three main types: (1) capacitively coupled plasmas or CCPs, also called reactive ion etchers (RIEs); (2) inductively coupled plasmas (ICPs), also called transformer coupled plasmas (TCPs); and (3) helicon wave sources, which are new and can be called HWSs.

6.2
Capacitively Coupled Plasmas

The principal parts of a CCP are shown schematically in Fig. 6.1. In its simplest form, the rf voltage is applied across two parallel metal plates, generating an oscillating electric field between them. This field accelerates electrons, heating their thermal distribution to have enough high-energy electrons in the tail to cause an ionization avalanche. The density rises to an equilibrium value set by the rf power and the density of the neutral gas. The silicon wafer to be processed is attached to the normally grounded electrode by an

Advanced Plasma Technology. Edited by Riccardo d'Agostino, Pietro Favia, Yoshinobu Kawai, Hideo Ikegami, Noriyoshi Sato, and Farzaneh Arefi-Khonsari
Copyright © 2008 WILEY-VCH Verlag GmbH & Co. KGaA, Weinheim
ISBN: 978-3-527-40591-6

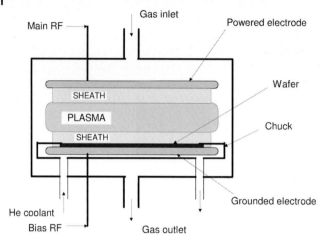

Fig. 6.1 Schematic of a capacitive discharge.

electrostatic chuck. The latter uses electrostatic charges to hold the wafer in contact with the electrode and also provides small channels for flow of helium to cool the wafer. To keep the plasma neutral, sheaths automatically form next to the electrodes providing an electric field (E-field) perpendicular to their surfaces. The potential drop in the sheath repels the fast-moving electrons so that they can escape no faster than the ions. At the same time, the sheath E-field accelerates the ions to bombard the surface and perform the beneficial functions mentioned above.

The sheath drop is of order $5\,KT_e$, where T_e is the electron temperature. For a 3 eV plasma, the ion energy is of order 15 eV. Since rf voltage is applied, the sheath drop and sheath thickness will oscillate at the rf frequency. Even if one electrode is grounded, the plasma potential will oscillate so as to make the two sheaths identical, but out of phase. The sheath oscillation will affect the ion energy distribution function (IEDF), depending on the transit time of the ions through the sheath. At low pressures, the IEDF at the wafer surface tends to be bimodal, with peaks at the maximum and minimum sheath drops, since a sine wave changes slowly at its extrema. CCPs are relatively inefficient ionizers and work best at high pressures and low densities. The sheath thickness, therefore, can become measurably large, of order millimeters. If the ion mean free path (mfp) for collisions with neutrals is smaller than the sheath thickness, the IEDF will be pressure-broadened. There are also other heating mechanisms. In resonant heating, some fast electrons can travel between the two sheaths without colliding, and those with just the right velocity can catch each sheath in its expanding phase, thus getting accelerated at each bounce. Such effects in classic CCPs are further described in textbooks [1,2].

6.2.1
Dual-Frequency CCPs

If ion bombardment energies larger than the normal sheath drop are desired, one can apply a second rf source, a *bias oscillator*, to the electrode bearing the substrate.

This is usually at a lower frequency, which has a larger effect on the massive ions. The time-averaged sheath drop will then increase by the following rectification effect. When the electrode is driven positive, a large electron current flows to it through the lowered Coulomb barrier, but when the electrode goes negative, there is no corresponding ion current because the ions are much slower. Unless the electrode emits electrons, it will accumulate a negative charge. Applying a large rf bias voltage to the substrate will, therefore, increase the dc (direct current) sheath drop even though the bias voltage is ac. It is not generally possible to apply a dc voltage directly, since parts of the wafer may be non-conducting. An rf voltage, however, will be conveyed capacitively through these insulating layers.

Bias power supplies have been used for many years, but recently the dual-frequency concept has found an important application in CCPs with extremely thin gaps. These new devices perform well in oxide etch; that is, in the etching of SiO_2, a difficult process since Si intrinsically etches faster than its oxide. The reason that thin-gap CCPs work is not yet understood, but interest in them has spawned computational studies which have advanced the science of CCPs in general. Figure 6.2 shows a schematic of this type of source. The electrodes are asymmetric; the wafer-bearing plate is smaller to enhance the sheath drop there. The high frequency produces the plasma, and the low frequency controls the ion distribution in the sheath. These devices are quite different from the original RIEs because they operate at high pressure (10–100 mtorr), and the gaps are very small (1–3 cm). With high rf bias voltage, the sheaths are quite thick and can occupy most of the volume, leaving only a small region of quasineutral plasma near the midplane. In this limit, what happens in the sheath controls the plasma production. Upon striking the substrate, those electrons that have penetrated the sheath produce more electrons by secondary emission. The emitted electrons are then accelerated toward the plasma by the sheath field. They ionize the neutral gas inside the sheath, since the ionization mfp can be smaller than the sheath thickness. The avalanche that creates the plasma then starts in the sheath.

Since the sheaths oscillate at two frequencies and their beats and harmonics, it is clear that extensive computer simulation is required to model the complicated

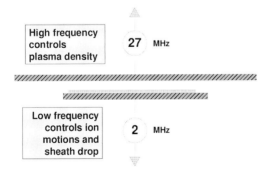

Fig. 6.2 Schematic of a thin-gap, dual-frequency CCP.

behavior in these collisional sheaths. We quote here just a few results from Lee's group at Pohang University in South Korea [3,5]. It is well known [1] in CCPs that the density increases with the square of the frequency. At constant power, therefore, the density increases, and the Debye length decreases, with frequency. The resulting change in sheath thickness is clearly demonstrated in Fig. 6.3. The IEDFs and EEDFs change not only with frequency but also with pressure. The pressure variation of IEDF is shown in Fig. 6.4, where it is seen that the typical bimodal distribution at low pressure is smoothed out by collisions at high pressure. Particle-in-cell simulations are invaluable in understanding complicated plasmas such as these. However, some aspects, such as why CCPs create less damage to oxide layers in etching, are still beyond the capabilities of theory. At this point, CCPs have been revived both as important manufacturing tools and as academically interesting subjects.

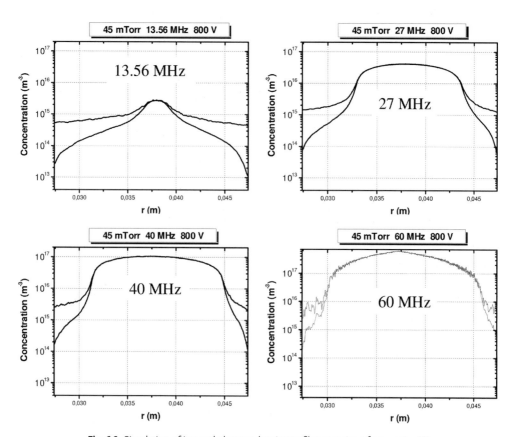

Fig. 6.3 Simulation of ion and electron density profiles at various frequencies [5].

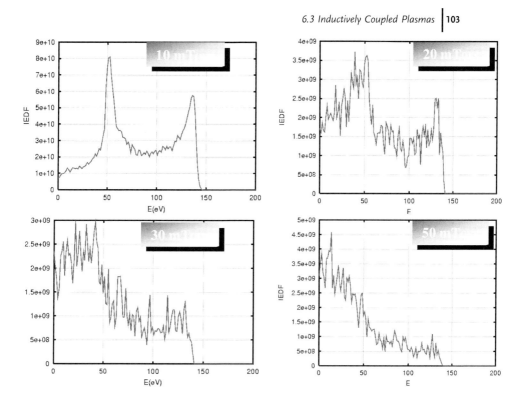

Fig. 6.4 Ion distributions at the substrate for different pressures [3].

6.3
Inductively Coupled Plasmas

6.3.1
General Description

Though simple and inexpensive, the original CCPs had a number of disadvantages, and a new generation of plasma sources was called for. For instance, the internal electrodes in CCPs introduced unnecessary impurities into the plasma. Until dual-frequency CCPs were introduced, there was a lack of control: changing the rf power changed both the plasma density and the sheath drop, and varying the pressure to do this would also change the chemistry. The high pressures also created a dust problem: negatively charged particulates of micrometer size or larger would form and be suspended above the substrate by the electric field, and these would collapse onto the wafer at plasma turn-off, thus destroying some of the chips. These problems are overcome in ICPs, which use an external coil ("antenna") to induce an electric field inside the chamber according to Faraday's law. The most common antenna shapes are illustrated in Fig. 6.5.

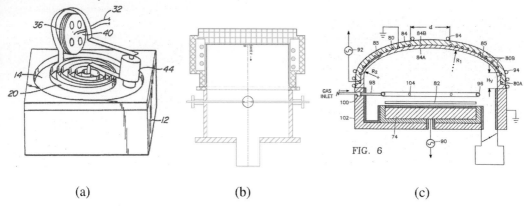

(a)　　　　　　　　　　(b)　　　　　　　　　　(c)

Fig. 6.5 Three types of ICP antennas: (a) planar coil; (b) cylindrical coil; (c) dome shaped.

Figure 6.5(a) shows the original Lam patent [6] for the TCP, whose antenna is a spiral coil, shaped like a stove-top heating element, separated from the plasma by a thick insulating plate. The diagram shows a variable transformer for impedance matching, giving the TCP its name. A capacitive automatch circuit is now standard. In Fig. 6.5(b), the antenna is a coil around the periphery of a cylindrical chamber. This type of antenna was successfully developed by the Plasma-Therm division of Unaxis. Figure 6.5(c) shows a patent from Applied Materials [7], the largest plasma source manufacturer. Its antenna is dome-shaped, combining the features of the previous two antennas, and has provision for adding a side coil also.

When an rf current is applied to a planar coil, an oscillating magnetic field (B-field) is created both above and below it. This generates a primarily azimuthal rf electric field. Inside the vacuum chamber, this E-field starts an electron avalanche which creates the plasma. Once the plasma is there, an electron current flows in a skin layer in the direction opposite to the current in the antenna and shields the plasma from the applied field. Thus, most of the rf energy is deposited within a skin depth of the surface. The plasma created there drifts downwards and decays while doing so. To be exposed to a large density, the substrate has to be placed not too far away from the top surface; but if it is too close, the plasma will be nonuniform because the discreteness of the antenna straps will be felt. There could be a problem with capacitive coupling if high voltages are applied at the ends of the antenna, and standing-wave effects can cause nonuniformity if the coil length is an appreciable fraction of the rf wavelength. These problems can be engineered away, and the TCP has been a very successful ICP.

To see how the design of the TCP coil could affect the plasma uniformity, one can compute the B-field lines induced. This is shown in Fig. 6.6, where the spirals have been approximated by circular rings. Figure 6.6(a) and (b) compare three rings with two rings: the field shapes are almost identical. Figure 6.6(a) and (c) compare copper

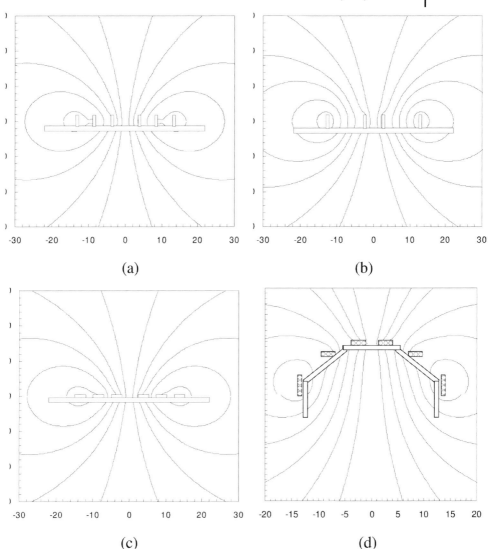

Fig. 6.6 Magnetic field patterns of ICP antennas: (a) three circular
rings; (b) two rings; (c) three rings lying flat; (d) rings arranged on a
dome.

strips standing on end with those lying flat. Though the field shapes are almost the
same, they are centered at the mid-height of the strips. Thus, flat conductors bring the
strong-field region closer to the plasma. Figure 6.6(d) shows the pattern created by an
antenna placed on a dome. It is quite different from the others; the lines diverge less
rapidly below the coils, possibly leading to more uniform ionization. This type of ICP
is manufactured by Applied Materials.

6.3.2
Anomalous Skin Depth

At first glance, the side-wound antenna of Fig. 6.5(b) should give poor plasma uniformity. Since the skin depth is of the order of a few centimeters, much less than the radius of the substrate to be processed, one would expect the density to be high only near the periphery. Actually, the opposite is true, and parameters can be adjusted to have excellent uniformity across the wafer. Radial profiles of $n(r)$, $T_e(r)$, and the rf field $B_z(r)$ are shown in Fig. 6.7. The rf field decays away from the wall with a skin depth of about 3 cm, and T_e peaks in the skin layer, as expected. However, the density actually peaks near the axis. How does plasma get created where there is no power deposition? This problem, called *anomalous skin depth*, has been known since the 1970s and has received much attention from theorists. Their favorite argument is that fast electrons accelerated in the skin layer can wander into the interior of the discharge by their thermal motions. These theories, which have been summarized [9], are usually linear, kinetic, and in Cartesian geometry. By following the orbits of electrons during many rf cycles, we have shown [8] that this fortuitous effect is caused by a combination of cylindrical geometry and the nonlinear Lorentz force $\mathbf{F}_L = -e\mathbf{v} \times \mathbf{B}_{rf}$. Figure 6.8 shows two orbits of an electron starting at rest in the skin layer as the rf field varies over four cycles. One orbit is computed including $\mathbf{F}_L$, and the other without it. The electron is accelerated in the azimuthal E-field and is reflected specularly from the Debye sheath at the wall. It makes glancing collisions with the wall and does not wander into the interior of the discharge until it has slowed down below ionizing velocities. The Lorentz force due to $v_\theta \times B_z$ is in the radial

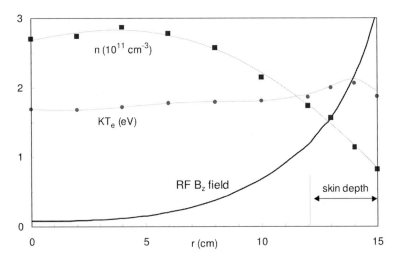

Fig. 6.7 Measurements in a side-fed ICP of plasma density, electron temperature, and rf field strength vs. radius [8].

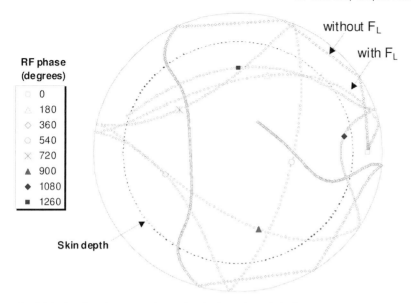

RF phase (degrees)

☐ 0
△ 180
◇ 360
◌ 540
✕ 720
▲ 900
◆ 1080
■ 1260

without F_L

with F_L

Skin depth

Fig. 6.8 Paths of an electron starting in the skin layer, through the first four rf cycles, with and without the Lorentz force.

direction and causes the electrons to collide with the wall at steeper angles, bringing them more rapidly into the center, while they still have ionizing energies. When F_L is included, $n(r)$ actually peaks near the axis, in agreement with Fig. 6.7, due to two effects: (a) orbits crossing the discharge tend to raise the density in the interior because of the smaller volume there, and (b) electrons created near the center spend a long time there before reaching the skin layer where they can gain energy. The puzzle of anomalous skin depth would appear to be solved, but further computations including particle motion in the axial direction need to be done. In any case, how ICPs actually work is not a simple problem.

6.3.3
Magnetized ICPs

In the field patterns of Fig. 6.6, it is clear that half the magnetic field energy appears above the antenna and is not utilized. This field is partly canceled by the skin current in the plasma, but the reduction is not large because the skin current is farther away and is diffuse. To channel that energy into the plasma, Colpo and co-workers [10] have modified ICPs by covering the antenna with a magnetic material. The mechanism is sketched in Fig. 6.9. The magnetization vector **H** is unaffected by the permeability μ and is the same in both parts of the diagram. When a high-μ ferrite material is added to the antenna, the field $\mathbf{B} = \mu\mathbf{H}$ is greatly enhanced, effectively capturing the magnetic energy which is normally lost and injecting it back into the plasma. Further details can be found in Chapter 2.

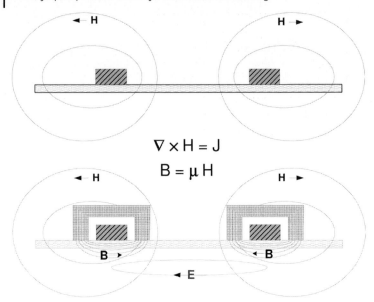

$$\nabla \times H = J$$
$$B = \mu H$$

Fig. 6.9 Schematic of magnetic field lines of an ICP antenna before and after modification with ferromagnetic material.

To cover large areas, ICP antennas can consist of parallel rods connected in various series–parallel combinations. The device of Meziani and co-workers [10,11] is shown in Fig. 6.10. It is found that the magnetic cover increases not only the rf field but also the plasma uniformity. Lee *et al.* [12–15] have also used various serpentine antennas, with permanent magnets placed in pairs above the antenna (Fig. 6.11). A physical picture of the resulting fields and their effect has not been given, but these configurations have been modeled in detail by Park *et al.* [16]. Though ICPs are standard in the semiconductor industry, they are being further developed for large-area display applications by the use of magnetic materials.

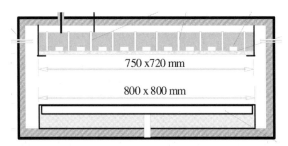

Fig. 6.10 Processing chamber with a magnetized array of linear ICP antennas [11].

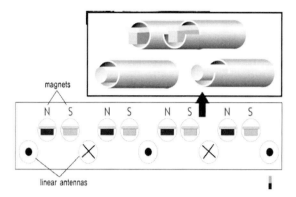

Fig. 6.11 ICP enhanced with permanent magnets above the antenna legs, all encased in quartz tubes (G.Y. Yeom, personal communication).

6.4
Helicon Wave Sources

6.4.1
General Description

This type of plasma source was discovered by Boswell [17] in 1970, and its wave nature was verified in 1984 [18]. As in an ICP, an antenna, an rf power source, and a matching circuit are used; but a dc magnetic field is added. In the presence of this B_0-field, the antenna launches circularly polarized *helicon* waves, related to "whistler" waves in the ionosphere, along $\mathbf{B}_0$. For reasons not known for over a decade, these waves are very efficient ionizers, producing plasma densities well over $10^{19}\,\mathrm{m}^{-3}$ with only a kilowatt of rf power. Helicon sources differ from CCPs and ICPs in several essential ways. First, they are more complicated because of the dc B-field. Second, they can generate plasma densities an order of magnitude higher than previous devices with the same power. And third, they were studied and understood before widespread acceptance by industry.

Figure 6.12 shows a typical apparatus for studying the propagation of helicon waves and the nature of the plasmas created. A commercial helicon source [19] is shown in Fig. 6.13. This device uses two ring antennas with opposite currents, and the magnetic field shape is controlled with the current ratio in two coils, one enclosing the other, also carrying opposite currents. These $m = 0$ antennas, where m is the azimuthal symmetry number, are less common than $m = 1$ antennas, of which two are shown in Fig. 6.14. The Nagoya III antenna is symmetric and launches both right-hand (RH) and left-hand (LH) circularly polarized waves in both directions. The HH antenna is a half-wavelength long and is meant to match the helicity of the helicon wave. It launches RH waves in one direction and LH waves in the other, the directions reversing with the $\mathbf{B}_0$ field. The most efficient coupling is with bifilar antennas (two HH antennas 90° apart in azimuth, also phased 90° apart in time, giving a field that rotates with the helicon wave [20]).

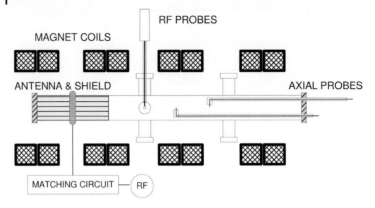

Fig. 6.12 Typical experimental setup for studying helicon waves.

6.4.2
Unusual Features

Many challenging problems have arisen in the behavior of helicon discharges, and these have been solved one by one. When either the rf power or the magnetic field is increased after breakdown, the plasma density increases not continuously but by

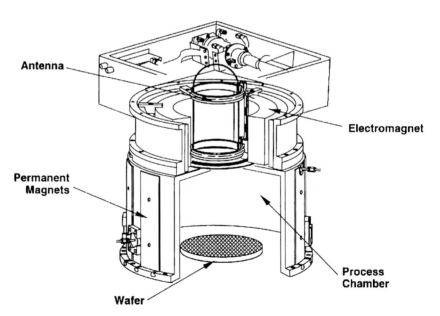

Fig. 6.13 The PMT MØRI device, a commercial reactor with two coplanar magnetic coils at the source and with multi-dipole confinement with permanent magnets in the processing chamber. The cut-away box at the top is the matching circuit [19].

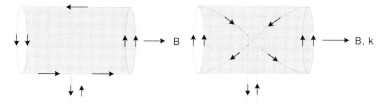

Fig. 6.14 Two common helicon antennas: a Nagoya Type III antenna (left) and a half-helical (HH) antenna (right).

discrete jumps. At low power, the coupling is capacitive, and the discharge is a CCP. As the power is raised, the plasma jumps into ICP operation as inductive coupling takes hold. When the conditions for propagation of helicon waves are met, there is a large jump into the lowest helicon mode, and the peak density can be 20 times higher than in ICP mode. There may be further jumps into higher-order radial modes. A second observation is that the density peaks, not under the antenna, but many centimeters downstream from it, as shown in Fig. 6.15. There are three possible reasons for this. First, as seen from the T_e curve, the temperature decays downstream because of inelastic collisions (line radiation and ionization). Second, plasma is ejected from the antenna region with a drift velocity comparable to the ion acoustic

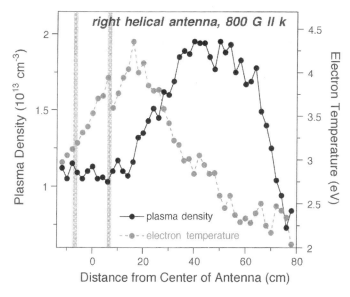

Fig. 6.15 Plasma density $n(z)$ and temperature $KT_e(z)$, where z is the direction of $\mathbf{B}_0$. The antenna is located between the vertical bars.

velocity c_s. This is just the Bohm criterion for sheath formation. There is no sheath here, but the criterion still has to be met if there are very few ions created downstream which can travel back to the antenna. Finally, there may actually be a little ionization downstream due to parametric instabilities, an effect recently verified in experiment. The result is that the helicon discharge is an ideal "remote" source, in which the substrate can be located in a desirable region of high density and low T_e, far from the high fields near the antenna.

A second, more important, problem is: What causes the high ionization efficiency of HWSs compared with ICPs? There is no difference in confinement, since the B-fields of 50–1000 Gs (5–100 mT) normally used are not sufficient to confine the ions, and the electrons are not confined axially. Hence, the difference must be in the way in which rf energy is absorbed. In ICPs, collisional absorption converts the electron energy gained in the skin layer into a general rise in KT_e, and the tail of this Maxwellian distribution does the ionization. Since helicon waves travel along $\mathbf{B}_0$ with velocities comparable to those of 100 eV electrons, could they not trap electrons and accelerate them by Landau damping? This mechanism was suggested by Chen [21], and several groups have indeed detected the fast electrons indicative of this process. However, these electrons were too few in number to account for the increased ionization, and this hypothesis was later disproved [22]. Meanwhile Shamrai *et al.* [23] suggested a new absorption mechanism; namely mode conversion to Trivelpiece–Gould (TG) modes at the boundary. The TG mode, essentially an electrostatic electron cyclotron wave in a cylinder, is needed to satisfy the radial boundary conditions. The helicon wave itself is weakly damped by collisions, but it transfers its energy to the TG wave, which is rapidly damped as it propagates slowly inward from the boundary. Computations by Arnush [24] have confirmed the dominance of this absorption process. TG modes are difficult to detect, however, because they only occur in a thin layer at the wall; however, by using a low B-field to widen this layer and developing an rf current probe, Blackwell et al. [25] verified the existence of this mechanism.

An efficient absorption mechanism increases the plasma resistance R_p, and therefore a greater fraction of the rf energy is deposited in R_p rather than in the parasitic resistances R_c in the matching circuit and connections. If $R_p/R_c \gg 1$, there would be no advantage of higher R_p/R_c. In ICPs with $n \leq 10^{18}\,\mathrm{m}^{-3}$, however, R_p/R_c is small enough that increasing it by operating in the helicon mode would deposit more energy into the plasma. The large densities $\geq 10^{19}\,\mathrm{m}^{-3}$ in the helicon's "Big Blue Mode" are a different matter. The density is high and fully ionized only in a central core; the more uniform deposition with TG modes at the edge is not seen. We believe that in this case there is an ionization instability, in which neutrals are depleted near the axis, and this allows T_e to rise and the ionization rate to grow exponentially.

In free space, the whistler wave is known to propagate only when it is RH polarized. Helicons, however, are in a bounded medium, and it is easily shown that both RH and LH polarizations are possible. It was unexpected that only the RH mode is strongly excited in practice; the LH mode hardly exists. A helical antenna is therefore highly directional and launches helicon waves only in the direction dictated by the sign of its helicity and the direction of $\mathbf{B}_0$. Computations confirm this, but the physical explanation of this effect is not simple. The LH wave has a somewhat smaller amplitude at the edge than the RH wave does. Perhaps this causes the coupling to the TG mode to be much weaker.

A final puzzle we can mention is that of the low field peak: the density is found to have a small peak at low B-fields of the order of 10–100 Gs (1–10 mT), whereas it should increase linearly with $\mathbf{B}_0$. Computations have shown [26] that this peak is caused by constructive interference by the helicon wave reflected from a back plate and occurs only with bidirectional antennas. This effect can be used to design more economical helicon reactors using low fields. Most of these advances in under-standing were made with simple geometries and uniform B-fields. To model a realistic reactor such as that shown in Fig. 6.13 would require extensive computer simulations. Several of these have been done, and these have shown that features such as the downstream density peak and TG modes actually play a role even in complex geometries.

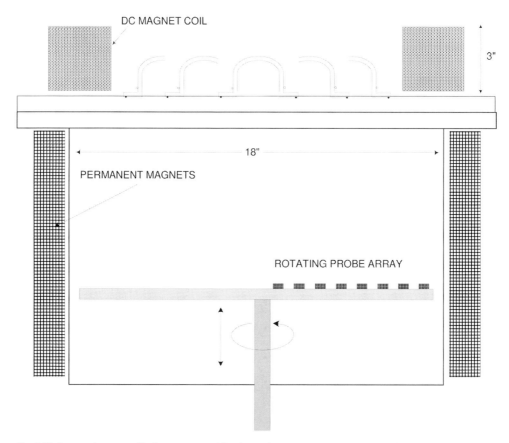

Fig. 6.16 Seven-tube array of helicon sources with a large electromagnet.

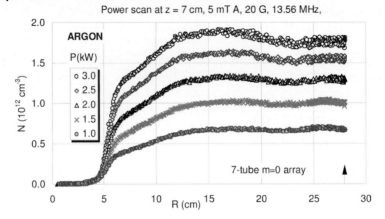

Fig. 6.17 Density profiles $n(r)$ at various rf powers in argon, 7 cm below the distributed source of Fig. 6.16.

6.4.3
Extended Helicon Sources

As in the case of ICPs, helicon sources can also be extended to cover large areas. This can be done with serpentine antennas [27] or with multiple small tubes. Figure 6.16 shows a distributed source arrayed with one tube surrounded by six others [28]. Each tube is very short, with a simple $m = 0$ antenna; and a single large magnet coil surrounds the array. Figure 6.17 shows density profiles at various rf powers. With a total power of 3 kW, a plasma uniform to $\pm 3\%$ can be created 7 cm below the sources with a density of nearly $10^{18}\,\mathrm{cm}^{-3}$ over a 400 mm diameter substrate.

In summary, the study of plasma sources has not only played an essential role in the production of semiconductor chips but has also provided challenging problems for the academic community to solve. The standard CCP and ICP sources are being extended and modified for new applications, and the new helicon sources are a promising prospect for the next generation of etching and deposition reactors.

References

1 Lieberman, M.A. and Lichtenberg, A.J. (2005) *Principles of Plasma Discharges and Materials Processing*, 2nd edn , Wiley-Interscience, Hoboken, NJ.

2 Chen, F.F. and Chang, J.P. (2003) *Principles of Plasma Processing*, Kluwer Academic/Plenum Publishers, New York.

3 Lee, J.K., Babaeva, N.Yu., Kim, H.C., Manuilenko, O.V. and Shon, J.W.

(2004) *IEEE Trans. Plasma Sci.*, **32**, 47.

4 Babaeva, N.Yu., Lee, J.K. and Shon, J.W. (2005) *J. Phys. D: Appl. Phys.*, **38**, 287.

5 Lee, J.K., Manuilenko, O.V., Babaeva, N.Yu., Kim, H.C. and Shon, J.W. (2005) *Plasma Sources Sci. Technol.*, **14**, 89.

6 US Patent 4,948,458, Lam Research (1990).

7 US Patent 4,948,458, Applied Materials (1993).

8 Evans, J.D. and Chen, F.F. (2001) *Phys. Rev. Lett.*, **86**, 5502.

9 Kolobov, V.I. and Economou, D.J. (1997) *Plasma Sources Sci. Technol.*, **6**, R1. For example.

10 Meziani, T., Colpo, P. and Rossi, F. (2001) *Plasma Sources Sci. Technol.*, **10**, 276.

11 Colpo, P., Meziani, T. and Rossi, F. (2005) *J. Vac. Sci. Technol. A*, **23**, 270.

12 Lee, Y.J., Han, H.R. and Yeom, G.Y. (2000) *Surf. Coat. Technol.*, **133**, 612.

13 Lee, Y.J., Kim, K.N., Song, B.K. and Yeom, G.Y. (2002) *Mater. Sci. Semicond. Process.*, **5**, 419.

14 Lee, Y.J., Kim, K.N., Song, B.K. and Yeom, G.Y. (2003) *Thin Solid Films*, **435**, 275.

15 Kim, K.N., Lee, Y.J., Kyong, S.J. and Yeom, G.Y. (2004) *Surf. Coat. Technol.*, **177**, 752.

16 Park, S.E., Cho, B.U., Lee, J.K., Lee, Y.J. and Yeom, G.Y. (2003) *IEEE Trans. Plasma Sci.*, **31**, 628.

17 Boswell, R.W. (1970) *Phys. Lett. A*, **33**, 457.

18 Boswell, R.W. (1984) *Plasma Phys. Control. Fusion*, **26**, 1147.

19 Tynan, G.R., Bailey, A.D., III, Campbell, G.A., Charatan, R., deChambrierA., Gibson, G., Hemker, D.J., Jones, K., Kuthi, A., Lee, C., Shoji, T. and Wilcoxson, M. (1997) *J. Vac. Sci. Technol. A*, **15**, 2885.

20 Miljak, D.G. and Chen, F.F. (1998) *Plasma Sources Sci. Technol.*, **7**, 61.

21 Chen, F.F. (1991) *Plasma Phys. Control. Fusion*, **33**, 339.

22 Blackwell, D.D. and Chen, F.F. (2001) *Plasma Sources Sci. Technol.*, **10**, 226.

23 Shamrai, K.P. and Sharanov, V.B. (1995) *Plasma Phys. Control. Fusion*, **36**, 1015. (1996) *Plasma Sources Sci. Technol*, **5**, 43.

24 Arnush, D. (2000) *Phys. Plasmas*, **7**, 3042.

25 Blackwell, D.D., Madziwa, T.G., Arnush, D. and Chen, F.F. (2002) *Phys. Rev. Lett.*, **88**, 145002.

26 Chen, F.F. (2003) *Phys. Plasmas*, **10**, 2586.

27 Jewett, R.F.Jr., (1995) PhD thesis, University of New Mexico.

28 Chen, F.F., Evans, J.D. and Tynan, G.R. (2001) *Plasma Sources Sci. Technol.*, **10**, 236.

7
Advanced Plasma Diagnostics for Thin-Film Deposition
R. Engeln, M.C.M. van de Sanden, W.M.M. Kessels, M. Creatore, and D.C. Schram

An ever-increasing amount of diagnostics is available to study plasma and obtain quantities like atomic and molecular densities, ion and electron densities, temperature of electrons and heavy particles, and their velocities, to just name a few. But also surface diagnostics that can unravel mechanisms leading to deposition or etching have become available, also to the non-laser specialist, and are now being introduced in the field of plasma physics. In this chapter we will focus on *laser-based* diagnostics that have been introduced in the field of plasma physics. The basic principles of the diagnostics will be explained and some examples will be discussed where these techniques were successfully applied. For a more comprehensive understanding of the techniques the reader will be referred to textbooks.

7.1
Introduction

The high temperature of electrons within plasma makes plasma chemically reactive. As a result of this high specific reactivity (in other words the chemical influence per particle) a very fast chemical conversion or modification is possible with relatively few particles. Many applications such as plasma deposition, etching and surface modification are based on this high chemical reactivity. Plasmas can break apart all molecules and this enables the synthesis of materials, which would otherwise be impossible. The ability of plasmas to radiate has been exploited in lighting and gas lasers. The conductivity of plasmas has been applied in power switches and their high energy density can be employed to weld, cut and melt materials.

However, usually plasmas are of low or moderate density and therefore the amount of treatable material per unit time is not very large. This makes working with plasmas relatively expensive. Therefore a good knowledge of the most important processes is necessary to keep the process costs within an acceptable range. Also the particles that are responsible for, for example, the deposition or etching process need to be known in order to be able to optimize their production. But also to extend the

Advanced Plasma Technology. Edited by Riccardo d'Agostino, Pietro Favia, Yoshinobu Kawai, Hideo Ikegami, Noriyoshi Sato, and Farzaneh Arefi-Khonsari
Copyright © 2008 WILEY-VCH Verlag GmbH & Co. KGaA, Weinheim
ISBN: 978-3-527-40591-6

technological possibilities of plasma, better insights in the production mechanisms of the particles is indispensable.

Many diagnostics are available to the (plasma) researcher and it is up to the researcher to find the right diagnostic that will deliver the information being sought. In this chapter we will focus on a selection of *laser-based* diagnostics that have shown to be applicable to the study of plasma. Some techniques that will be described are already being used for many years in plasma physics, while others have recently been introduced. Although the focus will be on gas-phase diagnostics, some surface diagnostics will be introduced as well.

7.2
Diagnostics Available to the (Plasma) Physicist

A large variety of gas-phase and surface diagnostics are available to the plasma physicist. Non-optical techniques like Langmuir probes and mass spectrometry are very successfully applied in the study of plasma [1]. But also optical techniques, which can be divided into passive and active spectroscopic techniques, have shown their merits. In *Plasma Spectroscopy* by Griem the working principle of several passive optical techniques is discussed [2]. Demtröder discusses many laser-based active spectroscopic techniques and their applications [3]. Most of the laser-based diagnostics were first applied under clean and well-determined conditions, and used to measure, for example, absorption cross-sections. But, while lasers were becoming more and more available also to non-laser specialists, these techniques are now also shown to be applicable in more harsh circumstances, like plasmas and flames.

With active spectroscopic gas-phase diagnostics like optical emission, actinometry, absorption and scattering spectroscopy, quantities like atomic and molecular densities, ion and electron densities, temperature of electrons and heavy particles, and their velocities, can be determined. Surface diagnostics, like (spectroscopic) ellipsometry and attenuated total internal reflection spectroscopy, give information on, for example, roughness of the surface, growth rate, and optical properties of the thin film. If applicable in the harsh environment of plasma, second harmonic generation (SHG) and sum-frequency generation (SFG) spectroscopy could unravel mechanisms leading to deposition or etching.

7.3
Optical Diagnostics

7.3.1
Thomson–Rayleigh and Raman Scattering

In most of the applications where plasma is used the motivation behind its use is the creation of excitation, light, and the dissociation of molecules. These processes all begin with the production of electrons, i.e. ionization usually accompanied by excitation and thus light production. The most important parameter that characterizes the plasma is therefore the electron density. It determines the conductivity of the plasma, the excitation and light emission, the production of radicals, and thus the chemical reactivity. Another important parameter is the electron temperature. In most technological plasmas this temperature is of the order of 1 to 4 eV (1 eV = 11 600 K).

Several techniques are available to determine both the electron density and temperature, i.e. Langmuir probes, Stark broadening, interferometry and line and continuum emission [4], to name just a few. All have their advantages and disadvantages. The use of Langmuir probes is cheap; however, it cannot always be applied, due to its intrusive character. Also, as the probe is in fact measuring the disturbance by the probe on the charged species, a model is necessary to extract the electron density and temperature from the measurement. The Stark broadening, interferometry and line and continuum emission measurements are non-intrusive, but, as they are all *line-of-sight* measurements, Abel inversion is needed to obtain spatial resolution. Thomson scattering is a non-intrusive optical diagnostic with which spatially resolved electron densities and temperatures can be readily obtained.

Thomson scattering is based on the scattering of light from free electrons. The amount of scattered photons is linear proportional to the electron density and laser power. Due to the movement of the electrons, the scattered light is Doppler broadened with respect to the excitation source. As the electrons have a much higher velocity than the heavy particles, the light that scatters of the free electrons shows a much broader scattering feature than the scattering originating from the heavy particles, i.e. the Rayleigh scattering feature. Although the electron scattering cross-section is much smaller than the heavy particle scattering cross-section and the density of the heavy particles is usually much higher than the electron density, the large difference between the Doppler broadening allows distinguishing between the electron and heavy particle scattered signal in frequency space. Interpolation of the Thomson scattering feature in the central part of the spectrum allows for the determination of the Rayleigh scattering signal. In addition to the Thomson and Rayleigh scattering components, a third component is always present, i.e. the stray light component arising from windows and surfaces of the chamber. The width is determined by the laser line width. The basic components of a Thomson–Rayleigh setup are shown in Fig. 7.1(b).

The total Thomson scattered light intensity is directly proportional to the *electron density*. This means that if the sensitivity of the system is calibrated, the electron density can be determined from the area under the Thomson spectrum. The calibration can easily be performed by measuring the Rayleigh scattered signal from a known amount of gas in the plasma chamber [5]. The *electron temperature* can be determined from the Doppler width of the Thomson scattered spectrum. The width is determined by the velocity distribution function of the scattering electrons. It should be mentioned that the velocity distribution function that is actually measured, is the one-dimensional velocity distribution function in a direction determined by the relative directions of the incident laser and the detection axes [5]. When the

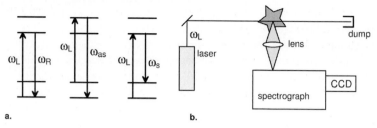

Fig. 7.1 (a) Energy level diagrams showing, from left to right, Rayleigh scattering, anti-Stokes scattering and Stokes Raman scattering. (b) Schematic view of a setup that can be used to perform Thomson–Rayleigh or Raman scattering experiments.

distribution function is Maxwellian, the electron temperature, $\hat{T}_e$ (in eV), is determined by

$$\hat{T}_e = \left(\frac{\Delta\nu_{\text{Th}}}{4\nu_0\sin(\theta/2)}\right)^2 \times \frac{m_e c^2}{2e} \tag{7.1}$$

Here, $\Delta\nu_{\text{Th}}$ is the full width at half maximum of the Thomson scattering spectrum, ν_0 the laser frequency, θ the scattering angle, and m_e, e, and c are the electron mass, the electron charge, and the speed of light, respectively.

As the Thomson scattering signal is proportional to the laser power, high-power, and often pulsed, lasers are used in these experiments. The advantage of a pulsed laser is the fact that the measured emission of the plasma can be decreased relatively with respect to the measured scattered radiation. However, care has to be taken not to use too high powers for the radiation, since that can lead to excitation and/or dissociation of the gasses in the plasma, thus disturbing the system under investigation.

The strength of the Rayleigh scattering signal is proportional to the total density of the heavy particles in the plasma. From this scattering signal no direct information can be obtained about what kind of species are in the plasma. However, using the difference in depolarization ratio between different species, one can in some cases get more information about the species in the scattering medium.

Thomson and Rayleigh scattering are elastic scattering processes. The inelastic scattering process, in which the molecule is in a different state after the scattering process, is called Raman scattering. When the molecule is after the scattering process in a higher (lower) state, i.e. the scattered light has a longer (shorter) wavelength, the scattering is called (anti-)*Stokes* Raman scattering (see Fig. 7.1(a)). This technique has not often been employed in the study of plasma, because of the small cross-section for Raman scattering. However, the advantage is that every molecule has a Raman spectrum, and can thus in principle be detected. This is especially interesting for homonuclear diatomic species like H_2, N_2, and O_2, which are very difficult to detect otherwise. As in the case of Thomson–Rayleigh scattering, a laser is used as a light source. As the Raman scattering cross-section becomes bigger at shorter wavelength (the reason that a cloudless sky appears blue), light in the blue part of the spectrum is preferably used. A laser that is very often used in these studies is the Ar ion laser, which has strong emissions at 488 and 514 nm.

A main drawback of the spontaneous scattering processes discussed above is the fact that the scattered light is emitted in all directions, which makes it very often difficult to detect in environments that strongly radiate. A technique in which the signal is created in a laser-like beam is the so-called coherent anti-Stokes Raman scattering (CARS) technique. During a CARS experiment two collinear laser beams with frequencies ω_1 and ω_2 ($\omega_1 > \omega_2$) are focused into the sample. The two waves are mixed via the nonlinear polarization of the sample. When $\omega_1 - \omega_2$ equals the frequency of a Raman-active transition of the medium, an anti-Stokes (as) and Stokes (s) wave at the frequencies $\omega_{as} = 2\omega_1 - \omega_2$ and $\omega_s = 2\omega_2 - \omega_1$ are generated [6]. Spatial filtering allows for very efficient reduction of the background radiation without loss of CARS signal [6,7].

7.3.2
Laser-Induced Fluorescence

In order to optimize operational characteristics of plasma in terms of, for example, deposition, etching or surface modification, densities of the most important radicals have to be determined. Several techniques have successfully been applied in determining densities and fluxes of radicals in processing plasmas. One of these experimental techniques is called laser-induced fluorescence (LIF). During a LIF experiment a light source, very often a laser, is used to excite species to an excited state and the fluorescence emitted by the excited state is detected. The number of photons, N_{fl}, emitted at a wavelength λ_{jk} from a volume V after excitation of atoms or molecules from an initial state i to a final state f, with a laser with intensity I_l, is given by

$$N_{fl}(\lambda_{jk}) = \sigma_{ij} I_l n_i V \frac{A_{jk}}{A_j + R} \tag{7.2}$$

Here, $\sigma_{ij} I_l n_i$ is the amount of photons absorbed per unit volume and time, with σ_{ij} the cross-section for absorption from state i to state j. A_{jk} is the Einstein coefficient for the transition from state j to k, which is responsible for the emitted fluorescence at λ_{jk}. A_j is determined by the fluorescence lifetime, τ_j, of state j, i.e. $A_j = 1/\tau_j$. R is the total loss rate due to other processes than fluorescence. At low pressures this loss process can usually be neglected.

During any experiment only part of the total emitted fluorescence can be collected. The measured LIF signal, S_{fl}, is then written as

$$S_{fl} = N_{fl} \frac{\Omega}{4\pi} Q \tag{7.3}$$

where Ω is the solid angle over which the fluorescence is detected. Q incorporates all losses due to the optics between the imaged LIF volume and the detector used to measure the LIF signal and the quantum efficiency of the detector [8]. Equations (7.2) and (7.3) show that the LIF signal is proportional to the density of the atoms or molecules in the lower state *i*.

The sensitivity of the technique stems from the fact that the detector recording the LIF signal is not detecting any LIF signal when the laser is not on resonance with a transition of the species under investigation; this is a so-called *zero-background* measurement. Also, the detection can be performed on a different wavelength as the excitation wavelength, i.e. *off-resonant* detection. In this way spurious scatter from optics in the beam path can be blocked by means of optical filters.

As in Thomson–Rayleigh scattering experiments, also here the emission of the plasma detected during the experiment can be reduced by gated detection, i.e. the detector only "looks" at the plasma when species, which were excited by the light source, fluoresce. For an absolute density measurement a calibration is necessary to determine, for example, V, Q, and Ω.

In order to be able to detect a molecule by means of laser-induced fluorescence, the molecule should possess at least an excited state that is preferably single-photon accessible and is at an energy that corresponds to a wavelength that can be produced with the help of a dye laser. Roughly speaking, this means between 800 and 200 nm. Nd:YAG-pumped tunable dye lasers readily deliver light at wavelengths with which ground state densities of molecular radicals can be determined. Also, the fluorescence lifetime of the upper state should not be too long, since otherwise the excited particle might have moved out of the detection volume or de-excited via collisions.

For the detection of the ground state of several atoms and small molecules the first prerequisite is not fulfilled. For example, in the case where atomic ground state densities have to be determined, the wavelength for excitation is very often in the vacuum ultraviolet (VUV) part of the spectrum, and other excitation schemes have to be used. One of the schemes that have been successfully applied in determining ground state densities of atomic hydrogen, nitrogen, and oxygen is two-photon absorption laser-induced fluorescence (TALIF) [9–14]. The excitation from the ground state is performed with two photons in the UV, and the fluorescence is detected in the (infra)red. The detected fluorescence is proportional to the ground state density, like in the normal LIF detection scheme, but the dependence on light intensity is quadratic. This makes calibration a tedious exercise.

In the case where the density of molecular hydrogen in its electronic ground state needs to be determined, CARS has been successfully applied [6,7]. However, the populations in the higher ro-vibrational states ($v > 3$) are too low to be detected with this technique, and LIF can be used, but VUV light is necessary to excite the molecule to the first electronically excited state [15,16]. Also the LIF signal is emitted in the VUV, which makes a more elaborate experimental setup necessary.

7.3.3
Absorption Techniques

Absorption techniques are based on the measurement of the intensity decrease of a light beam that passes through a medium. One can show that the intensity decrease as function of path length through the medium can be written as

$$\frac{I}{I_0} = \exp(-n\sigma l) \tag{7.4}$$

This is the so-called Lambert–Beer law. I is the intensity after passing through the medium, I_0 the intensity before passing through the medium, n the density of the absorbing specie, σ the absorption cross-section, and l the length over which the medium absorbs. The advantage over LIF spectroscopy is immediately clear: when the cross-section of the absorbing medium is known, the density can be directly determined from the ratio of I and I_0, and no calibration is necessary. However, in contrast to LIF, the absorption measurement is not a *zero-background* measurement. One has to record small changes, i.e. $\Delta I = I_0 - I$, on a large signal I_0. The absorption measurement is also a so-called *line-of-sight* measurement, which means that spatial resolution cannot be obtained from one measurement.

There are many different absorption detection schemes reported in the literature, all having their advantages and disadvantages. One can make a rough distinction between (a) schemes in which broadband light sources are used and (b) schemes in which narrow band light sources are used. When a broadband light source is used in an absorption experiment the light can be analyzed by means of a monochromator, which disperses the light, or with a Fourier transform (FT) spectrometer, in which the intensity of the light after passing a Michelson interferometer is measured. When a photomultiplier is used as the detector behind a monochromator, every wavelength has to be recorded one after the other. However, nowadays very often a CCD camera is used, which allows recording a range of wavelengths at the same time. In a FT spectrometer a collimated beam from a light source is divided into two by a beamsplitter and sent to two mirrors. These mirrors reflect the beams back along the same paths to the beamsplitter, where they interfere. If the optical path difference between the two beams is zero or a multiple of the wavelength of the light then the light beams will constructively interfere and the output will be bright, but if the optical path difference is an odd multiple of half the wavelength of the light then the light beams will destructively interfere and the output will be dark. In one arm the light is reflected to the beamsplitter after traveling a fixed distance, while in the other arm the light is reflected to the beamsplitter from a mirror of which the position is changed during the experiment. After interference of both beams on the beam splitter the beam is directed through the sample and the intensity of the light is recorded as function of the optical path length difference introduced by the moving mirror, i.e. a so-called interferogram is recorded. The Fourier transform of this interferogram shows the intensity of the light at every wavelength, as in the case of the monochromator. During an experiment two interferograms are recorded, i.e. one with and one without sample in the beam. The ratio of the Fourier transforms of the interferograms shows the absorption spectrum (see also Eq. (7.4)). The advantage of the FT spectrometer is the fact that during the experiment the detector is measuring all wavelengths at the same time, and thus the total intensity of the light source (this is called multiplexing), while in the monochromator the detector only measures the intensity at a certain wavelength. Certainly in the infrared part of the spectrum, where detectors are less sensitive, this is an important advantage. This is why in the early days FT spectrometers were mainly used to record spectra in the (far) infrared. Due to the fact that it is easier to record spectra with a resolution of about $0.1 \, \text{cm}^{-1}$ with a FT spectrometer than with a monochromator, FT spectrometers are nowadays also used to record spectra in the visible and even the UV part of the spectrum.

In the case where narrow band tunable lasers are used in absorption spectrometers, the laser itself acts as the frequency selective element. An absorption spectrum is recorded by tuning the laser over the absorption feature. The intensity is recorded in front and behind the sample, and using Eq. (7.4) the absorption can be deduced. If the cross-section of the transition is known, the density can directly be determined. This is in sharp contrast to the previously discussed LIF technique, where a calibration is always necessary.

Absorption spectroscopy has been successfully applied both with continuous wave (cw) and pulsed lasers. With cw tunable lasers many different techniques have been introduced, all aimed at increasing the sensitivity of the technique. One is particularly interesting to mention, since it is very often easy to implement. In that case the intensity of the laser is modulated, e.g. by means of a chopper, and the signal is recorded with a lock-in amplifier at the modulation frequency. In this way any background light from the sample (which is not modulated) or electronic signals at other frequencies than the modulation frequency is suppressed, which can lead to a sensitivity increase of an order of magnitude.

In the case where pulsed lasers are used, another very successful experimental scheme has been developed. Pulsed lasers have inherently large pulse-to-pulse intensity fluctuations, which make it very difficult to detect small changes on the intensity. In 1988 O'Keefe and Deacon [17] demonstrated a new direct absorption spectroscopic technique that can be performed with a pulsed light source and that has a significantly higher sensitivity than obtainable in "conventional" pulsed absorption spectroscopy. This so-called cavity ring down (CRD) technique is based upon the measurement of the rate of absorption rather than the magnitude of absorption of a light pulse confined in a closed optical cavity with a high Q-factor. The advantage over normal absorption spectroscopy results from (i) the intrinsic insensitivity of the CRD technique to light source intensity fluctuations, and (ii) the extremely long effective path lengths (many kilometers) that can be realized in stable optical cavities. In a typical CRD experiment a short light pulse is coupled into a stable optical cavity, formed by two highly reflecting planoconcave mirrors. The fraction of light that enters the cavity on one side rings back and forth many times between the two mirrors. The time behavior of the light intensity inside the cavity can be monitored by the small fraction of light that is transmitted through the other mirror (see Fig. 7.2). If the only loss factor in the cavity is the reflectivity loss of the mirrors, one can show that the light intensity inside the cavity decays exponentially with time with a decay constant τ, the "ring down time",

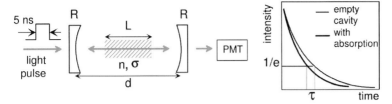

Fig. 7.2 Principle of pulsed cavity ring down spectroscopy.

given by

$$\tau = \frac{d}{c|\ln R|} \tag{7.5}$$

Here d is the optical path length between the mirrors, c is the speed of light, and R is the reflectivity of the mirrors. If R is close to unity, the approximation

$$\tau = \frac{d}{c(1 - R)} \tag{7.6}$$

can be made.

If there is additional loss inside the cavity due to the presence of absorbing and light-scattering species, the light intensity inside the cavity will still decay exponentially in time provided the absorption follows Beer's law. The ring down time can now be written as

$$\tau(\nu) = \frac{d}{c\left(1 - R + \sum \sigma_i(\nu) \times \int_0^L N_i(x)dx\right)} \tag{7.7}$$

and the sum is over all light scattering and absorbing species with frequency-dependent cross-sections $\sigma_i(\nu)$ and a line-integrated number density $\int_0^L N_i(x)dx$. The product of the frequency-dependent absorption cross-section with the number density $N_i(x)$ is commonly expressed as the absorption coefficient $\kappa_i(\nu, x)$. In an experiment it is most convenient to record $1/[c\tau(\nu)]$ as a function of frequency, i.e. to record the total cavity loss as a function of frequency, as this is directly proportional to the absorption coefficient, apart from an offset, which is mainly determined by the finite reflectivity of the mirrors.

To date, several absorption detection schemes based on the CRD principle have been reported and reviewed [18–20]. One that is worth mentioning is the so-called time-resolved cavity ring down detection scheme (τ-CRDS) [21,22]. It is based on a conventional CRD measurement of the density evolution of a particle after a temporary change has been introduced to the steady-state density. The measurement gives information on both the gas-phase kinetics and gas–surface interactions, leading to, for example, film growth.

7.3.4
Surface Diagnostics

There exist several diagnostics that give information (on processes) on the surface, like (spectroscopic) ellipsometry, attenuated total reflection (ATR) spectroscopy, sum frequency generation (SFG), and second harmonic generation (SHG). Ellipsometry and ATR spectroscopy have been successfully applied for many years in the study of depositing and etching plasmas. For example, hydrogen-containing vibrations such as SiH_x in a-Si:H thin films can be detected via infrared absorption spectroscopy. Most often this is used *ex situ* on films deposited on an infrared transparent substrate using a Fourier transform infrared (FTIR) spectrometer. Infrared spectroscopy, however, can also be applied *in situ* and real time during film growth when the infrared beam is reflected from the substrate. However, this single reflection yields usually a very poor sensitivity and is certainly not sensitive enough the measure the absorption of, for example, a single monolayer of surface species. The sensitivity can be enhanced using the principle of total internal reflection in an attenuated total reflection (ATR) crystal [23]. In an ATR crystal, the light undergoes several (typically 20–40) total internal reflections at the top surface of the crystal. This implies that a film deposited at the ATR crystal can be probed several times by the evanescent wave. ATR-FTIR has therefore been applied to detect hydrogen *in situ* by means of the SiH_x stretching mode in the bulk of a-Si:H films. Depending on the refractive indices of the ATR crystal and film, the total reflections can either take place at the ATR–film interface or at the film–vacuum interface.

SFG and SHG are techniques that have shown their possibilities in the field of surface science, and are now being introduced in the field of plasma physics. Both techniques are nonlinear optical techniques for investigation of molecules at surfaces and interfaces and are sensitive to sub-monolayers and show high spatial, temporal, and spectral resolution. In SFG studies, a pulsed tuneable infrared laser beam (ω_1) is mixed at an interface with a visible beam (ω_2) to produce a sum frequency signal (Ω) (see Fig. 7.3). Active vibration modes of molecules at the interface contribute

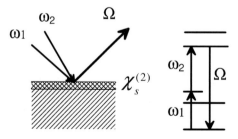

Fig. 7.3 Schematic of the interaction of the beams in a SFG and SHG ($\omega_1 = \omega_2$) experiment.

resonantly to the SFG signal, while in SHG ($\omega_1 = \omega_2$) electronic states are responsible for the second harmonic signal. The theory of nonlinear phenomena shows that the second-order nonlinear susceptibility, $\overleftrightarrow{\chi}_s^{(2)}$, is responsible for the effects like sum frequency and second harmonic frequency generation. $\overleftrightarrow{\chi}_s^{(2)}$ is a third-rank tensor that represents the second-order nonlinear susceptibility. The symmetry argument rules that the second order nonlinear polarization in media with inversion symmetry (like bulk isotropic media) is always equal to zero ($\overleftrightarrow{\chi}_s^{(2)} = 0$), and SFG (SHG) signals may be registered from only one area, i.e. where the symmetry is broken and $\overleftrightarrow{\chi}_s^{(2)} \neq 0$.

While SHG requires one monochromatic light beam and generates a signal at the doubled frequency, SFG photons carry away the sum of the energies of a visible photon and an infrared photon. For both techniques the efficiency is enhanced by resonance of any of the involved wavelengths with electronic or vibrational transitions. SHG is employed to study electronic states, while SFG can also access vibrational transitions located in the infrared.

7.4
Applications

7.4.1
Thomson–Rayleigh Scattering and Raman Scattering

Very often the spurious scattering, i.e. the stray light, from windows in the path of the laser beam and walls near the excitation volume determines the sensitivity of a Thomson scattering setup. Or in the case where electron densities need to be measured at atmospheric pressures, Rayleigh scattering from heavy particles causes extremely intense signals, swamping the Thomson scattering signal.

In case one wants to measure, for example, the electron density and temperature in a low-pressure positive argon column, a special Thomson scattering experimental design is necessary. In [24] electron density and temperature measurements are reported in a low-pressure argon mercury discharge lamp. The tube of this discharge lamp is only 26 mm in diameter; the total pressure is 5 mbar, which is merely argon with a small amount of mercury (between 0.14 and 1.7 Pa). In the standard Thomson scattering configuration the Thomson scattering signal would de drowned in the stray light, the intensity of which is very high due to the proximity of the tube wall. Bakker et al. use an excimer pumped tunable dye laser of which the amplified spontaneous emission is reduced by means of a specially designed filter [25]. The laser light is focused into the discharge tube and the scattered light is detected by means of two lenses, which image the laser focus on the entrance slit of a spectrograph, which is equipped with an intensified charge-coupled device (I-CCD). Between the two lenses a sodium vapor absorption cell is placed. When the laser is tuned to 589 nm, the sodium vapor absorbs the stray light and the Rayleigh scattering. In the fit of the recorded Thomson scattering spectrum they did not use the wavelength region between 587.8 and 590.9 nm since in that region the absorption of both sodium lines distorted the spectrum. In [24] the radial dependence of the electron density (between 1.5×10^{18} and $1.5 \times 10^{17} \, m^{-3}$) and temperature (around 1.2 eV) as function of three different discharge currents and up to 3 mm from the discharge tube wall is reported.

De Regt et al.[26] report Thomson scattering measurements on atmospheric inductively coupled plasma (ICP). This plasma is created by a high-frequency field of 100 MHz in a coil within a quartz tube. The tube has an inner diameter of 18 mm, and three separately controlled flows of argon are applied through the torch (total flow of $21 \, L \, min^{-1}$). The power input was about 1.2 kW. The Thomson scattering experimental setup is similar to the setup used by van de Sanden et al.[27], who used the setup to measure electron densities and temperatures in an expanding thermal plasma. The scattered light is detected by means of a photodiode array in combination with a holographic grating, resulting in a high spectral resolution. At atmospheric pressures, due to blooming of the photodiode at the Rayleigh wavelength, the Rayleigh scattering part of the spectrum cannot accurately be discerned from the Thomson scattering signal. De Regt et al. developed a procedure that uses Raman scattering from nitrogen molecules to recover the Rayleigh scattering intensity. As the Raman scattering is measured at large Stokes and anti-Stokes shifts (as compared to no shift for the Rayleigh and stray light scattering), the calibration is not influenced by stray light.

7.4.2
Laser-Induced Fluorescence

Luque et al.[28] have determined the gas-phase number densities distributions of CH, C_2 and C_3 in a diamond depositing dc plasma arc-jet by means of LIF spectroscopy. In the wavelength region between 425 and 438 nm they recorded transition of the $C_2(d–a)$, $C_3(A–X)$, $CH(A–X)$ and $CH(B–X)$ bands. This region can be conveniently covered with one dye, i.e. Coumarin 120. The LIF signals are turned into absolute densities by, for example, determining the collisional quenching via time-resolved LIF measurements and by performing Rayleigh measurements to determine collection efficiency and wavelength dependence of the LIF detection system. They show radial number density profiles of all three species at positions close to the exit of the dc plasma source, in the middle of the plasma plume, and close to the substrate. The distributions show distinct structures; while the CH and C_2 concentrations are maximal on the centerline of the plasma plume, the C_3 is distributed in an annular shape. They also conclude that due to the low concentrations of C_2 and C_3 in the plasma plume, these species are not likely important in diamond growth.

The hydrogen atom is in many applications an important radical. For example, during deposition of amorphous silicon the hydrogen atom recombines with hydrogen atoms at the surface during growth. Also in the gas-phase plasma chemistry of the dc arc-jet the H radical plays an important role [28]. To determine the ground state density of H radicals and their velocity, TALIF has been applied in an Ar–H_2 thermal plasma expansion [8,13]. In the same plasma expansion the density of the hydrogen molecule in the electronic ground state has also been determined by means of LIF in the VUV region of the spectrum [15].

In order to determine the flux of radicals impinging on a growth surface, Doppler LIF spectroscopy can be used to determine the velocity of particles. In this application of LIF the fluorescence is recorded as a function of the excitation frequency of the atom or molecule. Due to the velocity of the absorbing particles, the absorbing frequency will be Doppler shifted with respect to "non-moving" particles. In [29] the velocity of Ar metastable atoms is measured in an expanding thermal plasma. The laser used to excite the Ar metastables is directed counter propagating the expansion direction, and the LIF signal is collected under 90° with respect to the expansion axis (see Fig. 7.4(a)). To increase the sensitivity and reduce the influence of the background emission of the plasma, the cw light beam of the diode laser is intensity modulated, and the fluorescence signal is recorded via a lock-in amplifier. The LIF signal is detected while the laser is scanned over one of the Ar[4s–4p] transitions. Simultaneously, the absorption in an Ar lamp and the transmission of a Fabry–Perot interferometer is recorded for respectively absolute and relative frequency calibration (see Fig. 7.4(b)). The velocity of the metastables is determined as a function of the distance from the exit of the plasma source. When for example C_2H_2 is injected into the plasma expansion, radicals like C, CH, C_2, C_2H, etc., are formed via reactions with Ar ions [29]. With the velocities measured with the Doppler LIF method and assuming that the radicals move with the same velocity as the metastables, the flux of radicals impinging on the growth surface can be determined.

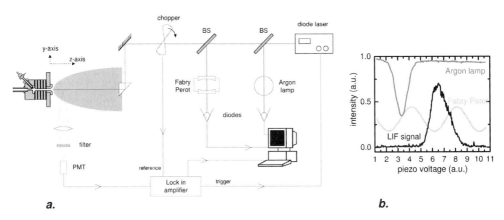

Fig. 7.4 (a) Experimental Doppler LIF setup for the determination of the velocity of Ar metastables in an expanding thermal argon plasma (from [29]). (b) A typical measurement showing the LIF signal measured with a photomultiplier tube (PMT), the absorption in an argon lamp, and the transmission signal of a Fabry–Perot interferometer.

7.4.3
Absorption Spectroscopy

Bulcourt et al.[30] have used broadband absorption spectroscopy between 220 and 300 nm to record the spectrum of the CF_2 radical in a plasma etch reactor. They determined in different regions of the plasma absolute densities of the radical, as well as rotational temperatures and vibrational distributions. Outside the active plasma region the density of the radical was measured to be 10 times lower than in the active plasma volume. The vibrational distribution in the "hot" active part of the plasma could not be fitted with a single Boltzmann distribution. However, the sum of two Boltzmann distributions with $T_{vib} = 300$ K and $T_{vib} = 1000$ K could satisfactorily describe the measured vibrational distribution. They argue that this could be explained assuming that CF_2 molecules are initially produced in high vibrational states, and are partially relaxed in the plasma by collisions. Under similar circumstances the CF radical has also been detected [31].

The density of many atoms, molecules, and radicals has been determined by narrow band tunable laser absorption techniques. In the infrared, the group of Röpcke has detected a large variety of species in several different plasmas with tunable diode laser absorption spectroscopy (TDLAS). The infrared multicomponent acquisition system (IRMA) has been used to study, using time-resolved TDLAS, the time dependence of the conversion of methane to the methyl radical, CH_3, and three stable C-2 hydrocarbons under static discharge conditions [32]. Also species like CH_4, CH_3OH, C_2H_2, C_2H_4, C_2H_6, NH_3, HCN, CH_2O, and C_2N_2 have been detected in an $H_2/Ar/N_2$ microwave plasma in which several percent of methane and methanol are added [33]. In [34] TDLAS at 16.5 μm and broadband UV absorption spectroscopy at 216 nm have been used to detect the ground state density of the methyl radical in two different microwave plasmas.

The CRD technique is nowadays commonly used for the study of all kinds of plasma. Radicals like C, CH, C_2, and H have all been measured with pulsed CRD spectroscopy in thermal plasma expansions. In one of the experiments, the plasma is created in a cascaded arc, which is connected via a nozzle to a vacuum vessel. The pressure in the arc is about 500 mbar, while the pressure in the vessel is typically between 20 and 200 Pa (0.2–2 mbar). Due to the large pressure difference the thermal plasma expands supersonically into the vessel. At a certain distance from the nozzle, a stationary shock is formed. After the shock the plasma flows subsonically into the background. For the deposition of amorphous carbon, C_2H_2 is injected around the shock region, and radical densities at different distances from the exit of the arc as a function of parameters like plasma current and C_2H_2 flow rate have been determined by pulsed CRD spectroscopy [35,36]. These results have been used to develop a model that describes the plasma chemistry in a thermal $Ar/H_2/C_2H_2$ plasma expansion [35] and an $Ar/H_2/CH_4$ plasma expansion [37]. In the latter experiment, CRD has also been used to determine absolute densities of the stable specie C_2H_2 formed in the plasma vessel [38].

For basic understanding and modeling of plasma deposition processes, information on the density as well as the surface reactivity of the plasma species is essential. Often the surface reaction probability β of the species has been obtained indirectly [39] or under process conditions different from the actual plasma deposition process, e.g. from a molecular beam scattering experiment [40] or by time-resolved density measurements in an afterglow plasma [41,42]. The chemistry in the plasma expansion used to deposit amorphous silicon has been elucidated by the results of CRD measurements on Si, SiH, SiH_2, SiH_3, and H [22,43,44]. In [22] time-resolved cavity ring down spectroscopy (τ-CRDS) is used to obtain β *during* plasma deposition: the method is used to map an increased radical density due to a pulsed rf bias to the substrate in addition to the continuously operated remote SiH_4 plasma. Although τ-CRDS has been employed previously to obtain gas-phase loss rates of radicals [45,46], in [22] the technique has been extended to measurements of the surface loss rates of the radicals. This yields simultaneously information on the surface reaction probability β and the density of the radicals under the specific plasma conditions, here particularly for the case of high rate deposition of hydrogenated amorphous silicon (a-Si:H) [47]. Using this method, it is shown that Si is mainly lost in the gas phase to SiH_4, whereas SiH_3 is only lost via diffusion to and reactions at the surface. Moreover, β values of Si and SiH_3 are determined and it is shown that β_{SiH3} is independent of the substrate temperature.

In the expanding thermal plasma (ETP) technique (Fig. 7.5(a)) a remote expanding $Ar/H_2/SiH_4$ plasma is created. To detect (low-density) radicals such as SiH_3 and Si the

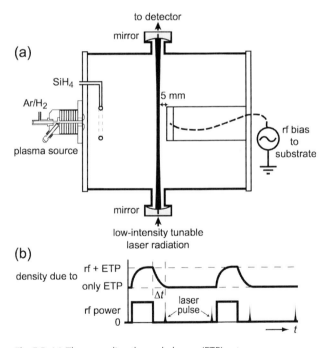

Fig. 7.5 (a) The expanding thermal plasma (ETP) setup equipped with the CRD spectroscopic setup and an rf power supply for pulsed bias voltage application to the substrate. (b) Schematic time diagram illustrating the modulation of the radical density and the synchronization of the CRDS laser pulses.

CRDS technique has previously been employed; SiH_3 has been identified at the $\tilde{A}^2A_1' \leftarrow \tilde{X}^2A_1$ broadband transition ranging from $\sim$200 to $\sim$260 nm [44], whereas Si radicals have been probed at the $4s\,^3P_{0,1,2} \leftarrow 3p^2\,^3P_{0,1,2}$ transition around 251 nm [48]. In τ-CRDS measurements, a minor periodic modulation of the radical densities is produced by applying 5 Hz, 2.5% duty cycle rf pulses to the substrate in addition to the continuously operating ETP. The additional absorption A_{rf} due to the radicals generated by the rf pulse is obtained from the difference in absorption at some point Δt in the rf afterglow and at a point long after the influence of the rf pulse has extinguished (see Fig. 7.5(b)). Every CRDS trace is handled separately by means of a "state-of-the-art" 100 MHz, 12-bit data acquisition system [49] and an averaged A_{rf} is obtained as a function of the time Δt in the afterglow of the rf pulse.

The result of a typical τ-CRDS measurement for Si and SiH_3 is shown in Fig. 7.6. A duty cycle of 2.5% has been carefully chosen in order to obtain a good signal-to-noise ratio in the additional Si and SiH_3 absorption, while possible powder formation due to the "anion confining" rf plasma sheath is suppressed. Figure 7.6 shows that both Si and SiH_3 decrease single exponentially, which is expected from the radicals' mass balance [50]. The corresponding loss rate τ^{-1} depends linearly on the gas-phase loss on the one hand and the loss due to diffusion to and reactions at the surface on the other hand [50]:

$$\tau^{-1} = k_r n_x + \frac{D}{\Lambda^2} \tag{7.8}$$

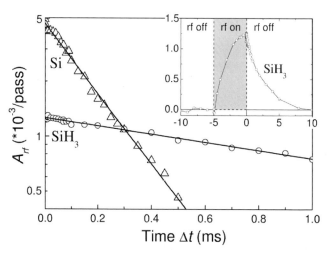

Fig. 7.6 Typical semi-logarithmic plot of the additional absorption A_{rf} of Si and SiH_3 during the rf afterglow showing a single exponential decay with a loss time of (0.226 ± 0.006) ms and (1.93 ± 0.05) ms for Si and SiH_3, respectively. Every data point is an average of 128 CRDS traces. The inset shows a linear plot of the A_{rf} of SiH_3 for the complete rf pulse of 5 ms.

In this equation, k_r is the gas-phase reaction rate with species x with density n_x, D is the diffusion coefficient for the specific radical in the $Ar/H_2/SiH_4$ mixture [51], and Λ is the effective diffusion length of the radical. The latter depends on diffusion geometry and on the radical's surface reaction probability β [50].

From Eq. (7.8) it is seen that the gas-phase loss processes need to be considered first before surface loss rates of the radicals can be deduced. For Si and SiH_3 in the ETP plasma the only candidate for a significant gas phase loss is SiH_4 [51]. Therefore the loss rate of Si and SiH_3 has been obtained as a function of the SiH_4 density keeping the pressure and thus the diffusion term in Eq. (7.8) nearly constant. The SiH_4 density has been calculated from the SiH_4 partial pressure using $T_{gas} = 1500$ K including a correction for the local SiH_4 consumption [52]. The loss rate of SiH_3 is fund to be independent of the SiH_4 density, which indicates no gas-phase loss of SiH_3, while the loss rate of Si increases linearly with the SiH_4 density. From the slope of this linear increase a reaction rate constant of $Si(^3P)$ with SiH_4 of $k_r = (3.0 \pm 1.3) \times 10^{-16}\,m^3\,s^{-1}$ is determined. This value corresponds well with literature values [22,52].

7.4.4
Surface Diagnostics

SHG is one of the diagnostic techniques that are used to detect dangling bonds. Dangling bonds have been proposed to play an important role at the surface in the growth mechanism of a-Si:H. It is also known that dangling bonds, i.e. free silicon bonds, in the bulk of a-Si:H determine the electronic quality of this material. Hydrogenated amorphous silicon (a-Si:H) films have important industrial applications, for example in thin-film transistors (TFT) for flat-panel displays. Furthermore a-Si:H is an important candidate for use in next-generation solar cells. In order to improve the quality of a-Si:H films it is necessary to improve the insight into the growth process. Whether a material has nonlinear optical properties and can therefore exhibit SHG depends on the symmetry in the structure of this material. The symmetry of the bulk of amorphous materials is such that no SHG can take place. However, at the surface this symmetry is broken and SHG is possible. As a consequence the technique of SHG is surface specific for amorphous materials.

The characteristics of SHG have already been exploited in surface science studies of c-Si. For Si, resonant optical transitions in the visible and near-infrared arise from the presence of surface states in the band gap of the Si material. The origin of these states is dangling bonds on the Si surface and a typical transition energy from the occupied to the unoccupied states is roughly 1.2 eV. The fact that the SHG signal generated at this energy is mainly due to surface dangling bonds on c-Si has been corroborated by the fact that the SHG signal almost completely disappears when hydrogen atoms are adsorbed on the c-Si surface [53]. This quenches the Si surface dangling bonds. In [53] results of an SHG experiment are shown carried out with a Nd:YAG laser with a photon energy of 1.17 eV (1064 nm) incident on an initially clean Si(111)-(7×7) surface. This surface is exposed to atomic hydrogen such that the H coverage of the surface increases. The SHG signal detected at 2.32 eV (532 nm), expressed in terms of the surface nonlinear susceptibility $\chi_s^{(2)}$, decreases when the H coverage of the surface increases and the dangling bond coverage decreases. The results show that for low H coverage, $\chi_s^{(2)}$ is directly proportional to the number of Si dangling bonds on the surface. The fact that $\chi_s^{(2)}$ increases again for higher H coverage is associated with the presence of non-resonant contributions to $\chi_s^{(2)}$. Due to the surface specificity of SHG the technique has extensively been used in *in situ* and real-time studies of the adsorption and desorption kinetics of atomic and molecular hydrogen on c-Si surfaces [53,54] as well as of the dynamics of surface diffusion of hydrogen on c-Si [55]. At the moment studies are undertaken at the Eindhoven University of Technology (see: http://www.phys.tue.nl/pmp) to explore the possibilities of SFG and SHG spectroscopy to study *in situ* plasma processes.

References

1 Kessels, W.M.M., Leewis, C.M., van de Sanden, M.C.M. and Schram, D.C. (1999) *J. Appl. Phys.*, **86**, 4029.

2 Griem, H.R. (1964) *Plasma Spectroscopy*, McGraw-Hill, New York.

3 Demtröder, W. (1981) in *Laser Spectroscopy: Basic Concepts and Instrumentation* (ed. F.P. Schäfer), Springer-Verlag, Berlin/Heidelberg/New York.

4 Meulenbroeks, R.F.G., Steenbakkers, M.F.M., Qing, Z., van de Sanden, M.C.M. and Schram, D.C. (1994) *Phys. Rev. E*, **49**, 2272.

5 Muraoka, K., Uchino, K. and Bowden, M.D. (1998) *Plasma Phys. Control. Fusion*, **40**, 1221.

6 Taran, J.-P.E. (1990) in *CARS Spectroscopy in Applied Laser Spectroscopy* (eds W. Demtröder and M. Inguscio), Plenum, New York.

7 Meulenbroeks, R.F.G., Engeln, R.A.H., van der Mullen, J.A.M. and Schram, D.C. (1996) *Phys. Rev. E*, **53**, 5207.

8 Boogaarts, M.G.H., Mazouffre, S., Brinkman, G.J., van der Heijden, H.W.P., Vankan, P., van der Mullen, J.A.M., Schram, D.C. and Döbele, H.F. (2002). *Rev. Sci. Instrum.*, **73**, 73.

9 Bokor, J., Freeman, R.R., White, J.C. and Storz, R.H. (1981) *Phys. Rev. A*, **24**, 612.

10 Bischel, W.K., Perry, B.E. and Crosley, D.R. (1981) *Chem. Phys. Lett.*, **82**, 85.

11 Alden, M., Edner, H., Grafstrom, P. and Svanberg, S. (1982) *Opt. Commun.*, **42**, 244.

12 Dimauro, L.F., Gottscho, R.A. and Miller, T.A. (1984) *J. Appl. Phys.*, **56**, 2007.

13 Mazouffre, S., Boogaarts, M.G.H., Bakker, I.S.J., Vankan, P., Engeln, R. and Schram, D.C. (2001) *Phys. Rev. E*, **64**, 016411.

14 Mazouffre, S., Foissac, C., Supiot, P., Vankan, P., Engeln, R., Schram, D.C. and Sadeghi, N. (2001) *Plasma Sources Sci. Technol.* **10**, 168.

15 Vankan, P., Heil, S.B.S., Mazouffre, S., Engeln, R., Schram, D.C. and Döbele, H.F. (2004) *Rev. Sci. Instrum.*, **75**, 996.

16 Mosbach, T., Katsch, H.-M. and Döbele, H.F. (2000) *Phys. Rev. Lett.*, **85**, 3420.

17 O'Keefe, A. and Deacon, D.A.G. (1988) *Rev. Sci. Instrum.*, **59**, 2544.

18 Berden, G., Peeters, R. and Meijer, G. (2000) *Int. Rev. Phys. Chem.*, **19**, 565.

19 Brown, S.S. (2003) *Chem. Rev.*, **103**, 5219.

20 Wheeler, M.D., Newman, S.M., Orr-Ewing, A.J. and Ashfold, M.N.R. (1998) *J. Chem. Soc. Faraday Trans.*, **94**, 337.

21 Atkinson, D.B. and Hudgens, J.W. (1997) *J. Phys. Chem. A*, **101**, 3901.

22 Hoefnagels, J.P.M., Stevens, A.A.E., Boogaarts, M.G.H., Kessels, W.M.M. and van de Sanden, M.C.M. (2002) *Chem. Phys. Lett.*, **360**, 189.

23 Chabal, Y.J. (1988) *Surf. Sci. Rep.*, **8**, 211.

24 Bakker, L.P. and Kroesen, G.M.W. (2000) *J. Appl. Phys.*, **88**, 3899.

25 Bakker, L.P., Freriks, J.M., de Hoog, F.J. and Kroesen, G.M.W. (2000) *Rev. Sci. Instrum.*, **71**, 2007.

26 de Regt, J.M., Engeln, R.A.H., de Groote, F.P.J., van der Mullen, J.A.M. and Schram, D.C. (1995) *Rev. Sci. Instrum.*, **66**, 3228.

27 van de Sanden, M.C.M., Janssen, G.M., de Regt, J.M., Schram, D.C., van der Mullen, J.A.M. and van der Sijde, B. (1992) *Rev. Sci. Instrum*, **63**, 3369.

28 Luque, J., Juchmann, W. and Jeffries, J.B. (1997) *J. Appl. Phys.*, **82**, 2072.

29 Engeln, R., Mazouffre, S., Vankan, P., Schram, D.C. and Sadeghi, N. (2001) *Plasma Sources Sci. Technol.*, **10**, 595.

30 Bulcourt, N., Booth, J.-P., Hudson, E.A., Luque, J., Mok, D.K.W., Lee, E.P., Chau, F.-T. and Dyke, J.M. (2003). *J. Chem. Phys.*, **120**, 9499.

31 Luque, J., Hudson, E.A., Booth, J.-P. (2003) *J. Chem. Phys.*, **118**, 622.

32 Röpcke, J., Mechold, L., Duten, X. and Rousseau, A. (2001) *J. Phys. D*, **34**, 2336.

33 Hempel, F., Davies, P.B., Loffhagen, D., Mechold, L. and Röpcke, J. (2003) *Plasma Sources Sci. Technol.*, **12**, S98.

34 McManus, J.B., Nelson, D., Zahniser, M., Mechold, L., Osiac, M., Röpcke, J. and Rousseau, A. (2003) *Rev. Sci. Instrum.*, **74**, 2709.

35 Benedict, J., Wisse, M., Woen, R.V., Engeln, R. and van de Sanden, M.C.M. (2003) *J. Appl. Phys.*, **94**, 6932.

36 Engeln, R., Letourneur, K.Y.G., Boogaarts, M.G.H., van de Sanden, M.C.M. and Schram, D.C. (1999) *Chem. Phys. Lett.*, **310**, 405.

37 Mankelevich, Yu.A., Suetin, N.V., Ashfold, M.N.R., Boxford, W.E., Orr-Ewing, A.J., Smith, J.A. and Wills, J.B. (2003). *Diamond Relat. Mater.*, **12**, 383.

38 Wills, J.B., Ashfold, M.N.R., Orr-Ewing, A.J., Mankelevich, Yu A. and Suetin, N.V. (2003) *Diamond Relat. Mater.*, **12**, 1346.

39 Kessels, W.M.M., van de Sanden, M.C.M., Severens, R.J. and Schram, D.C. (2000) *J. Appl. Phys.*, **87**, 3313, and references therein.

40 McCurdy, P.R., Bogart, K.H.A., Dalleska, N.F. and Fisher, E.R. (1997) *Rev. Sci. Instrum.*, **68**, 1684.

41 Perrin, J., Shiratani, M., Kae-Nune, P., Videlot, H., Jolly, J. and Guillon, J. (1998) *J. Vac. Sci. Technol. A*, **16**, 278.

42 Kae-Nune, P., Perrin, J., Guillon, J. and Jolly, J. (1995) *Plasma Sources Sci. Technol.*, **4**, 250.

43 Kessels, W.M.M., Leroux, A., Boogaarts, M.G.H., Hoefnagels, J.P.M., van de Sanden, M.C.M. and Schram, D.C. (2001) *J. Vac. Sci. Technol. A*, **19**, 467.

44 Boogaarts, M.G.H., Böcker, P.J., Kessels, W.M.M., Schram, D.C. and van de Sanden, M.C.M. (2000) *Chem. Phys. Lett.*, **326**, 400.

45 Atkinson, D.B. and Hudgens, J.W. (1997) *J. Phys. Chem. A*, **101**, 3901, and references therein.

46 Yalin, A.P., Zare, R.N., Laux, C.O. and Kruger, C.H. (2002) *Appl. Phys. Lett.*, **81**, 1409.

47 Kessels, W.M.M., Severens, R.J., Smets, A.H.M., Korevaar, B.A., Adriaenssens, G.J., Schram, D.C. and van de Sanden, M.C.M. (2001). *J. Appl. Phys.*, **89**, 2404.

48 Kessels, W.M.M., Hoefnagels, J.P.M., Boogaarts, M.G.H., Schram, D.C. and van de Sanden, M.C.M. (2001) *J. Appl. Phys.*, **89**, 2065.

49 Technical Laboratory Automation Group, Eindhoven University of Technology, Den Dolech 2, 5600 MB Eindhoven, The Netherlands.

50 Chantry, P.J. (1987) *J. Appl. Phys.*, **62**, 1141.

51 Perrin, J., Leroy, O. and Bordage, M.C. (1996) *Contrib. Plasma. Phys.*, **36**, 1.

52 Hoefnagels, J.P.M., Barrell, Y., Kessels, W.M.M. and van de Sanden, M.C.M. (2004) *J. Appl. Phys.*, **96**, 4094.

53 Höfer, U. (1996) *Appl. Phys. A*, **63**, 533.

54 Drr, M., Hu, Z., Biederman, A., Höfer, U. and Heinz, T.F. (2002) *Phys. Rev. Lett.*, **88**, 46104.

55 Raschke, M.B. and Höfer, U. (1999) *Phys. Rev. B*, **59**, 2783.

8
Plasma Processing of Polymers by a Low-Frequency Discharge with Asymmetrical Configuration of Electrodes

F. Arefi-Khonsari and M. Tatoulian

In this chapter surface modifications of polymer films treated by cold plasmas are reviewed. To illustrate these modifications, examples of plasma processing of polymers are given exclusively by using a low-frequency low-pressure reactor with asymmetrical configuration of electrodes. After briefly discussing the main processes involved in plasma polymerization, we end the chapter with some plasma deposition examples in the same reactor.

8.1
Introduction

Wet-chemical processes with solvents, oxidants such as chromates or permanganates, strong acids or bases, or the sodium–liquid ammonia etching for fluoropolymers [1] are becoming increasingly unacceptable because of environmental and safety considerations. Furthermore, wet chemical processes tend to have inherent problems of uniformity and reproducibility. The other issue of wet-chemical processes is that the chemicals can diffuse well below the surface and give rise to change of bulk properties such as mechanical properties. The plasma processing is a versatile dry process which can tailor polymers in different forms: webs, fibers, particles, etc., in order to modify their surface properties without changing their intrinsic bulk properties. The examples of technological applications of plasma processed polymers are numerous in the field of automobile industry, microelectronics, decoration or packaging, bioanalytical devices, biomedical applications, etc. [2–8].

For the plasma processing of polymers, one can find in the literature an extremely wide range of plasma devices both laboratory-scale experiments but also industrial reactors [9–11] proposed for surface treatment of polymers. Corona treatments in air were the first discharges which were used for the surface treatment of polymers [12,13] and have been present for the last 50 years in various polymer film production and processing industrial units. They are still widely used industrially due to the advantage of operating at atmospheric pressure and in air. In most corona

Advanced Plasma Technology. Edited by Riccardo d'Agostino, Pietro Favia, Yoshinobu Kawai, Hideo Ikegami, Noriyoshi Sato, and Farzaneh Arefi-Khonsari
Copyright © 2008 WILEY-VCH Verlag GmbH & Co. KGaA, Weinheim
ISBN: 978-3-527-40591-6

treatments in fact a dielectric barrier discharge (DBD) is used where the work piece on the transport rolls or high voltage electrode serves as a dielectric barrier. Different configurations of electrodes exist to treat not only polymer foils and fabrics, i.e. flat surfaces, but also three-dimensional substrates [14]. In many applications electrode assemblies of several parallel knife edges are used, and foils up to 10 m width are treated at a speed from 1 to $10\,\mathrm{m\,s^{-1}}$. This requires a discharge power ranging from 1 to $100\,\mathrm{kW}$, and operating frequencies vary in the range 10–70 kHz [13]. Recently DBDs using pulsed sources are very much developed for the surface modification purposes resulting in improved statistical distribution of the microdischarges across the surface, a prerequisite for more uniform treatments [15,16]. Such atmospheric discharges could lead to less homogeneous treatments subjected to more surface reorganization and therefore ageing effect of the surface properties as compared to low-pressure systems. However recently a wide range of atmospheric pressure discharges have been designed, developed, and tested with great success for different applications. These discharges just like low-pressure cold plasmas do not generate extensive heating in their surroundings, and as a result they are suitable for surface modification of polymers and processing of organic compounds [17–21].

However most of the research related to surface modification reactions involves low-pressure cold plasmas which are initiated and sustained by direct current (DC), radiofrequency (RF) or microwave (MW) power sources with or without an additional electric or magnetic field. Such low-temperature plasmas are usually characterized by pressures ranging between 10 to 1000 Pa. The commercial plasma systems used usually operate in the low-frequency (40–450 kHz), radio frequency (13.56 MHz or 27.12 MHz), or microwave (915 MHz or 2.45 GHz) ranges.

The first functionalization studies on polymers at low pressure were reported by Hollahan and Stafford in 1969 [22] who used plasmas of ammonia and mixtures of nitrogen and hydrogen in an inductively coupled reactor to bind heparin to plasma-treated polypropylene (PP). The treatment times used in this study ranged from 3 to 50 h, probably to investigate the modifications created on the surface by techniques not adapted to observe surface modifications.

Although electron collision processes define the chemistry taking place in different plasma reactors, even with similar electron energy distribution environments one does not reach the same results in terms of final surface modification obtained. There are more specific parameters, such as the electrode configuration, reactor geometries, the external properties, the position of the electrodes or substrate in the reactor, and also individual specificity of the processes which should be taken into consideration. That is why it is extremely important and necessary to monitor plasma parameters in each system in the presence of a particular polymer substrate in order to have a successful approach for the specific aimed application. That is to say that one cannot generalize a particular discharge to different substrates for a specific application since we shall see that the plasma–surface interactions are closely dependent on the chemical nature of the substrate.

In this chapter firstly the general interactions of plasma with a polymer surface will be presented; secondly examples of plasma processing of polymers will be given exclusively with a particular low-frequency discharge with asymmetrical configuration of electrodes.

8.2
Plasma Treatment of Polymers

8.2.1
Surface Activation

The interaction of cold plasmas with polymers involves both gas and surface reaction mechanisms. The gas-phase reactions in the discharge volume lead to the production of species including atoms, molecules, free radicals, ions of both polarity, excited species in different electronic, vibrational and rotational states, electrons, and photons. Depending on the energy and the reactivity of the plasma-created species with respect to each other and with respect to the surface, recombination/deposition processes on the surface or on the contrary etching or ablation of the polymer take place. In order to illustrate these interactions more clearly the different steps of functionalization (grafting) of polyolefins are shown below.

The reaction of the polymer with the plasma-created species leads to bond cleavages of the polymer backbone, i.e. C–C and/or C–H bonds. Since all surfaces exposed to a plasma establish a negative potential [23,24], the positive ions in the plasma play a significant role in such processes. The vacuum ultraviolet (VUV) radiations [25] also contribute to the production of free radicals. The activation step is shown below in the case of the simplest polyolefins i.e. polyethylene giving rise to alkyl radicals:

These free radicals which are characterized by short lifetimes [26] can then go through various chemical reactions involving plasma species (grafting or functionalization) *in situ* or result in crosslinking of the polymer layer as shown below.

8.2.2
Functionalization (Grafting) Reactions

The carbon free radicals formed can react with the molecular and atomic species. In the presence of an oxygen plasma, the main radicals formed are peroxy ones which can further react with the monomer molecule or condensed phase compounds such as water in the absence of plasma to give rise to more stable functionalities such as ketones or carboxylic groups.

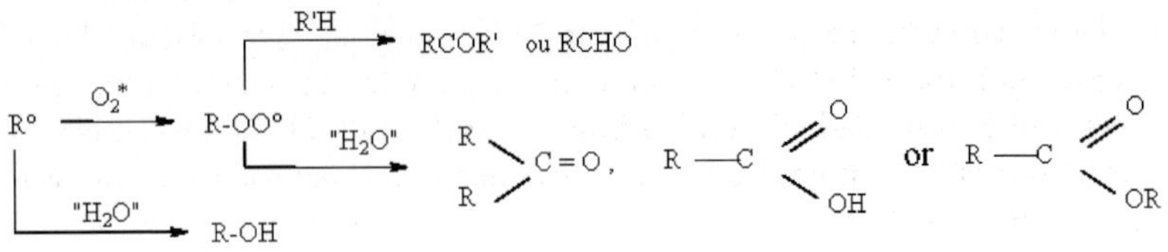

As can be seen by the different oxygen moieties which result from the reaction of the excited species and radicals obtained from the decomposition of oxygen, the plasma treatment of polymers is not a selective process.

The incorporation of nitrogen is also possible if plasma gases such as nitrogen, ammonia, or mixtures of nitrogen and hydrogen are used. In such cases the free radicals react with the reactive atomic and molecular species which are created in such plasmas.

$$R° \xrightarrow{N_2^+, NH_3^+, H°, NH_2°} R-NH_2 \quad \text{or} \quad R-NH-R'$$

Examples of other gases used for plasma treatment of polymers are carbon monoxide, nitrogen dioxide, fluorine, and water.

8.2.3
Crosslinking Reactions

Crosslinking reactions predominantly take place by interaction of radicals formed on polymer chains in plasmas of inert gases such as Ar, He, and Ne. The formation of crosslinks in the polymer is shown below in the case of isotactic polypropylene on which free radicals are formed readily due to the tertiary carbon bearing a methyl group:

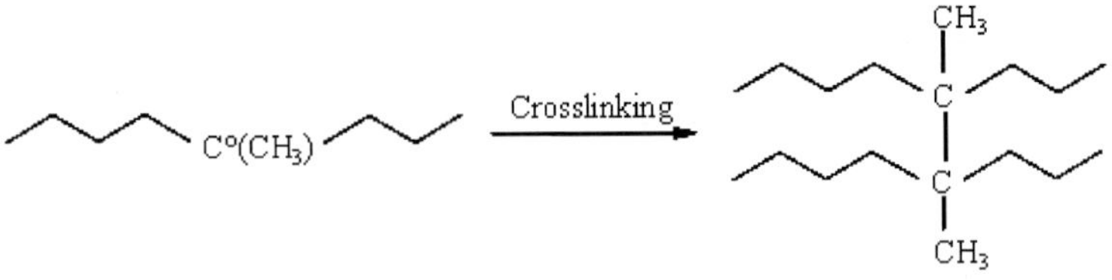

The thickness of the crosslinked layer reported in the literature for most of the polymers is generally in the range of 5 to 50 nm. However in the case of some polymers such as high-density polyethylene (HDPE) the crosslinked layer can increase up to a few micrometers. Generally low-power plasmas should be used in noble gases in order to obtain crosslinking. Plasma processing of polymers is extensively used to improve the adhesion characteristics of polymers. However it is not always sufficient to have an intimate contact between the adhesive and the polymer surface in order to obtain a strong adhesive force at the interface. Indeed the presence of polar groups at the interface increases the probability to bind the adhesive to the polymer either by hydrogen or covalent bonds depending on the chemical nature of the adhesive. This is true if and only if the polar groups are not placed on a layer which is mechanically weak. That is why Schonhorn and Hansen [27] developed the CASING (crosslinking by activated species of inert gases) process which consisted of bombarding the surface of PE by inert gases (Ar, He, H_2) to crosslink it superficially and to strengthen mechanically the interface in order to improve its adhesive characteristics. Such plasmas also remove low molecular weight fragments (LMWFs) by physical sputtering or convert them to higher molecular weight ones by crosslinking reactions. Consequently the weak boundary layer formed by the low molecular weight fragments is removed leading to greater adhesive forces at the interface. The authors have quantified the thickness of the crosslinked layer on PE and have obtained a precise correlation between this thickness and the measured adhesion of PE to aluminum. The method used was to dissolve the PE after treatment with helium in a solvent (toluene), and to separate the gel obtained which corresponded to the insoluble crosslinked phase. They showed that a treatment ranging from 5 to 15 min in He gave rise to a crosslinked layer whose thickness varied between 50 and 100 nm (Fig. 8.1), and this corresponded to the range where the maximum joint strength was measured at the interface. In other words, in the case of PE, a crosslinked layer around 50 to 100 nm thick conferred to the assembly a maximum adhesive force. Other authors also attribute the high cohesive force measured at the interface in the case of PE to the presence of this crosslinked layer [28].

The active species in a He discharge which could be responsible for the surface crosslinking are He metastables, ions [29,30] or VUV radiation [15]. An exhaustive study based on modeling and experiments has shown that in RF discharges, particularly helium ions play an important role in the surface modification of the polymer surface [31].

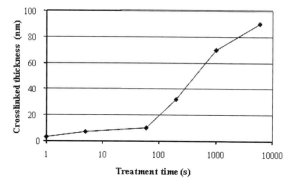

Fig. 8.1 Estimation of the thickness of the crosslinked layer of PE as a function of the treatment time in a He plasma (RF plasma: 100 W; 100 Pa, $f = 15$ MHz) [27].

8.2.4
Surface Etching (Ablation) Reactions

Surface etching reactions favor the formation of weak boundary layers. In plasma treatments, tough conditions such as high power density, low pressure and long treatment times give rise to the creation of LMWFs by bond cleavage and decarboxylation (see below) which can dramatically affect the mechanical properties of the assembly at the interface.

8.2.4.1   Decarboxylation

Ränby et al. [32] have shown the competition between oxidation and decarboxylation which is caused by UV excitation of the carbonyl groups present on the surface to triplet states. The delocalized electrons of –COO groups trap the UV radiation, and can in this way raise the excitation level of the molecule to cause cleavage of the R–C bonds and therefore lead to decarboxylation. Bond breaking, as explained above, can lead to the formation of a weak boundary layer. Therefore the decarboxylation reactions will compete with the oxidation reactions shown in Section 8.2.2, according to the reaction shown below:

This reaction is slow as compared to the oxidation reactions shown above (Section 8.2.2) and constitutes the rate limiting step of the surface oxidation reactions which levels off in all plasma treatments. This explains the reason why for example the oxygen uptake of plasma-treated polymers determined by X-ray photoelectron spectroscopy (XPS) reaches plateau levels.

8.2.4.2   β-Scission

If a radical is stabilized by a methyl group, a β-scission can take place, i.e. an electron pair, which is at a β-position with respect to the isolated electron, is separated. The tertiary carbon radicals give rise to a considerable number of β-scissions, resulting in several bond cleavages in the polymer backbone as well as the formation of double bonds:

One can note that, on the one hand, crosslinking reactions and the formation of double bond are synergic reactions. On the other hand, the vinyl functions formed are very sensitive to oxidation and depending on the surrounding environment can lead to ketone, aldehyde, and carboxylic functions.

8.2.4.3   Plasma Cleaning/Etching Effect

One can take advantage of this ablation effect of plasmas to eliminate the contamination layers existing on the surface of polymers. This layer is due to polymer processing aids and functional additives such as antistatic agents, lubricants (mineral oil, polyolefin waxes, fatty acid esters, etc.), antioxidants (phenols, amines, mercaptans, etc.), and light-protecting agents. Furthermore the large molecular weight polydispersity of the polymer chains is also a problem, and the presence of remnants of the initial monomers and solvents can never be excluded. Such compounds can be present over a thickness of 1 to 10 nm and can constitute a weak boundary layer responsible for a great number of adhesion failures in different polymer assemblies with other materials [33]. Many years ago Schonhorn *et al.* showed that most of the oxidizing plasmas which are used to improve adhesion of polymers to metals remove this contamination layer by the ablation reactions mentioned above. Since then, depending on the intended applications, different plasmas have been used to obtain a strongly adhesive joint between the polymer and the different coatings.

Polymer surfaces can be oxidized easily by O_2, N_2, N_2O, CF_4, and other low-temperature plasmas [34]. Plasma cleaning is an economical process if the build up of the contamination layer is not too massive, i.e. well below 1 μm (e.g. polymeric material). If not (e.g. metallic parts particularly in automobile industry), the etching process competes with the crosslinking of the hydrocarbon layer with the result of dramatically decreased etch rates. In these cases good results can be achieved if excessive contamination is removed by wet cleaning procedures prior to the plasma process [34]. Krüger *et al.* [35] used low-pressure plasma processes for cleaning the surface of glass, metals, carbon. and polymers from contaminants such as air pollutants, fingerprints, coupling agents, oxide layers, slip agents, light stabilizers, additive enrichment at the surface, etc. Depending on the type of contamination, different plasmas were proposed: inert gases such as Ar to remove different contamination layers by sputtering, oxygen plasmas to oxidize organic contaminants, and hydrogen plasmas to reduce inorganic contaminants such as oxides or sulfides. Examples were given for purification of glass or glass fiber surfaces and the consequences for optical properties or adhesion in epoxy–polymer matrix–carbon fiber composites [35].

The investment for a plasma cleaning installation is often higher than for solvent or water-based baths. However additional investment such as installations for safety and waste treatment, or running cost for chemicals and energy (drying) can add considerably to the expenses of liquid bath cleaning such that plasma cleaning is the most cost-effective solution [34].

So, as mentioned above, plasma processing of polymers can give rise to weight loss which one tries to minimize in the case of plasma grafting and to control in the case of etching of polymers [36].

Effect of the Chemical Structure of the Polymer   In the case of plasma processing of different polymers with oxygen, it has been reported that polymers with aromatic rings resist better degradation, while those bearing ether or other oxygen-containing groups are more fragile [37,38]. In the case of polyolefins, the degradation effect of polymers by an oxygen plasma has been investigated using mass spectrometry in order to determine the degradation products of the polymer, namely CO_2, H_2, CO, and H_2O, and the polymers studied were classified as follows: PP (polypropylene) > HDPE > LDPE (low-density polyethylene) > PTFE (polytetra-fluoroethylene).

PTFE, known to be the most inert polymer, has been used as a reference. The reason why PP is more easily degraded is, as mentioned above, because it possesses a tertiary carbon bearing a methyl group with an isolated hydrogen. The latter favors the formation of free radicals which are precursors for functionalization., and under severe plasma conditions (low pressure, high power and long treatment times) can give rise to polymer degradation (ablation).

Functionalization and Etching in Inert Gases   Surface treatment of the different polymers (even those that do not contain any oxygen) in helium plasmas also give rise to weight loss which is much lower (20 times less than in an oxygen plasma at the same pressure of 15 Pa) [39]. This is caused by physical sputtering (much less pronounced in He as compared to Ar plasma, because of the much higher momentum of Ar) and/or etching by active species resulting from the decomposition of residual air. Indeed residual nitrogen and oxygen even present at concentrations lower than 1% [40] react efficiently with the helium metastables and ions to give rise to excited atomic and molecular species and charged species [27]. Generally, in pure oxygen for example, electronic collision mechanisms, i.e. direct excitation from the ground state and dissociative excitation of molecular species, take place, giving rise to excited atomic oxygen (observed by optical emission spectroscopy at 777.4 nm). Both of these processes occur at electron energies greater than a threshold value (about 11 eV for the direct excitation from the ground state and about 18 eV for the dissociative excitation of molecular oxygen). But in a He discharge where oxygen is present, as an impurity or added in the gaseous mixture, the presence of He in the discharge renders very efficient the Penning reactions [29], fast transfer ($v \approx 10^{-12}$ s), which require practically no additional energy [42]. Such reactions are shown below:

$$He^m + O_2 \rightarrow He + O_2^*$$

$$He^m + O_2 \rightarrow He + O_2^+ + e$$

$$He^m + O_2 \rightarrow He + O^+ + O^* + e$$

$$He^+ + O_2 \rightarrow He^+ + O + O^*$$

$$He^+ + O_2 \rightarrow He + O^+ + O^*$$

Indeed in our previous works we have clearly identified by optical emission spectroscopy (OES) the presence of O^* at 777.4 and 844.6 nm and N_2^+ at 391.4 nm in helium discharge with oxygen present as residual gas (0.3% determined by mass spectrometry) [41].

Therefore the presence of oxygen on plasma-treated surfaces in inert gases can be explained by the reaction of the surface free radicals with the excited molecular and atomic oxygen species produced by the reactions above and/or by post-plasma oxidation initiated by reactions between remaining radicals and in-diffusing atmospheric oxygen.

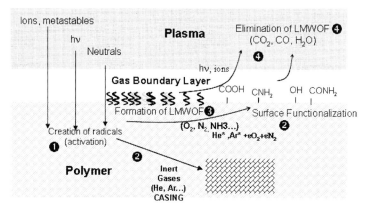

○❶ Activation step❷ reaction with inert gases and/or functionalisation (depending on the plasma gas)❸formation of LMWF and LMWOF ❹Elimination of LMWOF

Fig. 8.2 Plasma–polymer interactions during surface treatment of polymers: 1, activation step; 2, reaction with inert gases and/or functionalization (depending on the plasma gas); 3, formation of LMWFs (Low Molecular Weight Fragments) and LMWOFs (Low Molecular Weight Oxidized Fragments); 4, elimination of LMWOFs.

In the case where nitrogen is present as an impurity (residual air) or added in the gaseous mixture with helium, the Penning reactions are very efficient as well and can give rise to ionization or dissociation according to the following reactions which also explain the reason why N_2^+ excited species have been detected in our discharge [43]:

$$He^+ + N_2 \rightarrow He + N_2^+$$

$$He^m + N_2 \rightarrow He + N_2^+ + e-$$

$$N_2 + e- \rightarrow N_2^+ + 2e \text{ threshold energy} : 18.6\,eV$$

$$He^m + N_2 \rightarrow He + 2N$$

Figure 8.2 summarizes schematically the different steps which can take place during plasma modification of polymers and which we have explained in the above sections.

8.3
Surface Treatment of Polymers in a Low-Frequency, Low-Pressure Reactor
With Asymmetrical Configuration of Electrodes (ACE)

In this chapter, in order to illustrate the different plasma–polymer interactions given above, we will concentrate on one type of discharge which we have used, i.e. a low-pressure, low-frequency discharge with Asymmetrical Configuration of Electrodes (ACE). The details of the reactor and the experimental setup used for the plasma treatment of polymer films have been described elsewhere [44]. It consists mainly of a bell jar glass reactor with an asymmetrical configuration of electrodes. Low-frequency power (70 kHz), provided by an industrial generator from STT-France, was capacitively coupled to the hollow blade-type electrode through which the gases were introduced. The polymer film ($22 \times 22\,cm^2$) was rolled on the grounded cylinder which rotates in front of the high voltage electrode. The asymmetrical configuration of electrodes (corona type) allowed, for the first time, investigation of the surface modification on polymers submitted to a low-pressure plasma for very short treatment times (milliseconds to a few seconds) [45]. The other advantage of this reactor is that due to the specific electrode configuration it can be transferred to production scale and on-line treatment of polymers quite easily. Other authors have also shown that plasma surface modifications could be very effective with short residence times ($t < 1\,s$) [47].

This reactor (Fig. 8.3) has been used for surface treatment of polymers as well as plasma polymerization by introducing organic precursors (cf. Section 8.4). The treatment gas is introduced in the hollow electrode, all through its length (200 mm), which diffuses out in the interelectrode gap ($d = 8$–$10\,mm$). The cylinder turns in front of this electrode and the width of the plasma on the cylinder is around 8 mm in normal conditions. The rotation speed is normally fixed at 1 turn per second and for each turn the polymer is treated in usual conditions for 23 ms.

For plasma polymerization of organic precursors, where longer treatment times are required to deposit a coating on the polymer rolled on the cylinder, although the discharge is continuously on, the substrate turns in front of the electrode. The gaseous mixture which is introduced in the inter-electrode gap can surround the cylinder either by diffusion or by rotation of the cylinder. So in normal conditions (1 turn/s), the substrate is exposed to the discharge for 23 ms and is in the post discharge for the rest of the time.

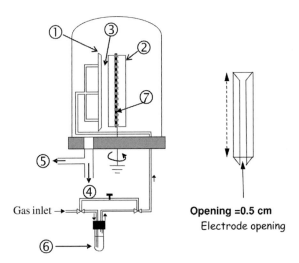

Opening =0.5 cm
Electrode opening

Fig. 8.3 Schematic of the low-frequency, low-pressure reactor with asymmetric configuration of electrodes: 1, hollow electrode; 2, grounded cylinder and sample holder (length 22 cm and diameter 7.0 cm); 3, plasma zone; 4, turbomolecular pump; 5, primary pump resistant to chemicals; 6, introduction of plasma gas for surface treatment or bubbling system for introduction of organic precursors; 7, discharge width on the cylinder (0.5 cm in normal conditions).

8.3.1
Surface Functionalization

Figure 8.4 shows the nitrogen uptake of a PP film treated in the same ACE reactor by using ammonia with two different excitation frequencies namely 70 kHz and 13.56 MHz. The first remark is that the nitrogen uptake of the surface takes place very quickly and that, at 70 kHz, the ammonia plasma gave rise to nitrogen uptake of the polypropylene surface which reached a plateau (N/C = 10%) for treatment times as short as 0.1 s. Indeed a low-frequency RF discharge has many qualitative features as a DC glow discharge. At low frequencies, where the ions can follow the electric field, the discharge will behave similarly to a DC discharge. At the pressure (30–150 Pa) and driving voltages (900–1000 V) studied, the ions reach the cathode with significant energy [48] and their energy distribution is rather broad. This ionic impact brings about in the case of normal glow discharges the emission of secondary electrons from the electrodes, which is considered to be the predominant mechanism to sustain the discharge. The ionic impact in the case of polymer-covered electrodes, i.e. in our case, will give rise to the formation of active sites which would accelerate the functionalization process for short treatment times. Furthermore in the low-frequency discharge, the cathode position alternates from one electrode to another each half-period; this explains the ion bombardment of the cylindrical electrode on which the polymer is placed. At higher frequencies (>500 kHz) the ions will no longer be able to follow the electric fields but will respond to the time-averaged fields. Therefore for a 13.56 MHz discharge the transition time of the ions in the sheath being much longer than the oscillation time of the electric field, the energy of ions narrows down.

This explains the difference of kinetics of nitrogen incorporation which is much higher with the low-frequency discharge (Fig. 8.4) However, for treatment times exceeding 1 s, the N 1s/C 1s ratio is higher in the 13.56 MHz discharge compared to the 70 kHz one (around 12–13% at 13.56 MHz). Fisher and co-workers have studied, using the imaging of radicals interacting with surfaces (IRIS) technique, the

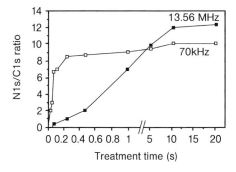

Fig. 8.4 Variation of the N 1s/C 1s from XPS as a function of the treatment time in the ACE reactor at two different frequencies (NH$_3$ plasma: $p = 100$ Pa; $Q = 40$ sccm; $P_w = 20$ W). From [52].

scattering of CF_2 or NH_2 radicals on a variety of substrates as well as polymers in various plasmas of fluorocarbons [49], NH_3 [49] or NH_3/SiH_4 [50]. Briefly IRIS combines the molecular beam technique and laser-induced fluorescence (LIF) to measure the steady-state surface reactivity of gas-phase species during plasma processing of a surface as well as the gas-phase density of species as a function of different plasma parameters. For reactivity measurements a substrate is rotated directly into the path of the molecular beam and LIF measurements are again collected. The difference between the spatial distributions with the surface in and out of the path of the molecular beam is used to measure radical–surface reactivity [50]. Fisher and co-workers have concluded that ion bombardment contributes significantly to the production of CF_2 and NH_2 radicals from the surface, which are measured by LIF in the gas phase. These results explain the rapid leveling off of the fluorine or nitrogen uptake of the surface as well as the lower levels obtained in the case of the low-frequency discharge (Fig. 8.4).

Besides ammonia [51], surface treatment of polypropylene has been carried out in this reactor in plasmas of nitrogen as well as mixtures of helium and ammonia [52], in order to obtain stable adhesive properties to aluminum [14]. Due to the energetic character of our discharge, it has been shown that the layer adhesion reaches a maximum for very short treatment times [51,53]. Other authors have reached the same conclusion using ECR plasmas and/or ion beams [54] to enhance the adhesion strength of sputtered and evaporated copper layers to different polymers. They have concluded that the layer adhesion reaches a maximum at extremely low doses of pretreatment and therefore the application of available plasma and ion sources allows one to implement web speeds in a range of some meters per second.

In general, as mentioned above, in the case of surface treatment of polymers severe discharge conditions should be avoided. Other authors working with completely different systems have found the same type of results, i.e. experimental conditions should be optimized in order to avoid overtreatments. d'Agostino and co-workers used modulated RF discharges in ammonia and found high selectivity for primary amine functions in the N 1 s envelope (NH_2/N 1 s) for short treatment times [55]. Ohl and co-workers [56] investigated 13.56 MHz and 2.45 MHz continuous wave (CW) and pulsed discharges in ammonia and showed a high nitrogen uptake of polystyrene (PS) films at short treatment times and high selectivity for treatment times as short as 0.02 to 0.2 s.

8.3.2
Ablation Effect of an Ammonia Plasma During Grafting of Nitrogen Groups

As explained in Section 8.2.4 oxygen plasmas which are used for grafting of polymers are utilized also for etching of polymers employed as photoresists in microlithography. However, etching occurs in all plasmas, particularly under tough plasma conditions (high power density, low pressure and prolonged treatment times). That is why the weight loss of isotactic PP films obtained by gravimetric measurements is compared in O_2 and NH_3 plasmas in Fig. 8.5. The latter shows, first, that in the case of

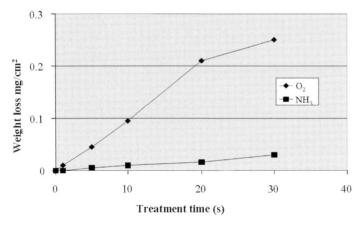

Fig. 8.5 Weight loss measured on NH_3 and O_2 plasma-treated polypropylene films as a function of treatment time (ammonia treatment: $p = 100$ Pa; $P = 5$ W, $Q = 100$ sccm; oxygen treatment: $p = 100$ Pa; $P = 3$ W, $Q = 100$ sccm). From [5].

an ammonia plasma the weight loss is almost linear with the treatment time and, second, that in the case of oxygen, for a treatment time of 30 s it is 8 times higher than that measured in ammonia. For a treatment time of 10 s the etching rate of PP in oxygen, obtained by gravimetric measurements, was $100\,\mu\mathrm{g\,cm}^{-2}$ (etching rate $= 10\,\mu\mathrm{g\,cm}^{-2}\,\mathrm{s}^{-1}$). Similar results were reported on the ablation of PE films by O_2 and NH_3 plasmas in the same reactor [57]. In this case the weight loss for a treatment time of 10 s was 5 times higher in oxygen ($25\,\mu\mathrm{g\,cm}^{-2}$) as compared to ammonia. The reason why PP is more degraded than PE (4 times as much) has been explained in Section 8.2.4.3 and is in agreement with what is reported in the literature [37].

Commercial, conventional polyolefins present undefined surfaces which are generally amorphous and contaminated with functional additives that migrate to the surface. That is why it is difficult to obtain a deep insight into the ablation effect of different plasmas on such polymers. We have used n-octadecyltrichlorosilane (OTS) self-assembled monolayers (SAMs) on oxidized silicon wafers as a model for polyethylene (PE) since their alkyl chains contain methylene groups exclusively (Fig. 8.6(b)). Similarly, hexatriacontane crystals ($C_{36}H_{74}$) have been used in the past [58] to study the effect of argon and oxygen RF plasmas on HDPE.

The degradation effect of ammonia plasma is generally not easy to quantify when the treatment times are short. However, ellipsometry offers the possibility to measure ablation on the OTS SAMs after 0.11 s of ammonia plasma treatment (Fig. 8.6(a)). The ablation process is linear with time and the slope yields an etching rate of $1.7\,\mathrm{nm\,s}^{-1}$. This is significantly smaller than the etching rate determined by gravimetric measurements on PE films under the same conditions, as also shown in Fig. 8.6. Typical values in this case are $5\,\mathrm{nm\,s}^{-1}$. Therefore this would mean that, if a surface cleaning is desired on PE, 1 s treatment time is enough for our experimental conditions. This effect is due to the semi-crystalline structure of the PE (LDPE provided by BASELL and ATOFINA and used as received) films which is less resistant

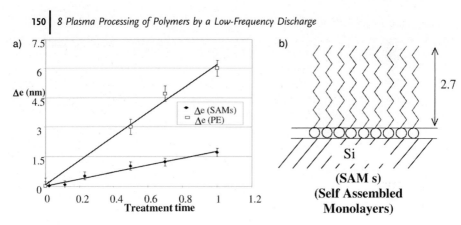

Fig. 8.6 (a) Ablation measured by ellipsometry for SAM and high precision balance for PE films, as a function of NH$_3$ plasma treatment time ($p = 100$ Pa; $P = 5$ W, $Q = 100$ sccm). (b) Schematic representation of the self-assembled monolayers of n-alkyltrichlorosilanes on silicon wafers. From [57].

to the plasma treatment than the well-defined crystalline SAM layers. Volatile LMWFs are more easily produced in the topmost layer of PE films during the bombardment by energetic incident plasma particles.

It has been shown in [57] that nitrogen-containing groups can be grafted by a NH$_3$ plasma onto both PE films and OTS SAMs. However, although the kinetics of incorporation of nitrogen groups is approximately the same in both cases and the saturation occurs after 0.5 s, the nitrogen content after similar plasma treatment conditions is two times higher for the PE films than for the SAM samples. For example, for a treatment time of 1 s in our experimental conditions the N/C ratio is 12% in the case of PE films whereas for SAMs it is only 6%. This is due on the one hand to the small layer thickness of the SAMs as compared to the 5 nm escape depth of the XPS photoelectrons and, on the other hand, to the high structural order of SAMs, with close-packed and oriented hydrocarbon chains (Fig. 8.6(b)), which favors the formation of carbon–carbon crosslinks: the radicals consumed in such process are no longer available for the nitrogen functionalization. Now the optimal treatment time for SAMs should correspond to a maximum of nitrogen containing moieties grafted to the surface with a minimum of ablation effect. If an ablated thickness of 10% could be acceptable, this yields an optimum treatment time of 0.11 s for OTS SAMs and with a N/C ratio close to 2.4%. Such a ratio corresponds to about one nitrogen function per alkyl chain, as measured independently by XPS on octadecylamine powders (C$_{18}$H$_{36}$NH$_2$) [57].

These results show that not only does the chemical structure of the polymer play an important role in plasma polymer interactions (as shown in Section 8.2.4.3), but the degree of the crystallinity of the polymer also determines the ablation rate of the polymer. Therefore in the case of a conventional semicrystalline polymer which can be presented by a structure composed of amorphous and crystalline zones, as shown

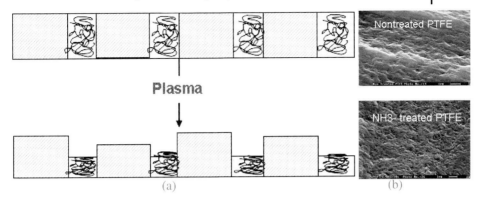

Fig. 8.7 (a) Schematic of the preferential ablation of a semicrystalline polymer. (b) SEM micrographs of nontreated and ammonia-treated PTFE.

schematically in Fig. 8.7(a), there will be a preferential ablation of the amorphous zones as compared to the crystalline ones. This will lead to an increase of the roughness of the treated polymers. This change in surface morphology can improve mechanical interlocking and can increase the area available for chemical or molecular interactions. The scanning electron microscopy (SEM) images in Fig. 8.7(b) also show the change of roughness of a semicrystalline PTFE after a plasma treatment in ammonia.

8.3.3
Acid–Base Properties

8.3.3.1 Introduction
As mentioned in Section 8.2, the surface treatment of polymers allows one to insert new functions in the surface layer which will change the polarity of the chemically inert surfaces to tailor them for a wide range of applications. However, plasma processing of polymers aiming at improving the macroscopic properties such as adhesion to metals or to other organic coatings or biomolecules requires a thorough understanding of the interfacial phenomena.

We discussed in Section 8.3.2 the plasma treatment of PP with two different gases, i.e. an oxidizing atmosphere such as oxygen and a reducing one which was ammonia.

For a long time, intermolecular interactions were classified as "polar" or "nonpolar" and it had become usual to discuss dipole–dipole interactions and the dipole–induced dipole interactions for molecules in the liquid or solid states or at liquid/solid interfaces. It was in the early 1990s that Fowkes and Whiteside pointed out the importance of acid–base interactions between the so called "polar" groups in liquids and solids and found that these interactions are quite independent of "polarity" as measured by dipole moments [59,60]. Therefore, Fowkes concluded that dipole–dipole interactions, in those cases, were negligibly small compared to acid–base and dispersion force interactions [61].

However, to validate this concept it is necessary to characterize such properties by adequate techniques and methods. The arsenal of techniques available to assess acid–base properties of polymers includes contact angle measurements [63,64], calorimetry, inverse gas chromatography, FTIR, NMR [59,62,65] and derivatization-aided XPS. The latter has been used to quantitatively evaluate the acid–base properties of polymers (in the solid state) by monitoring the chemical shift experienced by an adsorbed molecular probe such as chloroform, a Lewis acid [66]. Whitesides *et al.* [67] used "contact angle titrations", among other diagnostics, to investigate the behavior of acid and basic functional groups grafted by conventional methods at the surface of low-density polyethylene: the extent of ionization of the grafted groups was investigated, as well as the relationships between water contact angle (WCA) values and density of surface groups.

Scanning force microscopy (SFM) measurements using chemically modified tips were also used to determine the basicity or acidity of different plasma-modified polymers which contain ionizable functional groups. For example the pull-off forces of an OH-terminated tip on ammonia plasma-treated PP, and on a polymer obtained by plasma polymerization (cf. Section 8.4) of allylamine under different pH conditions were measured. The corresponding force titration curves, i.e. average pull-off force as a function of pH, were obtained [68,69]. For low pH values the plasma-polymerized allylamine film and ammonia plasma-treated PP samples, which both contain basic amino groups, showed negligible pull-off forces due to the protonation of such groups, while at $pH > 5$ the adhesion was quite pronounced. This work showed that the amino groups of the plasma-modified polymers were highly reactive and could be reversibly protonated or deprotoanted ($-NH_2 \leftrightarrow NH_3^+$) by variations in the pH environment. This has been used, for example, as a basis for immobilizing negatively charged DNA molecules on plasma-polymerized allylamine surfaces [70]. Furthermore it is shown that the pH-dependent force titration measurements allow one to map functional group distributions with a sub-50 nm resolution [68,69].

Among the different methods cited above we will focus on the contact angle titration technique which is a simple technique accessible to all and easy to use.

8.3.3.2 Contact Angle Titration Method

Hüttinger *et al.* [64] showed that by measuring the reversible work of adhesion between a solid and aqueous test liquids with different pH values, one can obtain the contact angle titration curves. Depending on the shape of these curves (Fig. 8.8) one can determine the neutral, acidic, basic, or amphoteric character of the surfaces depending on the trend of the total reversible solid–liquid work of adhesion obtained by the Young–Dupre equation ($W_{sl} = \gamma_w(1 + \cos \Theta)$, where $\gamma_w = 72.8$ mJ m^{-2}).

Ammonia Plasma-Treated PP [71,72] Contact angle measurements were determined on the surface of nontreated and plasma-treated PP by using test liquids with different pH values [63,64]. Aqueous acidic and basic solutions with pH values ranging from 1 to 14 were prepared with doubly distilled water ($\gamma_w = 72.8$ mJ m^{-2}, $\gamma_w^d = 21.6$ mJ m^{-2}, $\gamma_w^{ab} = 51.2$ mJ m^{-2}) and NaOH or HCl as the base or acid. The surface tension of the solutions was measured by the Wilhelmy technique. The latter

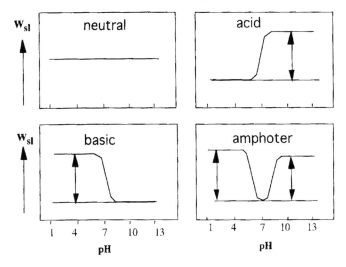

Fig. 8.8 Contact angle titration method [64].

showed that there were no differences between the surface tensions of water and those of the acidic and basis solutions. The contact angles of aqueous solutions at the surface were measured immediately after the plasma treatments.

The measurements carried out on ammonia plasma-treated PP for different treatment times are shown in Fig. 8.9. As can be seen in this figure, the behavior of nontreated PP is independent of the pH of the test liquids (Fig. 8.9(a)). In this case, the substrate being nonpolar, W_{sl} (the total solid–liquid reversible work of adhesion) which has been shown by Fowkes to be $W_{sl} = \gamma_L(1 + \cos\theta) = W_{sl}^d + W_{sl}^{ab}$ should be

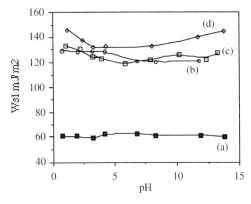

Fig. 8.9 Variation of the total work of adhesion with different test liquids for different treatment times: (a) untreated PP; (b) 0.7 s; (c) 5 s; (d) 30 s. Treatment conditions: gas, NH₃; pressure = 150–200 Pa; $Q = 150$ sccm; power = 7 W, $d = 1$ cm, $f = 70$ kHz. From [72].

equal to W_{sl}^d, since the acid–base interactions do not exist [73]. It is noteworthy that for treatment times in the range of 0.7–1 s (Fig. 8.9(b)), the surface shows a clear basic character, i.e. the total work of adhesion decreases from 129 mJ m^{-2} for acid pH values to 120 mJ m^{-2} for basic pH ones. By increasing the treatment times ($t \geq 5$ s), one can show an increase in the total work of adhesion. However, in these cases (Figs. 8.9(c) and (d)), the surface shows an amphoteric character with about the same total work of adhesion ($W_{sl} = 145$ mJ m^{-2} for $t = 30$ s) for the strongly basic and strongly acidic aqueous solutions.

On the other hand, peak fitting the N 1 s and C 1 s signals suggested that the grafted moieties on the surface were mainly amine and amide groups. [73]. These chemical functional groups confer a predominant basic character to the surface for treatment times of the order of 1 s. Minor oxygen functional groups were also detected (O/C $\leq 4\%$) on the ammonia plasma-treated surface [73]. The fact that the most basic surface is the one treated for treatment times less than a few second is in agreement with what has also been shown by other authors, i.e. NH$_2$ densities are promoted by short treatment times [55,56].

Besides the contact angle titration method, XPS has been used to quantitatively evaluate the basic character of ammonia plasma-treated surfaces by monitoring the chemical shift experienced by an adsorbed molecular probe such as chloroform, a Lewis acid [71,72].

Oxygen Plasma-Treated PP The results obtained with aqueous unbuffered solutions on the oxygen-treated PP surfaces were discussed in detail in a previous paper [73], and showed the acidic character of the treated surfaces. This property was enhanced with treatment times, due to an increase of the grafted acidic groups at the surfaces estimated by XPS. Indeed, the latter has shown firstly that the new functional groups which appear on the oxygen plasma-treated polymer are hydroxyl, carbonyl, ester and/or carboxylic acid groups [74,75] and secondly that the concentration of the grafted acidic groups increases with treatment time. DMSO was used as a basic Lewis probe in order to quantify the acidic character of oxygen plasma-treated PP [74,75].

These measurements clearly show that an ammonia plasma confers a basic character to the surface while a plasma treatment in oxygen gives rise to an acidic surface. This has been used for industrial printing applications by using aqueous-based acidic or basic inks. Indeed it has been shown that aqueous based acidic inks show a much better fixing to ammonia plasma-treated polymers and vice versa. Recently Favia *et al.* used the contact angle titration method to determine the acid–base properties of carbonaceous materials, in the form of flat graphite slabs, surface modified by mixtures of NH$_3$/O$_2$ RF plasmas [76]. In this way they tuned the acid–base properties of the carbonaceous material surfaces in a predictable way, as a function of the feed composition and of the power. This study was to investigate acid–base interactions of carbon black granules which are widely used in many applications such as reinforcing fillers in polymer matrices, as adsorbents and filters in many practical applications, e.g. for gases, vapors, liquids, etc.

8.3.4
Aging of Plasma-Treated Surfaces

The long-term stability of a plasma-treated polymer is crucial if the sample is not stored in a controlled environment or coated immediately for the intended application after treatment. It has been observed by many researchers that modified polymer surfaces are susceptible to aging effects, when exposed to a nonpolar medium such as air [77–79]. Very often, the surface rendered wettable by the plasma treatment is found to revert to a less wettable state with time. This process called ageing might result from a combination of two effects: (i) a thermodynamically driven reorientation of polar moieties away from the surface into the subsurface; (ii) a reaction of the surface with the atmospheric constituents such as oxygen, water vapor, and CO_2.

The surface reorientation or reconstruction is in response to the interfacial energy difference between the plasma polymer and its environment. This often leads to a substantial decrease in the surface density of functional groups [79]. Greisser et al. have reported a quantitative analysis of surface restructuring by distinguishing the groups on the surface as immobile and mobile polar groups [80]. However the internalization of the mobile polar groups inside the nonpolar polymer is not a thermodynamically favorable process, and the authors have hypothesized that these polar groups stabilize themselves by forming hydrogen-bonded dimers or micromicellar clusters in the subsurface region [78].

It has been shown that plasma modified surfaces go through a much pronounced surface reconstruction as compared to plasma polymers [79]. Indeed the high crosslink density of the plasma polymerized coatings can limit the surface restructuring, i.e. in the vicinity of crosslinks polar groups do not have sufficient mobility to respond to interfacial forces [78]. Therefore if one succeeds by plasma functionalization to crosslink the surface and to graft polar functionalities on top, then a stable "interphase" can be created between the polymer and the the the overlayer, which could be a metal, an oxide, a polymer coating, or a biomolecule.

The "interphase" defines a region intermediate to two contacting solids which is distinct in structure and properties from either of the two contacting phases. This crosslinked interphase will limit the mobility of the polar functions and also to some extent the migration of polymer additives to the surface [81].

It is well known that the plasma-treated surfaces react with the atmospheric constituents, what will cause a loss of desired functionality and/or surface rearrangement. Example of reactions of plasma-treated surfaces with atmospheric contaminants has been proposed by Gerenser [82] who showed that after exposing nitrogen plasma-modified PS to air, 4–5% of the total amount of nitrogen was lost. After air exposure, the integrated area under the peak due to imine species (which are clearly observed on nitrogen plasma-treated polymers) identified by XPS reported decreased by ∼60%. These results were consistent with the hydrolysis of imines to carbonyl functions via the following reactions [81]:

$$R-CH = NH + H_2O \rightarrow R-CH = O + NH_3$$

$$R-CH = NR' + H_2O \rightarrow R-CH = O + R'NH_2$$

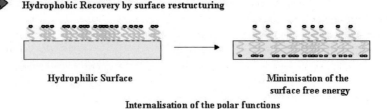

Hydrophobic Recovery by surface restructuring

Hydrophilic Surface

Minimisation of the
surface free energy

Internalisation of the polar functions

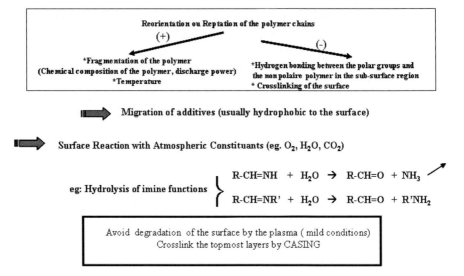

Migration of additives (usually hydrophobic to the surface)

Surface Reaction with Atmospheric Constituents (eg. O_2, H_2O, CO_2)

Fig. 8.10 Different phenomena involved in aging of plasma-
modified polymers stored in air and suggestions to limit surface
aging.

The first reaction can explain the loss of nitrogen, while the second reaction could contribute to the same effect if R' is a low molecular weight fragment that can volatilize in the vacuum of the XPS spectrometer or when exposed to air. If this is not the case, the second reaction becomes a rearrangement reaction without the loss of nitrogen.

Figure 8.10 summarizes the most important phenomena involved in the ageing of plasma treated polymers stored in air.

8.3.4.1 Aging of Ammonia Plasma-Treated PP

In the case of ammonia plasma-treated polypropylene indeed we have shown that the contact angle decreases drastically with time, i.e. the surface ages with time [80,81]. Therefore even though it is well known that the amine functions are oxidized by post-oxidation to amide function with time [79], it seems that similar to what reported on amine functionalized surfaces, the surface reorientation process is more efficient than post-oxidation. That is why to avoid this process we have tried first of all a two-step process: a helium pre-treatment in order to crosslink the surface and to obtain a dense skin by the CASING process explained in Section 8.2.3 followed by an ammonia plasma [84] to functionalize the reinforced layer. The reason why the inert gas helium was used instead of argon which is less expensive and more commonly used was to limit the sputtering of the polymer. Indeed as explained in Section 8.2.4.3.2 argon ions being much bigger than helium ones, have a much higher momentum which can therefore give rise to physical sputtering of the surface by bombardment and favor the formation of a weak boundary layer.

This crosslinked interphase as mentioned above will also limit the migration of low molecular weight polymer additives to migrate to the surface, limiting in this way the aging phenomenon. Then we have used plasmas of mixtures of ammonia and helium to crosslink and functionalize PP at the same time [52,83].

8.3.4.2 Stability of PP Treated in Plasmas of Mixtures of He + NH$_3$ for Improved Adhesion to Aluminum

XPS results showed that comparable N/C ratios with that of pure ammonia were obtained when PP films were treated in helium-rich mixtures (~2% ammonia) [51,83]. The contact angle measurements have confirmed the efficiency of the nitrogen incorporation for small amounts of ammonia introduced in a helium discharge which was comparable to that obtained in pure ammonia [83,85,86]. The aging of the surface was investigated first by the variation of WCA. The measurements were made during 70 days after the treatment and were correlated to the XPS results, i.e. (O + N)/C ratio [83,86]. A good stability of the wettability was obtained for samples treated in gaseous mixtures which contain less than 5% of ammonia in the discharge. Thus, we could estimate the variation of Θ_{water} (determined by $[(\Theta_t - \Theta_0)/\Theta_0)Q_0]$, in which Q_0 represents the contact angle at time 0) for 1% of ammonia in the mixture to be about 22%, while it is around 45% for pure ammonia discharge. Since the (O + N)/C ratio is practically constant over the whole composition of the mixture (~20%), the reorganization of the surface can not be a matter of the number of polar groups grafted to PP for different He + NH3 mixtures. In this way, the gain in the stability observed above with only a few percentages of ammonia in the plasma (<5%) can be explained by the lower mobility of the surface polar groups as we increase the crosslink density of the topmost layers of He + NH3 plasma treated PP. This crosslinked layer could also limit the process of diffusion of low molecular weight polymer additives up to the surface, limiting in this way the hydrophobic recovery of the surface [90].

However the intended application in this study was the adhesion improvement of PP to aluminum, and the wettability is only one of the theories of adhesion i.e. the thermodynamic theory [91].

After the plasma treatment in the low frequency low pressure reactor, we deposited *in situ* a thin aluminum coating (20–50 nm) by thermal evaporation at a base pressure of 10^{-3} Pa. In order to study the aging effect on the metallized plasma treated PP films, metallization of pretreated PP films was carried out either in-situ or on the aged samples in the same chamber [51,87,88]. To evaluate the improvement in adhesion, we used a U form peel test specially adapted to flexible substrates [87,88]. After the peel test, the percentage of the metal peeled off was determined by an image processing system on the peeled off films [87].

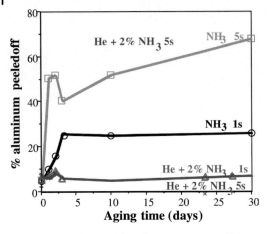

Fig. 8.11 Degradation of the adhesive properties of Al–PP joints with aging time for two different treatment times ($P = 100$ Pa; $Q = 100$ sccm; $P_w = 6.5$ W). From [52].

The stability of the adhesive properties of aluminum to plasma-treated PP studied by the peel test and the measure of the % of the metal peeled off has been reported in Fig. 8.11. For both treatment durations in ammonia plasma, the variation of the Al peeled off with the aging time depicts an increase between the freshly treated *in situ* metallized PP and a metallized aged sample. Nevertheless, one can note that for longer treatment times, the variation of the percentage of the metal peeled off is more pronounced (26% NH_3 1 s $\rightarrow$ 70% NH_3 5 s) [51,83,88]. This is explained by the formation of a weak boundary layer for over-treated PP films. Treatments carried out in He + 2% NH_3 plasma mixtures show a stabilization of the adhesion with the aging time and the over-treatment. These results are in good agreement with the stress–strain measurements, which show a cohesive strengthening of the polymer-surface treated in plasmas of He + 2% NH_3 mixtures with respect to pure ammonia [83,86,90].

8.4
Plasma Polymerization

The second way to modify the properties of different surfaces including polymers is plasma polymerization. While in surface treatment of polymers usually simple monatomic or diatomic molecules are used, and there is no deposition, in the case of plasma polymerization any kind of organic molecule can be polymerized even methane that does not contain any particular functional groups. The plasma polymerization on different substrates offers the advantage of fundamentally modifying surfaces by depositing nanometric scale polymers with different functionalities. Films deposited by plasma polymerization are free of pinholes and remnants of initiator or solvents used in conventional polymerization. Excellent reviews exist which present the complex elementary mechanisms which take place in a plasma reactor [2]. While the structure of the conventional plasma polymer depends on the precursor and has a well discernable structure, a wide range of polymers with different structures can be obtained from the same organic precursor by changing simply the operating conditions of the reactor. Therefore, if for example styrene is introduced in a plasma reactor, one cannot obtain a polymer like the conventional PE but a wide range of polymers from diamond like carbon to unsaturated polymer powders.

We have shown in the first part of the paper that the plasma treatment of polymers is tightly dependent on its chemical structure. In the case of plasma polymerization, it is often intuitively assumed that plasma polymers can be deposited on any substrate. However the deposition process is an interactive process of gas phase species with the top surface of the substrate material, and the nature and the extent of such interactions are very important factors of plasma polymerization. That is why; the plasma-deposited polymer boundary layer is composed not only of radicals resulting from the decomposition of the precursor but also gaseous species obtained from the plasma–surface interactions. In fact without a certain degree of interaction, an acceptable adhesion to the plasma polymer cannot be obtained. Such interactions exist at the beginning of the process, i.e. on the uncovered substrate material and continue on the plasma deposited polymer layer during the growth of the polymer, leading to bond scission i.e. ablation of the polymer at the same time as deposition. Yasuda has defined such a mechanism as CAP (competition between ablation and polymerization) [2].

In a plasma there is a considerable fragmentation of the starting precursor and a wide range of functional groups are incorporated into the deposit. Yasuda and co-workers showed that the deposition rate of a polymer depends on the operational composite parameter W/FM (W: discharge power; F: monomer flow rate; M: molecular weight of the monomer) in $J\,kg^{-1}$, i.e. energy injected per monomer mass [2,92].

In order to reduce the fragmentation of the precursor high powers should be avoided and one should work at low W/FM ratios, i.e. in the power deficient region where the monomer is sufficient to consume the power. In this region the deposition rate increases linearly with the parameter W/FM. In the monomer deficient region corresponding to high W/FM ratios, the power supplied is more than sufficient to fragment the monomer, and further increase in power will not give rise to an increase of the deposition rate. In this region the deposition rate reaches a plateau and can also decrease for higher values of W/FM [2,92]. So in the monomer sufficient region there is a higher probability to have less fragmentation of the polymer and to deposit thin films with a high degree of functionality retention [93]. Pulsed plasma deposition processes offer the possibility to treat surfaces which are sensitive to ion bombardment, UV radiation and temperature but also to limit to a great extent the fragmentation of the precursor in order to have a high degree of retention of the desired functionality [94,95].

Most of the plasma polymerization studies at low pressure have been carried out in 13.56 MHz RF discharges. There are very limited papers which report on the plasma polymerization of organic precursors with low frequency excitation sources (in the kilohertz range) [96].

In the next two sections we shall present two examples of plasma polymerization in our low-frequency ACE reactor in which as we mentioned earlier the ion bombardment plays an important role.

8.4.1
Influence of the Chemical Composition of the Substrate on the Plasma Polymerization of a Mixture of $CF_4 + H_2$

Fluorocarbon plasmas have been used extensively in the microelectronics industry to etch Si and SiO_2 films since the 1970s. Plasma polymerization of fluorocarbons has been largely studied since the early 1980s for their low dielectric constants, high thermal stability, and high chemical resistance [97]. Since F atoms and CF_x radicals have been demonstrated to be the main etching and deposition precursor species respectively, the fluorine to radicals F/CF_x density ratio in the plasma has shown to be characteristic of their ability to etch or cause polymerization [98,99]. Kay and Dilks [99] and Yasuda *et al.* [100] noted that in discharges in perfluorocarbons such as CF_4 and C_2F_6, the dominant process is etching rather than polymerization, but when a reducing agent (e.g. H_2, C_2F_4) was present in the discharge then polymerization predominated.

d'Agostino and co-workers studied by OES the effect of the addition of hydrogen, a scavenger of fluorine atoms, to $C_nF_{(2n+2)}$ discharges [101,102]. In all cases the fluorine atom concentration fell to a minimum at a hydrogen percentage varying between 10 and 18% and the CF and CF_2 radical concentrations increased with the percentage of hydrogen.

In order to show that the plasma–surface interactions play an important role on plasma polymerization processes, we have studied in the same ACE reactor (cf. Section 8.3.1), the polymerization of $CF_4 + H_2$ mixtures on two different substrates: polyethylene (PE) as well as on a fluorine-containing polymer polyvinyldifluoride (PVDF) [44,103]. As mentioned in Section 8.3, the polymerizing precursor can be introduced in this high voltage blade-type reactor, with the polymer substrate (PE or PVDF) placed on the grounded cylinder. As mentioned in Section 8.3, the polymer is treated for 23 ms during each turn in front of the electrodes (1 s), and the rest of the time it is in the post-discharge. The treatment times mentioned in this section can be considered to be the total time the polymer has passed in the discharge.

Optical emission spectroscopy was used to characterize the excited species in the discharge and the spectra obtained for the $CF_4 + H_2$ mixtures showed the following features: two different continua centered approximately at 300 and 580 nm which were attributed to CF_2 ions and CF_3 radicals [104] and lines at 685.6, 703.7 and 739.8 nm corresponding to atomic fluorine. Furthermore, $H\alpha$ and $H\beta$ lines were detected at 656.2 and 486 nm, respectively. Because of the energetic character of our discharge no CF_2 radicals were detected; the latter is the predominant species present in an RF glow discharge in fluorocarbons [104]. This is in agreement with the fact that at low frequencies, where the ions can follow the electric field, the discharge will behave similarly to a DC discharge, and therefore the mean electron temperature in

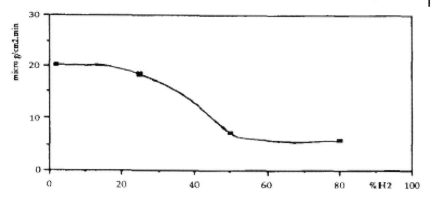

Fig. 8.12 Polymerization rate of $CF_4 + H_2 + 2\%$ Ar mixture vs. % H_2 in the feed on a PE substrate ($p = 150$ Pa; $Q = 200$ cm^{-1}; $p = 170$ W; $t = 2.7$ min). From [103].

the discharge is probably higher than the excitation threshold of CF_2 radicals [104]. By performing actinometry with Ar we could follow the fluorine atom concentration with the addition of hydrogen to CF_4, which went through a minimum at 2% of hydrogen in the mixture. The reason for this trend has been explained in detail elsewhere [104]. Briefly the minimum observed is due to the scavenging of fluorine by hydrogen to give rise to HF in the gas phase which is not an etching reagent and is pumped out [44,103,104]. The further increase in the fluorine atom concentration above 2% of hydrogen introduced in CF_4 is due to the change of the energetic character of the discharge which will favor the vibrational excitation of atomic and molecular hydrogen which will result in the decomposition of HF to F [104].

The polymerization rate, which was recorded by gravimetric measurements (μg cm^{-2} min^{-1}) on PE, was greatest with 2% hydrogen in the CF_4 discharge (Fig. 8.12).

XPS experiments performed on the deposited fluorocarbon films confirmed the emission spectroscopic results, i.e. the fluorine content (F/C ratio) was greatest for 2% hydrogen, which corresponded to the point where the relative concentration of excited fluorine atoms (known to be an etching agent) went through a minimum. This is in agreement with what reported by other authors [105], i.e. for low H_2 feed content, fluorocarbons with high F/C ratio and low crosslinking degree are deposited (small C–CF component of the C 1 s peak as compared to CF, CF_2 and CF_3 groups), while for hydrogen rich mixtures, the polymer is more crosslinked (C–CF peak increases while the other groups decrease), and is characterized by a lower fluorine content (Fig. 8.13) [103].

The above results show that the structure of the film was highly dependent on the hydrogen concentration in the feed. This is how one can tune the wettability of the PE on which the fluorocarbon was deposited in a wide range depending on the $CF_4 + H_2$ mixture used [103].

We have looked then at the variation of the polymerization rate of the fluorocarbon films deposited on PE with a mixture of $CF_4 + 2\%$ H_2 with time (Fig. 8.14). The

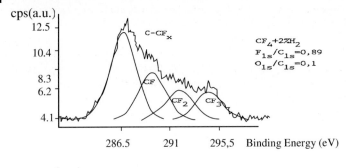

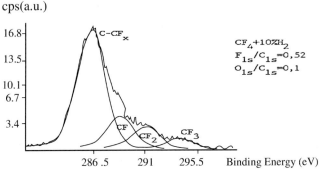

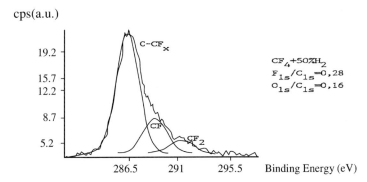

Fig. 8.13 Variation of C 1s photopeak of PE film treated with different percentages of H_2 in $CF_4 + H_2$ discharges ($p = 150$ Pa; $P = 17$ W; $Q = 200$ cm^3 min^{-1}; $t = 2.7$ min). From [103].

results show that as soon as the discharge was ignited a fluorocarbon polymer was formed on the PE substrate, while with longer treatment times when the fluorine content of the deposited polymer was increased (F : C = 1 : 1), the polymerization rate decreased; that is, etching (both physical and chemical) became competitive with polymer deposition. Figure 8.14 shows the plateau obtained for the polymerization rate (after 1.5 min) which was due to an equilibrium established between the polymerization and etching processes.

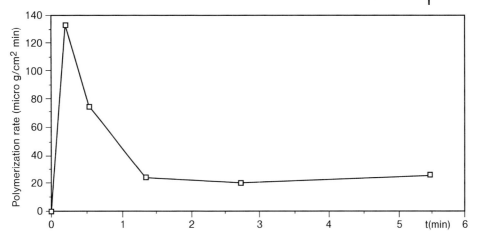

Fig. 8.14 Polymerization rate of $CF_4 + 2\% H_2 + 2\%$ Ar mixture as a function of treatment time on a PE substrate ($p = 150$ Pa; $Q = 200\,cm^3\,min^{-1}$; $p = 17$ W). From [103].

XPS results obtained with these samples showed that the F : C ratio of the deposit remained almost constant while the area of the peak at 286.6 eV attributed to C–CF$_x$ groups increased with time (Fig. 8.15), which shows a more crosslinked structure with time due to ion bombardment. We can note that for a treatment time of 0.23 min (time during which the polymer is actually in the discharge), the thickness of the polymer is less than 5 nm, the analytical depth of XPS, which is why a component at 285 eV is observed, which gradually disappears with time.

The PVDF substrate material was itself hydrofluorinated from the start, with an F : C ratio equal to unity, and therefore with the particular feed composition used ($CF_4 + 2\% H_2$), for short treatment times (less than 75 s), there was no polymer deposition but only etching of the substrate (Fig. 8.15). This is consistent with what was observed with the PE substrate using longer treatment times (more than 30 s) (Fig. 8.14). As soon as the fluorine content of the deposited polymer decreased, then the polymer could start to build up on the surface of the PVDF. However the small increase of the weight of the substrate could also be due to substitution of hydrogen atoms by fluorine ones, i.e. fluorination and not poly-merization.

Consequently, since we have shown that the fluorine constituent in the sub-strate clearly contributes to the growth mechanisms, the optimum hydrogen percentage in the feed (corresponding to the maximum polymer growth rate) will not be the same with a PVDF substrate as with a PE substrate. The polymerization rate on PE is greatest with a hydrogen fraction equal to 2% (Fig. 8.13), whereas with the fluorine-containing substrate PVDF, a higher concentration of H_2 is required. In the case of PVDF, the optimum hydrogen proportion is around 4% (Fig. 8.16).

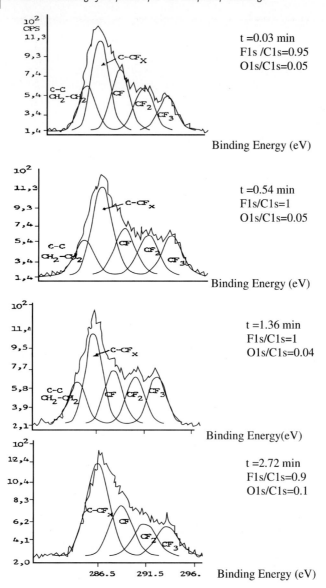

Fig. 8.15 Variation of C 1 s photoelectron peak of PE film treated for different treatment times in $CF_4 + 2\%$ H_2 discharges ($p = 150$ Pa; $P = 17$ W; $Q = 200$ cm^3 min^{-1}. From [103].

The above results show clearly that the deposition process is an interactive process of gas phase species with the top surface of the substrate material or the depositing polymer. Such gas–surface interactions can be more pronounced in the ACE reactor, as compared to 13.56 MHz discharges, in which ion bombardment plays an important role.

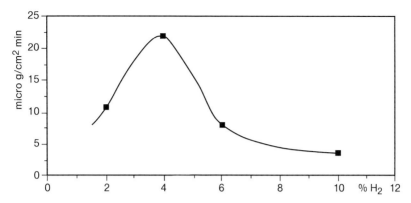

Fig. 8.16 Polymerization rate of $CF_4 + H2 + 2\%$ Ar mixture as a function of % H_2 in the feed on a PVDF substrate ($p = 150$ Pa; $Q = 200$ cm^3 min^{-1}; $p = 17$ W; $t = 2.7$ min). From [103].

8.4.2
Plasma Polymerization of Acrylic Acid

The plasma polymerization of acrylic acid, acetic acid, propionic acid has been an active area of research to produce surfaces with high densities of acid groups for "bio" applications [106–110]. Acid-rich surfaces have shown to absorb atmospheric moisture and to be very soluble in water. Therefore for such application, there is a strong need to obtain stable plasma polymerized acrylic acid coating, resistant to washing with water. A certain number of papers have dealt with the washing stability of PPAA in the past [113–114]. Different ways of stabilizing such polymers have been reported such as copolymerizing acrylic acid with hydrocarbons like 1,7-octadiene [109,111]. Some improvements in the stability of the coatings have been obtained. Alexander and Duc [113] obtained at higher powers a COOH content of the surface determined by XPS on TFE (TriFluoroEthanol)-labeled plasma-polymerized acrylic acid (PPAAA) coatings deposited on aluminum substrates less than 6% (of the total C 1 s peak) after rising in water. Detomaso *et al.* studied the increase in power and of duty cycle in a modulated plasma for the deposition of acrylic acid on PET and PS substrates: they obtained a surface containing 3–4% of COOH groups after washing [112]. Sciarratta *et al.* studied the stability of PPAA in a continuous and pulsed 13.56 MHz discharge. They found a maximum COOH retention about 5% after water washing [113]. All of these studies were carried out at 13.56 MHz.

We have also studied the plasma polymerization of acrylic acid with the ACE reactor in which, as explained in Section 8.3.1, due to the low excitation frequency (70 kHz), the ion bombardment should play an important role on the surface chemistry. Furthermore since as explained in Section 8.3.1, during every turn of 1s duration, the polymer spends a very short time (23 ms) in the discharge and the remaining time is in fact the post-discharge, the post graft polymerization which occurs when the polymer is not in the discharge zone should contribute to the deposition. The aim of this work is to produce a surface containing a high density of

COOH functions for the covalent immobilization of biomolecules onto the surface of polyethylene (PE). One should note that the treatment time mentioned in the figures could correspond to t_{on} if compared to a pulsed system.

The acrylic acid precursor was introduced in the hollow electrode by bubbling argon at a rate of $10 \, cm^3 \, min^{-1}$ in a 10 mL round-bottom flask containing the acrylic acid monomer (Sigma-Aldrich, 99% pure) which was heated at 45 °C. In such conditions, the flow rate of the acrylic acid was around $6 \, cm^3 \, min^{-1}$. The main chamber was evacuated by a TPH 170 (Balzers) turbomolecular pumping system and a base pressure of $10^{-3} \, Pa$ was then obtained.

The operating pressure was maintained at 0.3–0.4 mbar by a 2012 AC primary pump resistant to chemicals. Plasma emission was monitored through an optical fiber (diameter: 200 μm, incident angle = 47°). The radiation transmitted was analyzed by a SpectraPro-500i spectrophotometer (Acton Research Corporation) equipped with a 3600 and 1200 grooves mm^{-1} holographic grating for the analysis of the 300–800 nm spectral domain. The physicochemical characteristics of the PPAA coatings were analyzed by WCA, XPS and FTIR in transmission mode. The details of the techniques used have been presented elsewhere [115].

Advancing water contact angle (WCA) was measured for the plasma polymerized acrylic acid deposited on the PE films. The measured values varied from 90° for a nontreated PE film to an average of less than 10° for as-deposited PPAA coatings at different experimental conditions. After a thorough rinsing with water, the coating clearly loses a part of its hydrophilicity since the WCA increases to reach a value of 50–60° [115].

PPAA coatings were analyzed by XPS. Like most of the work reported in the literature, the C 1 s core level can be satisfactorily fitted by a combination of four distinct peaks: the peak at 285 eV corresponds to C–C and C–H moieties, the one at 286.4 eV to C–OH and C–O–C functional groups, the peak at 287.8 eV to C=O and O–C–O groups, and finally the component at 289.2 eV can be attributed to COOH and/or COOR groups [115]. Some authors have distinguished the contribution of the acid form from that of the ester by derivatization experiments with trifluoroethanol [116,117]. Their investigations showed that the contribution of the carboxylic acid functions correspond to more than 90% of the component at 289.2 eV [116]. Our XPS derivatization experiments with trifluoroethanol confirm that more than 95% of the peak at 289.2 eV corresponds to acid groups [119]. Therefore, we consider that this component of the C 1 s peak corresponds mainly to carboxylic acid groups. The role of the power on the stability of the coatings has been investigated by most of the authors depositing coatings from acrylic acid [111,112,116,118]. Figure 8.17 displays the variation in COOH retention before and after washing obtained by XPS on PPAA coatings deposited at different plasma powers. One can note that the COOH content of the as-deposited (unwashed) films decreased by increasing the power. But, in contrast, on the washed surfaces, the COOH retention increased at high powers, leading to more stable coating to washing with water.

These coatings were also analyzed by FTIR (Fig. 8.18) which allows one to characterize the whole thickness of the deposited PPAA coatings. The transmission bands visible in the FTIR spectrum of PPAA coatings deposited on the PE film can be

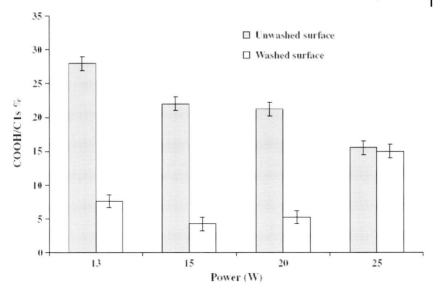

Fig. 8.17 Variation of the COOH content of PPAA coating as function of the plasma power ($f = 70\,kHz$, $Q_{argon} = 10\,sccm$, $Q_{acrylic\ acid} = 6\,sccm$, $P = 0.4\,mbar$, time $= 10\,s$). From [115].

assigned to the stretching vibration peak of C–H at $1450\,cm^{-1}$, a stretching broad band of OH in the range of 2900–$3300\,cm^{-1}$, and the stretching vibration peak of C=O at $1706\,cm^{-1}$. By increasing the power, the peak intensity at $1706\,cm^{-1}$ decreased in the PPAA coating which is consistent with a decrease of the carboxylic group content in the as-deposited PPAA coating. These results are in good agreement with the XPS analyses.

Finally, in order to have further information about the deposition process in our low-frequency ACE reactor, OES was used to study the fragmentation of acrylic acid in the plasma as a function of the plasma power. The OH ($A^2\Sigma^+ - X^2\Pi$ at 306.6 nm), the CO_2 ($^1B_2 - X^1\Sigma^+$ at 434.4 nm) and the CO ($B^1\Sigma - A^1\Pi$ at 519.8 nm) lines were followed in order to link these results with the monomer functionality. The intensity of these lines increases with the plasma power [115–119] which is consistent with a higher level of fragmentation for acrylic acid. Palumbo *et al.* [116] have shown that the ICO/IAr ratio in a 13.56 MHz discharge could be advantageously used as a process control parameter to monitor the retention of the carboxylic groups in the film. We have also found in our low-frequency ACE reactor that there exists also a good correlation between the ICO/IAr ratio which could be a representative of the density of CO species in the fundamental state, and the carboxylic groups content in the PPAA coatings as measured by XPS (Fig. 8.19). There is an inverse relationship between the CO density measured in the plasma and the carboxylic acid content in the PPAA coating. An increase in the power leads to an increase of the ICO/IAr ratio (Fig. 8.19), indicating a more fragmented acrylic

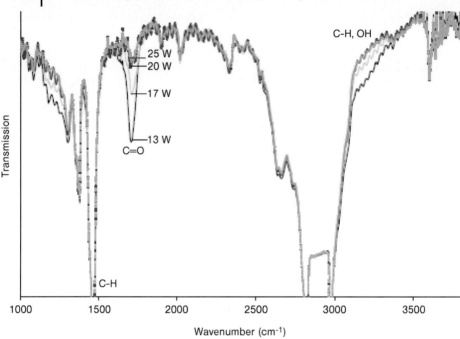

Fig. 8.18 FTIR spectra of PPAA deposited on PE films under various power conditions ($f = 70\,kHz$, $Q_{Argon} = 10\,sccm$, $Q_{Acrylic\ acid} = 6\,sccm$, $P = 0.4\,mbar$, $t = 10\,s$). From [115].

acid monomer in the gas phase; as a consequence, the content in carboxylic groups in the PPAA coating decreases. In other words, at higher powers due to the higher fragmentation of acrylic acid monomer in the discharge, the carboxylic groups instead of being incorporated in the films are decomposed in the form of CO species, which are pumped out of the reactor.

The above results show that the PPAA coatings deposited in the low-frequency ACE reactor at higher powers on PE films are quite stable to rinsing. Even though some of the coating and as a consequence carboxylic functions are washed away, the retention of the carboxylic groups is higher as compared to what has been reported in the literature. As mentioned above and in Section 8.3.1, even though it is a CW discharge since the cylinder rotates in front of the HV electrode, the discharge is somewhat similar to a pulsed DC discharge with a duty cycle equal to 2.3%. This discharge is however not a common pulsed system because the polymer rotates in and out of the discharge. When the polymer is in the discharge, plasma polymerization takes place and the bombardment of the cylindrical electrode on which the polymer is rolled with highly energetic ions gives rise to a higher range of production of nucleation sites on the starting substrate and on the growing film which in turn increases the deposition rate. When the polymer film is out of the discharge, mainly graft-polymerization on the surface activated polymer can occur.

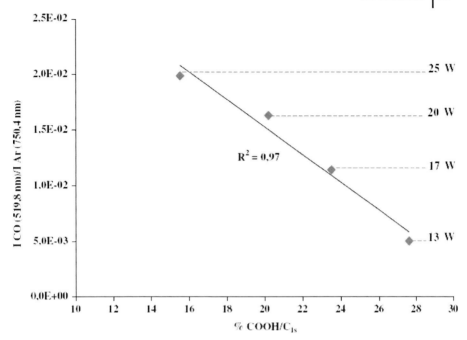

Fig. 8.19 Correlation between the ICO/IAr ratio (OES measurements) and the carboxylic content in the PPAA coating, as measured by XPS ($f = 70\,kHz$, $Q_{Argon} = 10\,sccm$, $Q_{Acrylic\ acid} = 6\,sccm$, $P = 0.4\,mbar$). From [115].

8.5
Conclusions

In this chapter the elemental reactions in plasma surface treatment of polymers have been reviewed. The main surface reactions: functionalization, crosslinking, etching/ablation take place in a complex synergy which depends on many parameters of the plasma system used (power density, electrode configuration, excitation frequency, reactor geometry, etc.) but also on the polymer (chemical composition, degree of crystallinity, etc.). That is why it is extremely important and necessary to monitor plasma parameters in each system in the presence of a particular polymer substrate in order to have a successful approach for the specific aimed application.

Plasma surface functionalization of polymers of different geometries (webs, fibers, powders, complex shapes) is now largely used in industry in particularly for improvement of their adhesive properties to different coatings. Plasma polymerization offers the possibility to increase the selectivity of the plasma process as compared to functionalization, to obtain specific functionalities on the surface. The decisive add-on value that allows one to obtain tailored products, the cost of which is much higher than the starting material, and/or which cannot be attained by other processes, makes low-pressure plasma polymerization a very promising technique to develop for large scale production in industry. The conventional wet processes are becoming increasingly unacceptable because of environmental and safety considerations. The advantages of plasma processing of polymers, being environmentally friendly and allowing one to produce high-performance materials compensates the initial equipment cost. There is literature [120] which compares the whole overall cost of a conventional process with that of a plasma one. In the case of the former usually additional investment such as installations for safety and waste treatment, or running cost for chemicals and energy (drying) can add considerably to the expenses of the overall costs.

In this chapter we have focused on the use of a particular cold plasma which was a low-frequency discharge with ACE to show through demonstrative examples the main surface reactions which take place in surface processing of polymers both by plasma treatment and plasma polymerization. We have shown that in this discharge, ion bombardment can play an important role in plasma–surface interactions. This can result in higher kinetics of insertion of functionalities during grafting or to the deposition of more crosslinked polymers during plasma deposition.

Acknowledgments

Our plasma research is supported by PhD student fellowships from the Ministère de la Recherche et de l'Enseignement Supérieur and European fundings such as IFCA (GSRD-CT-2002-00723) and ACTECO-EC project (515859-2).

References

1 Siperko, L.M. and Thomas, R.R. (1989) *J. Adhesion Sci. Technol.*, **3**, 157.

2 (a) Yasuda, H. (2003) *Nucl. Instrum. Methods Phys. Res., Sect. A*, **515**, 15; (b)Yasuda, H. and Yasuda, T. (2000) *J. Polym. Sci. A: Polym. Chem.*, **38**, 943.

3 Haïdopoulos, M., Turgeon, S., Laroche, G. and Mantovani, D. (2005)*Plasma Process. Polym.*, **2** (5), 424–440.

4 Denes, F.S. and Manolache, S. (2004) *Prog. Polym. Sci.*, **29**, 815–885.

5 Arefi-Khonsari, F., Tatoulian, M., Bretagnol, F., Bouloussa, O. and Rondelez, F. (2005) *Surf. Coat. Technol.*, **200**, 14–20.

6 Morstein, M., Karches, M., Bayer, C., Casanova, D. and Rudolf von Rohr, P. (2000) *Chem. Vapor Depos.*, **6**, 16–20.

7 Benrejeb, S., Tatoulian, M., Arefi-Khonsari, F., Fischer-Durand, N., Martel, A., Lawrence, J.F., Amouroux, J., LeGoffic, F. (1998) *Anal. Chim. Acta*, **376**, 133–138.

8 Hiratsuka, A. and Karube, I. (2000) *Electoanalysis*, **9**, 12.

9 Grunwald, H., Henrich, J., Krempel-Hesse, J., Dicken, W., Kunkel, S. and Ickes, G. (1997) *40th Annual Technical Conference Proceedings.*

10 Grünwald, H., Adam, R., Bartella, J., Jung, M., Dicken, W., Kunkel, S., Nauenburg, N., Gebele, T.,

Mitzlaff, S., Ickes, G., Patz, U. and Synder, J. (1999) *Surf. Coat. Technol.*, **111**, 287–296.

11 Boutroy, N. (2006) *Industrial workshop, Tenth International Conference on Plasma Surface Engineering, PSE 2006*, Abstracts, 90.

12 Goldman, M., Goldman, A. and Sigmond, R.S. (1985) *Pure Appl. Chem.*, **57**, 1353.

13 Uehara, T. (1999) in *Adhesion Promotion Techniques Technological Applications* (eds K.L. Mittal and A. Pizzi), Marcel Dekker, New York. 191–204.

14 Arefi-Khonsari, F., Tatoulian, M., Kurdi, J. and Amouroux, J. (2001) in *Progress in Plasma Processing of Materials* (ed. P. Fauchais), Begall House, New York/Wallingford. 457–471.

15 Linsley Hood, H.L. (1980) *Proceedings of Gaz Discharges and Their Applications* (GD 80), Edinburgh, 8–11 September 1980, 86–90.

16 (a) Eliasson, B., Egli, W. and Kogelschatz, U. (1996) *Pure Appl. Phys.*, **79** (8), 3877–3885; (b) Kogelschatz, U. (2002) *IEEE Trans. Plasma Sci.*, **30**, 1400.

17 Borcia, G., Anderson, C.A. and Brown, N.M.D. (2004) *Appl. Surf. Sci.*, **225** (1–4), 186–197.

18 Borcia, G., Anderson, C.A. and Brown, N.M.D. (2003) *Plasma Sources Sci. Technol.*, **12**, 335–344.

19 Zhu, X.D., Arefi-Khonsari, F., Petit-Etienne, C. and Tatoulian, M. (2005) *Plasma Process. Polym.*, **2**, 407–413.

20 Starostine, S., Aldea, E., deVries, H., Creatore, M. and van deSanden, M.C. (2007) *Plasma Process. Polym.*, **S1**, S440–S444.

21 Akdoğan, E., Çökeliler, D., Marcinauskas, L., Valatkevicius, P., Valincius, V. and Mutlu, M. (2006) *Surf. Coat. Technol.*, **201**, 2540–2546.

22 Hollahan, J.R. and Stafford, B.B. (1969) *J. Appl. Polym. Sci.*, **13**, 807–816.

23 Lichtenberg, M.A. and Lieberman, A.J. (1994) *Principles of Plasma Physics and Materials Processing*, Wiley, New York.

24 Grill, A. (1993) *Cold Plasma in Materials Fabrication*, IEEE Press, New York.

25 Holländer, A., Wilken, R. and Behnisch, J. (1999) *Surf. Coat. Technol.*, **788**, 116–119.

26 Kuzuya, M., Kondo, S.I., Sugito, M. and Yamashiro, T. (1998) *Macromolecules*, **31**, 3230.

27 Schonhorn, H. and Hansen, R.H. (1967) *J. Appl. Polym. Sci.*, **11**, 1461–1474.

28 Loh, In-Houng, Cohen, E. and Baddour, R.F. (1986) *J. Appl. Polym. Sci.*, **31**, 901.

29 Placinta, G., Arefi, F., Geoghiu, M., Amouroux, J. and Pop, G. (1997) *J. Appl. Polym. Sci.*, **66**, 1367–1375.

30 Arefi-Khonsari, F., Placinta, G., Amouroux, J. and Popa, G. (1998) *Eur. Phys. J. Appl. Phys.*, **4**, 193.

31 Amatanides, E. and Mataras, D., Chapter 4, Advanced Plasma Technology Book, Wiley 2007.

32 Ränby, B. and Rabek, J.F. (1976) *Singlet Oxygen Reactions with Synthetic polymers*, Stockholm Symposium, 2–4 September.

33 (a) Bikerman, J.J. (1968) *The Science of Adhesive Joints*, 2nd edn , Academic Press, New York, (b) Bikerman, J. (1961) *J. Appl. Chem.*, **11**, 81.

34 Grünwald, H. (1999) *14th international Symposium on Plasma Chemistry, Prague, Czech Republic, 2–6 August 1999, Symposium Proceedings*, 2749–2751.

35 Krüger, P., Knes, R. and Fredrich, J. (1999) *Surf. Coat. Technol.*, **112**, 220–240.

36 Egitto, F. and Matienzo, L.J. (1996) *Metallized Plastics: Fundamental and Applied Aspects V, 189th Meeting of the Electrochemical Society*, Los Angeles, CA, 5–10 May, 283–301.

37 Hansen, R.H., Pascale, J.V., DeBenedictis, T. and Rentzepis, P.M. (1965) *J. Polym. Sci.*, (Part A3), 2205.

38 Moss, S.J., Jolly, A.M. and Tighe, B.J. (1986) *Plasma Chem. Plasma Process*, **6** (4), 401.

39 Yasuda, H. *et al.* (1973) *J. Appl. Polym. Sci.*, **17**, 137.

40 Ricard, A. (1990) in *Plasma–Surface Interactions and Processing of Materials* (ed. O. Auciello), Kluwer Academic, Dordrecht.

41 Borcia, G., Arefi-Khonsari, F., Amouroux, J. and Popa, G. (1999) *Proceedings of the ISPC 14*, 2–6 August, Prague, Czech Republic, Vol. IV, 1815–1820.

42 Gordiets, B. and Ricard, A. (1993) *Plasma Sources Sci. Technol.*, **2**, 158–163.

43 Lindinger, W., Schemeltekopf, A.L. and Fehsenfeld, F.C. (1974) *J. Chem. Phys.*, **61**, 2690.

44 Arefi, F., Andre, V., Montazer-Rahmati, P. and Amouroux, J. (1992) *Pure Appl. Chem.*, **64**, 715–772.

45 Arefi, F., Tatoulian, M., André, V., Amouroux, J. and Lorang, G. (1992) in *Metallized Plastics 3: Fundamental and Applied Aspects* (ed. K.L. Mittal), Plenum Press, New York. 243–256.

46 Arefi, F., Tchoubineh, F., Andre, V., Montazer, P., Amouroux, J. and Goldman, M. (1988) *Plasma Surf. Eng.*, **2**, 679–686.

47 Foerch, R., McIntyre, N.S. and Hunter, D.H. (1991) *Kunststoffe*, **81**, 260.

48 Conti, S., Fridman, A., Grace, J.M. *et al.* (2001) *Exp. Thermal Fluid Sci.*, **24**, 79–91.

49 McCurdy, P.R., Butoi, C.I., Williams, K.L. and Fisher, E.R. (1999) *J. Phys. Chem. B*, **103**, 6919–6929.

50 Fisher, E.R. (2004) *Plasma Process. Polym.*, **1**, 13–27.

51 Arefi-Khonsari, F., Tatoulian, M., Kurdi, J., Ben Rejeb, S. and Amouroux, J. (1998) *J. Photopolym. Scie. Technol.*, **11**, 277–292.

52 Arefi-Khonsari, F., Kurdi, J., Tatoulian, M. and Amouroux, J. (2001) *Surf. Coat. Technol.*, 142–444, 437–448.

53 Khairallah, Y., Arefi, F., Amouroux, J., Leonard, D. and Bertrand, P. (1994) *J. Adhesion Sci. Technol.*, **8**, 363–381.

54 Milde, F., Goedicke, K. and Fahland, M., (1996) *Thin Solid Films*, **279**, 169–173.

55 Favia, P., Stendardo, V. and d'Agostino, R. (1996) *Plasmas Polym.*, **1**, 91.

56 Meyer-Plath, A., Schröder, K., Finke, B. and Ohl, A. (2003) *Vacuum*, **71**, 391–406.

57 Tatoulian, M., Moriere, F., Bouloussa, O., Arefi-Khonsari, F., Amouroux, J. and Rondelez, F. (2004) *Langmuir*, **20**, 10481–10489.

58 Clouet, F., Shi, M.K., Prat, R., Holl, Y., Marie, P., Leonard, D., DePuydt, Y., Bertrand, P., Dewez, J.L. and Doren, A. (1994) in *Plasma Surface Modification of Polymers* (eds M. Strobel, M. Lyons and C.K.L. Mittal), VSP, Utrecht. 65–97.

59 Fowkes, F.M. (1991) in *Acid–Base Interactions, Relevance to Adhesion Science and Technology* (eds K.L. Mittal and H.R. Jr Anderson), VSP, Utrecht.

60 Whiteside, G.M., Biebuyck, H.A., Folker, J.P. and Prime, K.L. (1991) in *Acid–Base Interactions, Relevance to Adhesion Science and Technology* (eds K.L. Mittal and H.R. Jr Anderson), VSP, Utrecht. 229–241.

61 Fowkes, F.M., Riddle, F.L., Pastore, W.E. and Weber, A.A. (1990) *Colloids Surf.* 43, 367–387.

62 (a) Fowkes, F.M. (1987) *J. Adhesion Sci. Technol*, **1**, 17–27; (b) Fowkes, F.M. (1983) in *Adhesion and Adsorption of Polymers vol. 2* (ed. Lieng-Huang Lee), Plenum Press, New York, 583.

63 Whitesides, G.M. and Laibinis, P.E. (1987) *Langmuir*, **3**, 62–76.

64 Hüttinger, K.J., Hôhmann-Wein, S. and Krekel, G. (1992) *J. Adhesion Sci. Technol.*, **6**, 317.

65 Fowkes, F.M., Kacsinski, M.B. and Dwight, D.W. (1991) *Langmuir*, **7**, 2464.

66 Chehimi, M.M. (1991) *J. Mater. Sci. Lett.*, **10**, 908.

67 Kaczinski, M.B. and Dwight, D.W. (1993) *J. Adhes. Sci. Technol.*, **7**, 165.

68 Schönherr, H., vanOs, M.T., Hruska, Z., Kurdi, J., Förch, R., Arefi-Khonsari, F., Knoll, W. and Vancso, G.J. (2000) *Chem. Commun.*, 1303–1304.

69 Vancso, G.J., Schonherr, H., vanOs, M.T., Zdenek, H., Kurdi, J., Forch, R., Arefi-Khonsari, F. and Knoll, W. (2000) *Polym. Prepr. (Am. Chem. Soc. Div. Polym. Chem.)*, **41** (2), 1416–1417.

70 Fôrch, R., Zhang, Z. and Knoll, W. (2005) *Plasma Process. Polym*, **2**, 351–372.

71 Shahidzadeh-Ahmadi, N., Chehimi, M.M., Arefi-Khonsari, F., Amouroux, J. and Delamar, M. (1996) *Plasmas Polym.*, **1**, 27–45.

72 Arefi-Khonsari, F., Tatoulian, M., Shahidzadeh, N. and Amouroux, J. (1997) in *Plasma Processing of Polymers* (ed R. d'Agostino *et al.*), Kluwer Academic, Dordrecht., 165–207.

73 Shahidzadeh-Ahmadi, N., Arefi-Khonsari, F. and Amouroux, J. (1995) *J. Mater. Chem.*, **5** (2), 229–236.

74 Shahidzadeh-Ahamadi, N., Chehimi, M.M., Arefi-Khonsari, F., Foulon-Belkacemi, N., Amouroux, J. and Delamar, M. (1995) *Colloids Surf. A: Physicochem. Eng. Aspects*, **105**, 277–289.

75 Shahidzadeh-Ahamadi, N., Arefi-Khonsari, F., Chehimi, M.M. and Amouroux, J. (1996) *Surf. Sci.*, **352–354**, 888–892.

76 Favia, P., DeVietro, N., DiMundo, R., Fracassi, F. and d'Agostino, R. (2006) *Plasma Process. Polym.*, **3**, 66–74.

77 Morra, M., Occhiello, E. and Garbassi, F. (1991) in *Polymer–Solid Interfaces* (eds J.J. Pireaux, P., Bertrand and J.L. Brédas), Galliard Ltd, Great Yarmouth, Norfolk, Great Britain, 407–428.

78 Griesser, H.J., Da, Y., Hughes, A.E., Gengenbach, T.R. and Mau, A.W.H. (1991) *Langmuir*, **7**, 2484–2491.

79 Siow, K.S., Britcher, L., Kumar, S. and Griesser, H.J. (2006) *Plasma Process. Polym.*, **3**, 392–418.

80 Chatelier, R.C., Xie, X., Gengenbach, T.R. and Greisser, H.J. (1995) *Langmuir*, **11**, 2576.

81 Wertheimer, M.R., Martinu, L. and Klember-Sapieha, J.E. (1999) in *Adhesion Promotion, Techniques* (eds K.L. Mittal and A. Pizzi), Marcel Dekker, New York, 139.

82 Gerenser, L.J. (1994) in *Plasma Surface Modification of Polymers* (eds M. Strobel, C. Lyons and K.L. Mittal), VSP, Utrecht, 43–64.

83 Kurdi, J., Arefi-Khonsari, F., Tatoulian, M. and Amouroux, J. (1998) in *Metallized Plastics 5 & 6, Fundamental and Applied Aspects* (ed. K.L. Mittal), VSP, Utrecht. 295–319.

84 Tatoulian, M., Arefi-Khonsari, F., Mabille-Rouger, I., Amouroux, J., Gheorgiu, M. and Bouchier, D. (1995) *J. Adhesion Sci. Technol.*, **9**, 923–934.

85 Kurdi, J., Ardelean, H., Marcus, P., Jonnard, P. and Arefi-Khonsari, F. (2002) *Appl. Surf. Sci.*, **189**, 119–128.

86 Kurdi, J., Tatoulian, M., Amouroux, J. and Arefi-Khonsari, F. (1999) *Proceedings of ISPC 14*, Prague, Czech Republic, 2–6 August 1999, IV, 1773–1778.

87 Arefi, F., Tatoulian, M., André, V., Amouroux, J. and Lorang, G. (1992) in *Metallized Plastics 3: Fundamental and Applied Aspects* (ed. K.L. Mittal), Plenum Press, New York, 243–256.

88 Tatoulian, M., Arefi-Khonsari, F., Shahidzadeh-Ahmadi, N. and Amouroux, J. (1995) *Int. J. Adhesion Adhesives*, **15**, 177–184.

89 Klemberg-Sapieha, J.E., Martinu, L., Kûttel, O.M. and Wertheimer, M.R. (1991) in *Metallized Plastics 2: Fundamental and Applied Aspects* (ed. K.L. Mittal), Plenum Press, New York, 315–329.

90 Kurdi, J. (2000) PhD thesis, Pierre Marie Curie University.

91 Schulz, J. and Nardin, M. (1999) in *Adhesion Promotion*, Techniques (eds K.L. Mittal and A. Pizzi), Marcel Dekker, New York, 1.

92 Yeh, Y.S., Shu, I.N. and Yasuda, H. (1987) *J. Appl. Polym. Sci. Appl. Polym. Symp.*, **42**, 1–26.

93 O'Toole, L., Beck, A.J. and Short, R.D. (1996) *Macromolecles*, **29**, 5172.

94 Savage, C.R. and Timmons, R.B. (1991) *Abstr. Pap. Am. Chem. Soc.*, **201**, 53.

95 Hynes, A.M., Shenton, M. and Badyal, J.P.S. (1996) *Macromolecules*, **29**, 4220–4225.

96 Yasuda, H.K. (2005) *Plasma Process. Polym.*, **2**, 293–304.

97 Takahashi, K., Mitamua, T., Ono, K., Setsuhara, Y., Itoh, A. and Tachibana, K. (2003) *Appl. Phys. Lett.*, **82**, 2476.

98 d'Agostino, R., deBenedictis, D. and Cramarossa, F. (1984) *Plasma Chem. Plasma Process*, **4** (1), 1.

99 Kay, E. and Dilks, A. (1981) *Thin Solid Films*, **78**, 309.

100 Yasuda, H. (1978) in *Thin Film Process* (eds J.L. Vessen and W. Kern), Academic Press, New York, 361.

101 d'Agostino, R., Cramarossa, F., Colaprico, V. and d'Ettole, R. (1983) *J. Appl. Phys.*, **54** (3), 1284.

102 d'Agostino, R. (1997) in *Plasma Processing of Polymers* (eds R. d'Agostino, P. Favia and F. Fracassi), *NATO ASI Series E:346*, Kluwer Academic.

103 Montazer Rahmati, P., Arefi, F. and Amouroux, J. (1991) *Surf. Coat. Technol.*, **45**, 369–378.

104 Montazer Rahmati, P., Arefi, F., Amouroux, J. and Ricard, A. (1989) *Proc. 9th ISPC (IUPAC)*, Pugnochiuso, Italy, 1195.

105 Lamendola, R., Favia, P. and d'Agostino, R. (1992) *Plasma Sources Sci. Technol.*, **1**, 256.

106 Gupta, B., Plummer, C., Bisson, I., Fery, P. and Hilborn, J. (2002) *Biomaterials*, **23** (3), 863–871.

107 Whittle, J.D., Bullett, N.A., Short, R.D., Ian Douglas, C.W., Hollander, A.P. and Davies, J. (2002) *J. Mater.*, **12**, 2726–2732.

108 De bartolo, L., Morelli, S., Lopez, L.C., Giorno, L., Campana, C., Salerno, S., Rende, M., Favia, P., Detomaso, L., Gristina, R., d'Agostino, R. and Drioli, E. (2005) *Biomaterials*, **26**, 4432–4441.

109 Daw, R., Candan, S., Beck, A.J., Devlin, A.J., Brook, I.M., MacNeil, S., Dawson, R.A. and Short, R.D. (1998) *Biomaterials*, **19**, 1717–1725.

110 Muguruma, H. and Karube, I. (1999) *Trends Anal. Chem.*, **18** (1), 62–68.

111 Alexander, M.R. and Duc, T.M. (1999) *Polymer*, **40**, 5479–5488.

112 Detomaso, L., Gristina, R., Senesi, G.S., d'Agostini, R. and Favia, P. (2005) *Biomaterials*, **26**, 3831.

113 Sciarratta, V., Vohrer, U., Hegemann, D., Muller, M. and Oehr, C. (2003) *Surf. Coat. Technol.*, 174–175, 805–810.

114 Betz, N., Begue, J., Goncalves, M., Gionnet, K., Déléris, G. and LeMoël, A. (2003) *Nucl. Instrum. Meth. Phys. Res. B*, **208**, 434–441.

115 Jafari, R., Tatoulian, M., Morscheidt, W. and Arefi-Khonsari, F. (2006) *React. Funct. Polym.*, **66**, 1757–1765.

116 Palumbo, F., Favia, P., Rinaldi, A., Vulpoi, M. and d'Agostino, R. (1999) *Plasmas Polym.*, **4**, 133–145.

117 Alexander, M.R. and Duc, T.M. (1998) *J. Mater. Chem.*, **8** (4), 937–943.

118 Candan, S., Beck, A.J., O'Toole, L. and Short, R.D. (1998) *J. Vac. Sci. Technol.*, **A16**, 1702.

119 European project report n°29, IFCA (Immunoprobes for Food Contamination Analysis) (2005).

120 Yasuda, H. and Matsuzawa, Y. (2005) *Plasma Process. Polym.*, **2**, 507–512.

9
Fundamentals on Plasma Deposition of Fluorocarbon Films
A. Milella, F. Palumbo, and R. d'Agostino

The first studies of fluorocarbon plasmas date back to the 1970s, when CF_4 and other feeds started to be utilized in microelectronics for dry etching process of silicon, SiO_2, and other materials. Since then, many points related to plasma and surface diagnostics, deposition kinetics, and applications of fluorocarbon coatings have been investigated [1–6].

The choice of the monomer is of primary importance for the process to be performed, since it is the source of reactive fragments and film precursors in the plasma. Volatile fluoroalkanes (C_nF_{2n+2}), fluoroalkenes, fluoroalkynes, and cyclic and aromatic fluorocarbon compounds can be utilized as monomers for plasma deposited fluorocarbon coatings. Compounds able to produce a high density of radicals, rather than atoms, should be preferred in plasma-enhanced chemical vapor deposition (PE-CVD), since F atoms and CF_x radicals have been demonstrated to be the main etching and deposition precursor species, respectively. Figure 9.1 shows some of the most utilized monomers from the literature.

Concerns for the environment can also drive the choice of the gas feed in plasma processing techniques. Recently, in fact, great research efforts have been spent to develop new effective etching and PE-CVD processes in microelectronics, with monomers allowing a reduced emission of greenhouse gases in the exhaust [7–9]. This is due to the evidence that most fluorocarbons show a global warming potential similar to or higher than that of CO_2.

9.1
Deposition of Fluorocarbon Films by Continuous Discharges

It is important to highlight that many of the concepts herein developed are of general validity and their applicability can be partially or totally extended to downstream and modulated plasma depositions.

In continuous discharge (CD) processes the discharge is kept continuously switched on and the substrate is directly exposed to the glow, thus it undergoes the direct interaction with neutral (chemical reactions) and ionic (positive ion

Advanced Plasma Technology. Edited by Riccardo d'Agostino, Pietro Favia, Yoshinobu Kawai, Hideo Ikegami, Noriyoshi Sato, and Farzaneh Arefi-Khonsari
Copyright © 2008 WILEY-VCH Verlag GmbH & Co. KGaA, Weinheim
ISBN: 978-3-527-40591-6

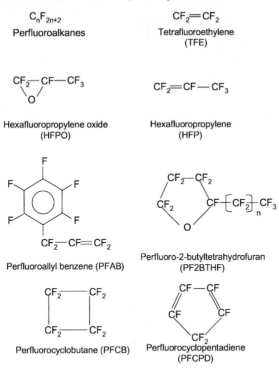

Fig. 9.1 Representative list of monomers studied in the literature.

bombardment) species generated in the plasma and radiation emitted. Both chemical reactions and ion bombardment play an important role in CD deposition kinetics, and lead to coatings with variable composition and crosslinking, where the composition and structure of the starting monomer is almost completely lost. A typical X-ray photoelectron spectroscopy (XPS) spectrum of a fluoropolymer deposited by CD PE-CVD is shown in Fig. 9.2 together with its chemical structure.

Generally, a maximum value of 1.6 for the F/C ratio on a film deposited in CD mode can be found, albeit d'Agostino *et al.* obtained a F/C ratio of 2 for a very thin film deposited from a C_2F_6–H_2 (50/50) discharge [1,10].

9.1.1
Active Species in Fluorocarbon Plasmas

As evident from plasma diagnostics [1,2,11–20], a fluorocarbon plasma is populated by CF_x ($1 \leq x \leq 3$) radicals, F and C atoms, and ions produced by the fragmentation of the monomer, as well as by heavier species originated by recombination reactions among different fragments and monomer molecules.

All active species are present in different excitation states, and their distribution highly depends on the experimental parameters of the discharge. The distribution of

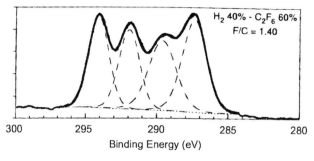
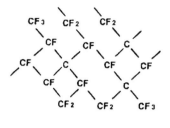

Fig. 9.2 Chemical structure and XPS C 1s spectra for a fluorocarbon coating plasma deposited in CD conditions [25].

species in the plasma and in its vicinity drives the interactions with the substrate along with its nature, temperature, ion bombardment, and position in the reactor, and controls structure and composition of the coating or of the etched/fluorinated layer.

Since F atoms and CF_x radicals have been demonstrated to be the main etching and deposition precursor species in CD fluorocarbon plasmas, the fluorine to radicals F/CF_x density ratio in the plasma plays a key role in describing the deposition of fluoropolymers [1,6,11].

Emitting species present in a plasma can be probed, in a non intrusive way, by optical emission spectroscopy (OES). Moreover actinometric optical emission spectroscopy (AOES) [1,6,14–16,21] is utilized to deduce semi-quantitative density trends of emitting species as a function of the experimental variables.

The decomposition of C_2F_4, along with those of other halocarbons, in a radio-frequency discharge was studied by d'Agostino *et al.* [18]. The simplified proposed scheme, shown in Fig. 9.3, reflects the high polymerizing properties of C_2F_4, and the lowest F/CF_x ratio found in the analyzed feed with respect to other fluorocarbon fed plasmas.

The presence of relatively high levels of CF_3 was taken into account with the two recombinative processes $CF_2 + CF \rightarrow CF_3 + C$ and/or $CF_2 + CF_2 \rightarrow CF_3 + CF$. The former reaction is endothermic by about 84 kJ mol^{-1} and can be probably activated by

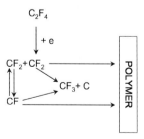

Fig. 9.3 Decomposition scheme for C_2F_4 [18].

the various energy-exchange processes occurring in the discharge, while the latter one, which requires about 167 kJ mol^{-1}, likely does not contribute to the formation of CF_3. The emission analysis has also allowed them to identify the continuum centered at 290 nm, which has been assigned to CF_2^+.

The relative densities of CF_x radicals over F atoms can be varied with the addition of reactive additives to the gas feed, such as O_2 or H_2, or by performing the discharge in presence of certain reactive materials. Adding O_2 or an oxidant molecule to a fluorocarbon plasma increases the density of F atoms, due to the reaction of CF_x radicals with oxygen atoms and excited oxygen molecules according to reaction (1); a consequence of reaction (9.1) and of the reduced rate of the radical–atom recombination is that the F/CF_x density ratio of the gas increases [1,2,6,11–13,17].

$$CF_x + O(O_2) \rightarrow CO(CO_2) + xF \tag{9.1}$$

H_2 can be added, instead, to lower the F/CF_x density ratio and favor the deposition of coatings with variable crosslinking degree and chemical composition [1,2,6, 11,17,19,22–24]. As shown for C_2F_6/H_2 CD discharges [1,17,22–25], in fact, hydrogen atoms scavenge (reaction (9.2)) F atoms in the plasma, and remove them from radicals (reaction (9.3)), leading to less F atoms and to radicals with a lower fluorination degree. HF formed in dry conditions does not contribute to surface reactions. Scavenging of F atoms can be achieved also when silicon, polymers, or certain metals are present in the reactor, due to the etching reactions (reaction (9.4), and similar ones).

$$H(H_2) + F \rightarrow HF + H \tag{9.2}$$

$$H(H_2) + CF_x \rightarrow CF_{x-1} + HF + (H) \tag{9.3}$$

$$Si + xF \rightarrow SiF_x \uparrow \tag{9.4}$$

C_2F_6/H_2 mixtures have been utilized, at variable ratios, to switch CD discharges from etching $(80\% \leq C_2F_6 \leq 100\%)$ [17] through the deposition of fluoropolymers $(20\% \leq C_2F_6 \leq 80\%)$ [17,19,22,25] to the deposition of fluorinated carbon films $(0 < C_2F_6 \leq 20\%)$ [23,24].

When highly fluorinated CF_x radicals dominate the distribution of active species (low H_2 feed content), fluoropolymers with high F/C ratio and low crosslinking degree are deposited, while less fluorinated radical distributions (high H_2 feed content) produce coatings with lower F/C ratio and higher crosslinking degree. The distribution of radicals in the plasma can thus be adjusted to obtain coatings with pre-determined F/C ratio and surface properties.

9.1.2
Effect of Ion Bombardment

Albeit the ionization degree (ions/neutrals) of low pressure plasmas almost never exceeds 10^{-5}, the contribution of positive ions must always be considered in plasma

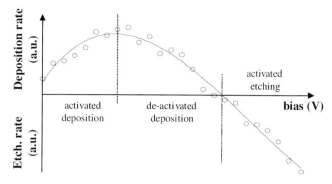

Fig. 9.4 Example of the bias-induced positive ion bombardment on the deposition/etching rate of plasma deposited fluoropolymers in CW RF glow discharges. Adapted from [29].

deposition/etching mechanisms. Due to the negative self bias potential that develops at plasma edges, between the surfaces exposed and the plasma itself, a positive ion bombardment is induced on any surface exposed to a low-pressure plasma, with intensity and energy depending on the geometry of the reactor and on the experimental parameters. In particular in a capacitively coupled parallel plate RF reactor the most intense bombardment develops at the surface of the smallest electrode according to a general law expressed by Koenig and Maisel [26,27]. Using a third electrode as sample holder allows an independent control of the ion bombardment by simply tuning the power delivered to the third electrode, thus the bias potential that develops there.

PE-CVD experiments performed in such a configuration have evidenced the effect of the ion bombardment on the deposition/etching rates, keeping the distribution of active species in the plasma (e.g. the F/CF_x ratio) constant [6,28,29]. Increasing the positive ion bombardment in the low energy (bias) range results, generally, in an increased deposition rate when the F/CF_x ratio is low (e.g. for feeds as C_2F_4), and in an higher etching rate when the F/CF_x ratio is high (e.g. for feeds as CF_4/O_2).

At a given F/CF_x it is possible to switch from deposition to etching by increasing the bias (see Fig. 9.4): an increased ion bombardment first activates surface to deposition, then further increase of the bias may depress the deposition rate of a fluoropolymer, due also to the competition of ion-activated etching and sputtering processes. Beyond a certain bias threshold, the etching of the plasma deposited layer can occur.

9.1.3
The Activated Growth Model

Homogeneous and heterogeneous reactions in CD PE-CVD processes of Teflon-like coatings were rationalized by d'Agostino [1,6,17,19], and the ion activated growth model (AGM) of deposition was proposed. AGM, which has been utilized also to rationalize other PE-CVD processes, like those from organosilicon monomers,

Fragmentation of the monomer in the plasma

$$\text{monomer} \rightarrow nCF_x \, (1 \leq x \leq 3) \quad \text{(a)}$$

Formation of addition compounds

$$nCF_x \leftrightarrows \text{addition compounds} \quad \text{(b)}$$

Ion activation of the coating/substrate

$$I^+ (\text{low energy}) + (\text{film})_n \rightarrow (\text{film})_n^* \quad \text{(c)}$$

Growth of the coating

$$CF_x \leftrightarrows CF_{x(\text{adsorbed})} \quad \text{(d)}$$

$$CF_{x(\text{adsorbed})} + (\text{film})_n^* \rightarrow (\text{film})_{n+1} \quad \text{(e)}$$

Fig. 9.5 Simplified scheme of the AGM.

combines the contribution of low-energy ion bombardment to activate the substrate surface, and of CF_x radicals formed in the glow discharge as building blocks of the coating. A simplified AGM scheme is shown in Fig. 9.5.

Reaction (a) describes schematically the production of the CF_x precursors from the monomer. If many F atoms are produced in this step, an etching route can be also established, which would compete with the deposition. The formation of heavier addition compounds in reaction (b) can be enhanced by increasing the production of radicals and decreasing their diffusion to the substrate, in other words at high powers and/or pressures. In this case, the polymer formation can occur in the plasma phase forming polymer nuclei. When these nuclei become micrometer sized they are called powders. Figure 9.6 shows a schematic of the different steps involved in the process of powder formation.

After agglomeration, the powders are negatively charged and float at plasma edge. This is the reason why they do not contribute to deposition. However, they fall onto the substrate when the discharge is switched off, if no contrivance is set [30].

The ion-activation step (c) depends on the energy of the ions bombarding the growing film (or the substrate, in the early deposition stages), thus on the related external parameters (pressure, power, bias, geometry). At low energy, the ions create surface defective sites (e.g. dangling bonds) that act as preferential chemisorption sites for the precursor CF_x radicals from the plasma. If the energy of the ions overcomes a certain threshold, however, the desorption of the precursors is induced. When the ion energy becomes very high the sputtering of the coating may also be activated. Moreover step (c) greatly influences the deposition rate, the composition and the crosslinking of the coating.

Steps (d) and (e) describe, respectively, the adsorption–desorption equilibrium of the CF_x radicals on the substrate (coating) and the following reaction of polymerization with the surface active sites originated by the ion bombardment. The temperature greatly affects both reactions, but in a different way. PE-CVD of fluoropolymers

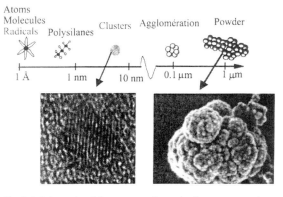

Fig. 9.6 Schematic of the genesis of powder formation in silane plasmas along with a scanning electron micrograph of a large powder and a high-resolution electron transmission electron micrograph of a nanocrystalline silicon particle [30].

are generally performed with substrates kept at room temperature; due to its $\Delta H < 0$ the equilibrium (d) is moved to the desorption side by the increase of temperature, while the polymerization kinetics (e) is favored. As an overall effect of the increased substrate temperature in the 25–100°C range, a reduced deposition rate is usually recorded [17,25] and a reduced F/C ratio in the deposited coating.

9.2
Afterglow Deposition of Fluorocarbon Films

In afterglow (AG) PE-CVD the substrates are positioned downstream, i.e. several centimeters away from the glow, in the direction of the feed flow, where no electrons are present, the density of ions is very low, and the ion bombardment of the substrate is negligible. In these conditions the interaction with the substrates of the monomer molecules, and of the unstable species (CF_x and other radicals, atoms) generated in the plasma, depends on the pressure, lifetime of the species, and on their residence time in the plasma reactor. Species with a long lifetime have the highest probability to interact with the substrate and, eventually to deposit a coating. The optimum plasma–substrate distance has to be found, for each single case, in order to deposit coatings with the required composition, crosslinking, and monomer structure retention degree, at an acceptable deposition rate. A short distance from the glow may result in coatings very similar to those deposited in the glow, or in substrate etching, depending on the density of species in that particular position of the reactor. Due to the absence of the ion-activation step and to the reduction of deposition precursors, in AG PE-CVD the rate of deposition is 1–3 orders of magnitude lower than in CD and modulated PE-CVD and the degree of structure retention of the monomer in the coating can be very high, due to a deposition mechanism which very likely has many aspects in common with a conventional polymerization. In certain conditions (high residence time of the species in the reactor), the deposition of coatings can occur also upstream the glow, in the so-called pre-glow positions [1].

Since in these processes the flow dynamics of the reactor can play a major role, AG PE-CVD processes are investigated mostly in tubular reactors, with many different plasma sources.

The deposition of very thin AG Teflon-like coatings with high retention of the monomer structure and F/C ratio very close to 2 has been performed with C_2F_4 [31–34], C_2F_6[25,34], C_2F_6/H_2 (80/20) [25,34], and C_3F_6O [35] feeds. The distance from the glow in each condition is the key parameter to select between cross-linked films very similar to those obtained by continuous mode, and AG Teflon-like coatings. The structure of AG coatings has been characterized by XPS, static secondary ion mass spectrometry (SIMS) and near edge X-ray absorption fine structure spectroscopy (NEXAFS), and it has been found that AG coatings are characterized by highly oriented $-CF_2-(CF_2)_n-CF_3$ chains grafted to the surface of the substrate (silicon, polymers), as shown in Fig. 9.7.

Depending on the deposition conditions, the orientation of the chains and the density of the (few) crosslinked sites can change. The deposition rate is very low, of the order of a few angstroms per minute; this can lead to a partial coverage of the substrate with the coating. Moreover, due probably to the low surface energy properties associated with terminal $-CF_3$ groups of the chains, this AG deposition process has been found self-limited: after a certain deposition time, in fact, the coating does not grow anymore, albeit the discharge is still on.

AG chain-oriented Teflon-like coatings display an hydrophobic character very close to that of PTFE (water contact angle, WCA ~115°); due to their peculiar structure, they show interesting adsorption–retention properties for proteins, that has stimulated a certain interest in the field of biomaterials.

In the case of AG coatings generated from C_2F_6 RF glow discharges, it is remarked that this kind of discharge produces, in the glow region, a high density of F atoms

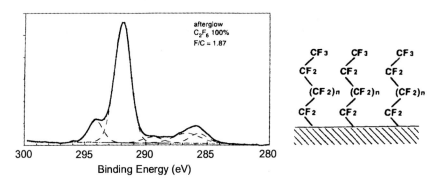

Fig. 9.7 Chemical structure and XPS C 1s spectra for a fluorocarbon coating plasma deposited in AG conditions [25].

[25] which induces the etching of any Si or polymer substrate rather than the deposition of the coatings. Then, it has been suggested that in AG positions, evidently, the distribution density of the active species becomes completely different from the glow, due to their different lifetime, or to different consumption rates with the walls of the reactor, and the deposition of very thin, highly fluorinated chain-oriented coatings occurs.

9.3
Deposition of Fluorocarbon Films by Modulated Glow Discharges

In modulated glow discharges the power input is delivered periodically to the plasma for micro- to milliseconds (t_{ON}), and switched off, or reduced to a certain fraction, during off time (t_{OFF}) [36–41]. The substrates are positioned in the region where the glow is activated, thus they experience ion bombardment and interact with unstable species from the plasma during t_{ON}, and with long-lived radicals/atoms and, in some cases, with unreacted monomer molecules during t_{OFF}, since ions and electrons extinguish being characterized by shorter lifetimes (see Fig. 9.8).

Modulated PE-CVD processes can be investigated as a function of the period ($t_{ON} + t_{OFF}$) and of the duty cycle (DC% $= 100 \times t_{ON}/t_{ON} + t_{OFF}$) of the discharge. Most of the experiments are investigated keeping the power at zero value during t_{OFF}.

Though modulated plasmas have been investigated since the 1970s [42,43], many papers have been produced in the last years on this subject [8,20,34,36–65], most of them dealing with plasma-deposited fluoropolymers from unsaturated or cyclic monomers.

With all the other parameters constant, the power delivered to the discharge and the duty cycle determines the degree of fragmentation of the monomer in the gas phase and, consequently, the degree of retention of the monomer structure in the coating.

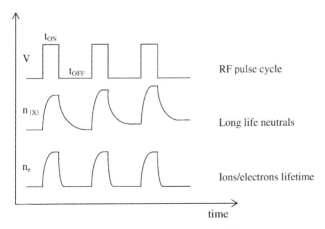

Fig. 9.8 Schematic of neutrals, ions, and electrons lifetimes.

The "effective" or "average" power delivered in MD can be roughly calculated as

$$W_{\text{eff}} = W \times \text{DC} \tag{9.5}$$

where W is the peak power value delivered by the power supply [62,63].

In this context the deposition rate is frequently expressed in terms of coating thickness per "unit input energy" or per pulse. Many experiments performed by different authors at constant t_{ON} and increasing t_{OFF} values showed that the coating can grow also during t_{OFF}. This is due to heterogeneous reactions of the radicals formed during t_{ON} with the substrate and eventually, of the monomer molecules with the activated substrate, depending on the reactivity of the monomer.

Various researchers [8,20,34,36–64] found that lowering the duty cycle results in a progressively less branched polymeric film structure with higher F/C ratio, which is more similar to polytetrafluoroethylene (PTFE). This was explained by a reduced extent of fragmentation of the monomer in the gas phase and a less energetic ion bombardment at the film surface during the on time. Ion bombardment may, in fact, lead to structural rearrangement, crosslinking, etching, or defluorination of the film, resulting in a more damaged polymer surface. Many authors deposited PTFE-like films from hexafluoropropylene oxide (HFPO, C_3F_6O) modulated glow discharges with a F/C ratio as high as 1.8–1.9 [38,60]. Moreover, Fisher and co-workers demonstrated that films deposited in modulated discharges fed with HFPO at lower duty cycle (i.e. 5%) contain chains perpendicular to the substrate surface [64].

Limb *et al.* [48] carried out a systematic study of film composition as a function of several discharge parameters, using HFPO as gas feed in modulated discharges. The main results can be summarized as follows. Increasing t_{OFF}, at constant t_{ON}, leads to an increase in the relative abundance of CF_2 groups in the deposited film and F/C ratio. At lower duty cycle, as the sample undergoes a DC averaged power, the resulting ion bombardment effects are reduced. Moreover it was found that the percentage of CF_2 in the film decreases with decreasing pressure. This indicates that greater gas-phase fragmentation and surface interaction defluorination may occur at lower pressure. At lowest flow rate the growth rate of the film is reduced and the amount of CF_2 strongly decreases. This is due to undesirable fragmentation reactions occurring to a larger degree when subsequent excitation periods occur before reactive precursors and fragments created by the previous pulse can exit the reactor, being the residence time much greater than period. A decrease in electrode spacing also leads to a decrease in CF_2 percentage in the film, along with the deposition rate per cycle. Surface morphology of films deposited from pulsed plasmas of HFPO, 1,1,2,2-tetrafluoroethane ($C_2H_2F_4$), and difluoromethane (CH_2F_2) reveals nodular growth (cauliflower-like appearance), with the size and distribution of the nodules dependent on the precursor, the degree of surface modification to which the growing film is exposed, and the substrate surface temperature (see Fig. 9.9).

The degree to which the growing surface will be subjected to surface modification is dependent on both t_{ON} and the film growth rate. Thus, at a fixed t_{ON}, Gleason and co-workers [53] quantified the extent of surface modification by defining the incident power/deposition rate per pulse cycle (J cycle nm^{-1}) as a function of t_{OFF}. At short

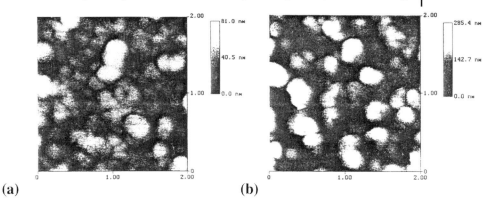

Fig. 9.9 AFM surface images of pulsed plasma films deposited from HFPO in different conditions: (a) 10/100; (b) 10/400 [54].

t_{OFF} the growing surface is subjected to the highest level of modification, leading to smoother surface, whereas increasing t_{OFF}, less modification occurs, resulting in a rougher surface. Moreover, heating the substrate at higher temperature during the deposition, results in an increased modification of the growing film, which becomes rougher. Changing the substrate from bare silicon to Al-coated silicon leads to an increase in surface roughness and in nodule size. This result has to be ascribed to the fingerlike structures on Al-coated silicon resulting in a surface roughness of about 7.48 nm compared to 0.53 nm of the bare silicon. Finally higher advancing contact angles were found on the rougher surface than on smoother one [54].

9.4
Deposition of Nanostructured Thin Films from Tetrafluoroethylene Glow Discharges

A systematic study of the morphology of fluorocarbon films deposited both from continuous and modulated discharges fed with tetrafluoroethylene (TFE) accomplished by means of atomic force microscopy (AFM) is reported in [66]. Figure 9.10 shows two-dimensional surface topography of the polished silicon substrate and of the films deposited in continuous plasmas at 5, 100, and 150 W, at a pressure of 200 mtorr.

It is evident that in all conditions the film surface is smooth and that no morphological features can be distinguished. The surface roughness ranges from 0.54 nm for the bare silicon to 1.19 nm for the film deposited at 150 W.

The AFM images of films deposited from continuous discharges at 100 W under varying pressure from 100 to 500 mtorr are displayed in Fig. 9.11.

At lower pressure the surface is smoother, with a very low value of roughness, 0.87 nm. At 500 mtorr, however, the morphology is characterized by bumps, some of them comprising larger agglomerates. The corresponding roughness increases to 10.94 nm.

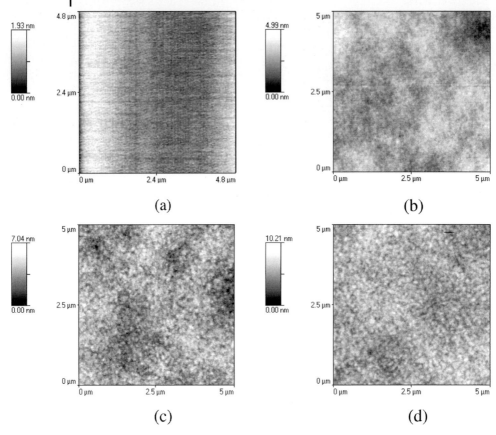

Fig. 9.10 Two-dimensional AFM topography of polished silicon (a) and of films deposited from TFE continuous plasmas at different input power: 5 W (b), 100 W (c), 150 W (d). Deposition time was 15 min. From [66].

The effect of duty cycle on film morphology at a fixed modulation period of 320 ms and deposition time of 90 min is shown in Fig. 9.12. The samples deposited at a duty cycle of 5% reveal a complex morphology characterized by ribbon-like structures, typically several micrometers long and hundreds of nanometers in width. These ribbons are randomly distributed over the entire surface and twisted in an intricate way. At higher magnification the image shows that the surface area between these structures is populated by islands of nanometric size, also randomly distributed. Higher magnification allows one to evaluate the diameters of such nuclei as ranging from 80 to 500 nm, with heights varying from 10 to 200 nm. The morphology of the film deposited at a 10% duty cycle (Fig. 9.12(b)) is drastically changed, since no ribbons appear, but only some grains with irregular shape, up to 2 μm long and with a maximum height of about 400 nm. Nuclei are still present, but with a lower spread in heights, from about 2 to 30 nm. The film corresponding to 20% of duty cycle (Fig. 9.12(c)) shows yet another kind of morphology: some larger bumps start to

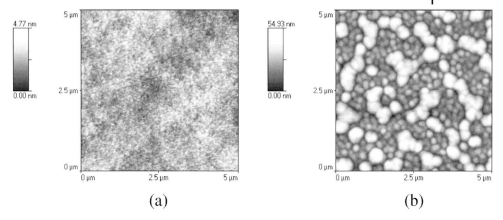

Fig. 9.11 Two-dimensional AFM topography of films deposited from TFE continuous plasmas at 100 mtorr (a) and 500 mtorr (b). Deposition time was 15 min. Input power and gas flow rate were fixed at 100 W and 6 sccm, respectively. From [66].

form, align and coalesce. They exhibit a more regular distribution of diameters, 400 to 600 nm, and heights in the range 70–90 nm. The remaining portion of surface is populated by bumps uniformly dispersed and with very similar diameters of about 200 nm, and maximum height of 20 nm. Further increasing the duty cycle up to 50% (Fig. 9.12(d)) does not change the morphology much over that of its 20% counterpart. However, in this case, the surface is very rich in bumps, 200 nm wide and with heights close to 10 nm.

Moreover, a smaller number of large aggregates with diameters around 400 nm and a maximum height of 30 nm can be observed. Thus, as the duty cycle increases, the height distribution becomes more uniform. At a duty cycle of 70% (Fig. 9.12(e)), the sample is so crowded with bumps that these collapse, lose their individual shape and lead to a more homogeneous surface. Finally, the continuous-mode deposition (100% DC, Fig. 9.12(f)) results in a flat surface with no evidence of bumps.

The effect of the modulation period on the topography of the deposited films is shown in Fig. 9.13.

Even though a duty cycle value of 5% is used, for a period of 40 ms (Fig. 9.13(a)) the process results in a featureless film with very low roughness (2.6 nm). Increasing the period to 80 ms (Fig. 9.13(b)) results in very few aggregates becoming visible and the surface roughness becomes 3.2 nm. At a period of 200 ms (Fig. 9.13(c)) the sample surface displays aggregates with almost ribbon-like appearance, but shorter than the ones observed at 320 ms (Fig. 9.13(d)). Moreover, with respect to the latter the bumps are reduced in number. The formation of short ribbons results in an increased roughness of 43.9 nm.

Another key parameter for understanding ribbon formation is the deposition time. This effect was thoroughly investigated in previous work [67], in which we basically reported that ribbon formation involves several steps, namely the formation of

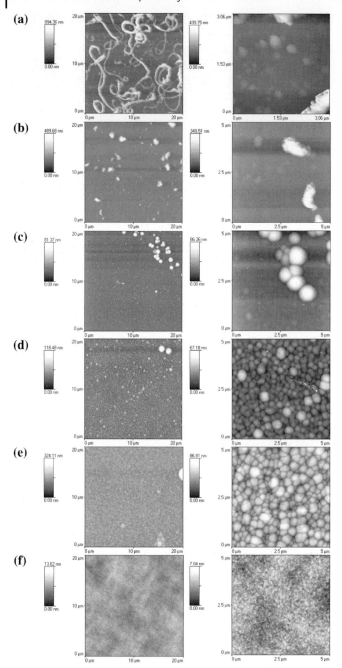

Fig. 9.12 Non contact AFM images ($20 \times 20\ \mu m^2$ and $5 \times 5\ \mu m^2$) of films deposited from modulated discharges at 320 ms period with different duty cycles: 5% (a), 10% (b), 20% (c), 50% (d), 70% (e), 100% (f). Deposition time was 90 min. Input power and pressure were fixed at 100 W and 200 mtorr, respectively. From [66].

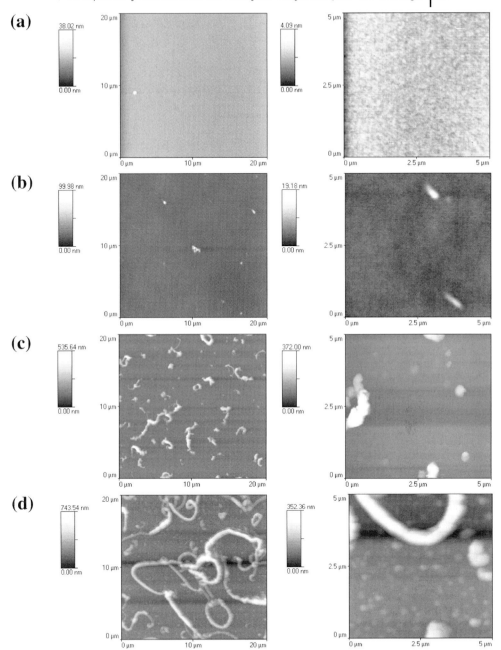

Fig. 9.13 Non contact AFM images ($20 \times 20\ \mu m^2$ and $5 \times 5\ \mu m^2$) of films deposited from modulated discharges at 5% duty cycle with different modulation periods: 40 ms (a), 80 ms (b), 200 ms (c), 320 ms (d). Deposition time was 90 min. Input power and pressure were fixed at 100 W and 200 mtorr, respectively. From [66].

nucleation centers at the very first stage of growth, alignment and subsequent attachment of further nuclei for the formation and development of the ribbons. Once the ribbons have grown enough in numbers and dimensions, the formation of further nucleation centers on the ribbon surfaces occurs. The shape and the dimensions of the morphological features have been confirmed by SEM analyses (not reported).

ToF/SIMS analysis evidenced higher structure retention in ribbon-like samples, whose fragmentation patterns (not shown) were similar to the one of bulk PTFE. This is in agreement with our findings by XPS, XRD and FTIR analyses, reported in previous papers [68,69]. Ribbon-like samples were found to be more ordered, partially crystalline, and fluorine-rich coatings.

On the basis of the results found on the compositional evolution of the plasma phase and for the structure and morphology of the deposited coatings, both in continuous and modulated regimes [66], the film deposition mechanism illustrated in Fig. 9.14 has been developed.

As previously described [68], at DC $\geq 10\%$ the depletion of monomer (reaction (a)), produces plenty of CF_2 radicals and modulated plasmas do not appear very different from continuous ones. When the density of radicals becomes high in the gas phase (high power, DC, and pressure), recombination reactions become significant, and lead to the formation of heavier fragments, C_yF_z, similar to the saturated and unsaturated molecules identified by means of FT-IRAS (route (d)). This reaction path can also proceed, mainly at higher pressure, to form particulates (route (e)); however, this can be ruled out in the regimes leading to ribbon-like structures, as FT-IRAS absorption analysis does not reveal any formation of powders. It should be considered that beside CF_2 other small radicals, CF_3, CF, and F atoms, can also be produced (route (b)). However, these latter species and C_yF_z radicals, under low duty cycle and input power, are likely formed to a much lesser extent, so that CF_2 remains

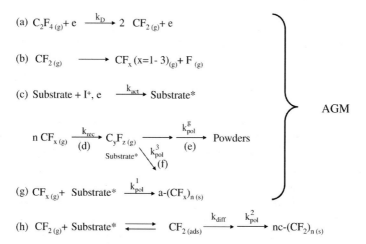

Fig. 9.14 Scheme of the deposition mechanism developed for TFE continuous and modulated plasmas. From [66].

the main film precursor. Reactions (g) and (h) represent two different film deposition pathways. In particular, (g) refers to the high-rate formation of amorphous, flat and crosslinked films, a-$(CF_x)_n$, occurring at high duty cycle and in continuous discharges according to the AGM of deposition [1].

In the low CF_2 radical concentration regime (route (h)), that is, with duty cycles smaller than 10%, the deposition rate and positive ion bombardment are drastically reduced and the few adsorbed CF_2 radicals have enough time to migrate on the surface to lower-energy sites, which become the nucleation centers for the formation of nanostructured crystalline PTFE-like ribbons, nc-$(CF_2)_n$.

According to the described mechanism, the changes in surface morphology seem to be mainly linked to:

- modulation itself, which provides an off time;
- migration of adsorbed radicals, mainly during this off time.

The schematic shown in Fig. 9.15 is an attempt to explain the evolution of surface morphology with duty cycle. Under the continuous deposition regime, the high and continuous flux of radicals and of ions ensures enough active sites for AGM, and many radicals readily stick to these sites. Surface migration cannot occur and only flat coatings are possible. At high pressure, the CF_2 and other small radical densities decrease, as found by AOES investigations [70], and heavier fragments (or even powders) are produced to a greater extent. The number of surface activated sites decreases, too, due to reduced ion bombardment; thus, radicals have appreciably more time to diffuse across the surface toward energetically favorable sites, where they can aggregate resulting in a rougher surface. However, in this regime it should be considered that heavier species, formed in the gas phase from association reactions of lighter species, and eventually powder granules, can contribute to film deposition (route (f)) and take part to the formation of the aggregates present on the film surface.

A similar granular and rough morphology was described by Silverstein and co-workers for the plasma deposition of fluorocarbon films in continuous hexafluoropropylene (C_3F_6) plasmas at higher pressures (750–1500 mtorr). In this gas-phase driven plasma polymerization, sub-micrometer particles formed by homogeneous nucleation are believed to deposit on the surface, where they undergo further polymerization and become incorporated into the structure of the coating [71,72].

At 5% duty cycle and long period, the number of surface activated sites is low, as is the radical concentration in the plasma phase. Moreover, adsorbed radicals have more time (long off time) to migrate on the surface and to reach activated sites before the arrival of other impinging radicals (beginning of the on time). When a nucleation center is formed, it slowly grows for the preferential attachment of the other few incoming radicals, very likely CF_2. This leads to the formation of PTFE-like ribbons, whose direction of propagation depends primarily on the direction in which a greater concentration of nuclei is formed. When ribbons start to form, they can grow by both nuclei coalescence of and by direct attachment of radicals diffusing along the surface.

Linear rod-like aggregates consisting of extended chains with the chain axis parallel to the long axis of the rods were observed in thin films obtained from conventional PTFE [73] and in PTFE dispersions [74]. It was also reported that, at

CONTINUOUS MODE DEPOSITION

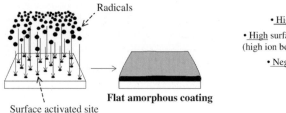

Radicals

• High radicals concentration

• High surface activated sites concentration (high ion bombardment; AGM is operative)

• Negligible surface migration

Flat amorphous coating

Surface activated site

MODULATED DEPOSITION AT LOW DC AND HIGH PERIOD

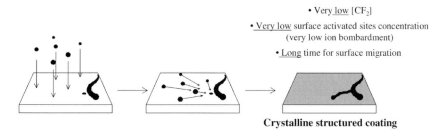

• Very low [CF₂]

• Very low surface activated sites concentration (very low ion bombardment)

• Long time for surface migration

Crystalline structured coating

MODULATED DEPOSITION AT HIGH DC AND HIGH PERIOD

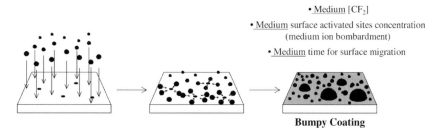

• Medium [CF₂]

• Medium surface activated sites concentration (medium ion bombardment)

• Medium time for surface migration

Bumpy Coating

Fig. 9.15 Schematized deposition mechanism in different regimes. From [66].

low polymerization rates, bulk PTFE polymerizes in chains of helical conformation [75]. Thus, at low duty cycle and long period, the deposition rate is low enough for the radicals to rearrange in ribbons, which appear to constitute a thermodynamically more stable form. However, when the duty cycle is increased, the process becomes kinetically driven, and the anisotropy is progressively lost. The number of active sites on the sample surface increases, together with the radical concentration. As a consequence, the probability that each active site is reached by radicals coming from all possible directions increases, too, and this results in the formation of bumps over the entire surface, in particular for duty cycles higher than 20%. Moreover, since the time for the radical migration decreases (shorter off time) and ion bombardment increases, the average heights of the nuclei as well as their height differences are lowered.

Surface morphologies similar to those obtained at high duty cycles in this work, have also been observed by Labelle and Gleason during the deposition of fluor-ocarbon films in modulated plasmas fed with hexafluoropropylene oxide, 1,1,2,2-tetrafluoroethane, and difluoromethane [54]. Lau *et al.* observed rod-like morphologies in addition to spherical grains during hot filament chemical vapor deposition (HFCVD) of hexafluoropropylene oxide, and they speculated that the formation of spherical grains occurs at high deposition rate as a result of gas-phase nucleation, while rod-like arrangements prevail in a slow deposition regime, when chain growth is favored over gas-phase nucleation [57]. Ribbon-like structured coatings deposited from modulated discharges fed with other perfluoroalkane monomers were also reported by other authors [76]. Similarly, the ribbons develop under limited duty cycle and input power conditions; moreover, evidence of the major role played by radical surface mobility in the occurrence of this unique morphology can be found therein.

At 5% duty cycle and decreased period, the CF_2 concentration becomes higher, ion bombardment is progressively more effective and the time left for radical diffusion is lower, which reduces both nuclei the formation of and growth. Going to lower periods (40 and 80 ms), only few aggregates (or none) are evidenced on a smooth grown film.

References

1 d'Agostino, R., Cramarossa, F., Fracassi, F. and Illuzzi, F. (1990) in *Plasma Deposition, Treatment and Etching of Polymers* (ed R. d'Agostino), Academic Press.

2 Kay, E., Coburn, J.W. and Dilks, A. (1980) in *Topics in Current Chemistry 94* (eds S. Veprek and M. Venugopalan), Springer, Verlag.

3 Yasuda, H. (1985) *Plasma Polymerization*, Academic Press.

4 Inagaki, N. (1996) *Plasma Surface Modification and Plasma Polymerization*, Technomic.

5 Biedermann, H. and Osada, Y. (1992) *Plasma Polymerization Processes*, Elsevier.

6 d'Agostino, R. (1997) in *Plasma Processing of Polymers* (eds R. d'Agostino, P. Favia and F. Fracassi), *NATO ASI Series, E: Appl. Sci*, **346**, Kluwer Academic.

7 Fracassi, F. and d'Agostino, R. (1999) *Plasmas Polym.*, **4**, 147.

8 Labelle, C.B., Karecki, S.M., Reif, R. and Gleason, K.K. (1999) *J. Vac. Sci. Technol. A*, **17**, 3419.

9 Shirafuji, T., Kamisawa, A., Shimasaki, T., Hayashi, Y. and Nishino, S. (2000) *Thin Solid Films*, **374**, 256.

10 d'Agostino, R., Cramarossa, F., Fracassi, F., DeSimoni, E., Sabbatini, L., Zambonin, P.G. and Capriccio, G. (1986) *Thin Solid Films*, **143**, 163.

11 Coburn, J.W. and Winters, H.F. (1979) *J. Vac. Sci. Technol.*, **16**, 391.

12 Flamm, D.L. and Donnelly, V.M. (1981) *Plasma Chem. Plasma Process.*, **1**, 317.

13 ManosD.M. and FlammD.L. (eds) (1989) in *Plasma Etching: An Introduction, Plasma–Materials Interaction Series*, Academic Press.

14 Favia, P. (1997) in *Plasma Processing of Polymers* (eds R. d'Agostino, P. Favia and F. Fracassi), *NATO ASI*

Series, E: Appl. Sci, **346**, 487, Kluwer Academic.

15 Coburn, J.W. and Chen, M. (1980) *J. Appl. Phys.,* **51**, 3134.

16 d'Agostino, R., Cramarossa, F., DeBenedictis, S. and Ferraro, G. (1981) *J. Appl. Phys.,* **52**, 1259.

17 d'Agostino, R., Cramarossa, F. and Illuzzi, F. (1987) *J. Appl. Phys.,* **61**, 2754.

18 d'Agostino, R., Cramarossa, F. and DeBenedictis, S. (1982) *Plasma Chem. Plasma Process,* **2**, 213.

19 d'Agostino, R., Favia, P. and Fracassi, F. (1990) *J. Polym. Sci. A: Polym. Chem.,* **28**, 3387.

20 Cruden, B.A., Gleason, K.K. and Sawin, H.H. (2001) *J. Appl. Phys.,* **89**, 915.

21 Favia, P., Creatore, M., Palumbo, F., Colaprico, V. and d'Agostino, R. (2001) *Surf. Coat. Technol.,* **142–144**, 1.

22 Truesdale, E.A. and Smolinsky, G. (1979) *J. Appl. Phys.,* **50**, 6594.

23 Lamendola, R., Favia, P. and d'Agostino, R. (1992) *Plasma Sour. Sci. Technol.,* **1**, 256.

24 d'Agostino, R., Lamendola, R., Favia, P. and Gicquel, A. (1994) *J. Vac. Sci. Technol. A,* **12**, 308.

25 Favia, P., Perez-Luna, V.H., Boland, T., Castner, D.G. and Ratner, B.D. (1996) *Plasmas Polym.,* **1**, 299.

26 Fracassi, F. (1997) in *Plasma Processing of Polymers* (eds R. d'Agostino,P. Favia and F. Fracassi), *NATO ASI Series, E: Appl. Sci,* **346**, 47, Kluwer Academic.

27 Koenig, H.R. and Maisel, L.I. (1970) *IBM J. Res. Dev.,* **14**, 168.

28 Fracassi, F. and Coburn, J.W. (1988) *J. Appl. Phys.,* **63**, 1758.

29 Fracassi, F., Occhiello, E. and Coburn, J.W. (1988) *J. Appl. Phys.,* **62**, 3980.

30 Cabarrocas, P.R., Morral, A.F., Lebib, S. and Poissant, Y. (2002) *Pure Appl. Chem.,* **74**, 359.

31 Kiaei, D., Hoffman, A.S., Ratner, B.D. and Horbett, T.A. (1988) *J. Appl. Polym. Sci.: Polym. Symp.,* **42**, 269.

32 Castner, D.G., Lewis, K.B., Fischer, D.A., Ratner, B.D. and Gland, J.L. (1993) *Langmuir,* **9**, 537.

33 Kiaei, D., Hoffman, A.S. and Horbett, T.A. (1992) *J. Biomater. Sci. Polym. Ed.,* **4**, 35.

34 Castner, D.G., Favia, P. and Ratner, B.D. (1996) in *Surface Modifications of Polymeric Biomaterials* (eds B.D. Ratner and D.G. Castner), Plenum Press, p. 45.

35 Butoi, C.I., Mackie, N.M., Gamble, L.J. and Castner, D.G. (1999) *Chem. Mater.,* **11**, 862.

36 Cicala, G., Losurdo, M., Capezzuto, P. and Bruno, G. (1992) *Plasma Sources Sci. Technol.,* **1**, 156.

37 Panchalingam, V., Poon, B., Huo, H.H., Savage, C.R., Timmons, R.B. and Eberhart, R.C. (1993) *J. Biomater. Sci. Polym. Ed.,* **5**, 131.

38 Limb, S.J., Lau, K.K.S., Edell, D.J., Gleason, E.F. and Gleason, K.K. (1999) *Plasmas Polym.,* **4**, 21.

39 Panchalingam, V., Chen, X., Savage, C.R., Timmons, R.B. and Eberhart, R.C. (1994) *J. Appl. Polym. Sci.: Appl. Polym. Symp.,* **54**, 123.

40 Coulson, S.R., Woodward, I.S., Badyal, J.P.S., Brewer, S.A. and Willis, C. (2000) *Langmuir,* **16**, 6287.

41 Favia, P., Cicala, G., Milella, A., Palumbo, F., Rossini, P. and d'Agostino, R. (July 2001) *Proc. 15th Int. Symp. on Plasma Chemistry, ISPC-15. Orleans, France, Vol, II,* 587.

42 Yasuda, H. and Hsu, T. (1977) *J. Polym. Sci., Polym. Chem. Ed.,* **15**, 81.

43 Vinzant, J.M., Shen, M. and Bell, A.T. (1979) *ACS Symp. Ser.,* **108**, 79.

44 Lau, K.K.S. and Gleason, K.K. (1999) *Mater. Res. Soc. Symp. Proc.,* **544**, 209.

45 Cruden, B., Chu, K., Gleason, K.K. and Sawin, H. (1999) *J. Electrochem. Soc.,* **146**, 4590.

46 Cruden, B., Chu, K., Gleason, K.K. and Sawin, H. (1999) *J. Electrochem. Soc.,* **146**, 4597.

47 Zabeida, O., Klemberg-Sapieha, J.E., Martinu, L. and Morton, D. (1999)

Mater. Res. Soc. Symp. Proc., **544**, 233.

48 Limb, S.J., Edell, D.J., Gleason, E.F. and Gleason, K.K. (1998) *J. Appl. Polym. Sci.*, **67**, 1489.

49 Labelle, B.C., Limb, S.J. and Gleason, K.K. (1997) *J. Appl. Phys.*, **82**, 1784.

50 Han, L.M., Timmons, R.B. and Lee, W.W. (2000) *J. Vac. Sci. Technol. B*, **18**, 799.

51 Mackie, N.M., Castner, D.G. and Fisher, E.R. (1998) *Langmuir*, **14**, 1227.

52 Coulson, S.R., Woodward, I., Badyal, J.P.S., Brewer, S.A. and Willis, C. (2000) *J. Phys. Chem. B*, **104**, 8836.

53 Labelle, C.B. and Gleason, K.K. (1999) *J. Vac. Sci. Technol. A*, **17**, 445.

54 Labelle, C.B. and Gleason, K.K. (1999) *J. Appl. Polym. Sci.*, **74**, 2439.

55 Panchalingam, V., Chen, X., Huo, H.H., Savage, C.R., Timmons, R.B. and Eberhart, R.C. (1993) *ASAIO J.*, **39**, M305.

56 Lau, K.K.S. (2000) PhD thesis, Massachusetts Institute of Technology.

57 Lau, K.K.S., Caulfield, J.A. and Gleason, K.K. (2000) *Chem. Mater.*, **12**, 3032.

58 Lau, K.K.S. and Gleason, K.K. (2001) *J. Phys. Chem. B*, **105**, 2303.

59 Lau, K.K.S., Caulfield, J.A. and Gleason, K.K. (2000) *J. Vac. Sci. Technol. A*, **18**, 2404.

60 Savage, C.R., Timmons, R.B. and Lin, J.W. (1991) *Chem. Mater.*, **3**, 575.

61 Wang, J.-H., Chen, J.-J. and Timmons, R.B. (1996) *Chem. Mater.*, **8**, 2212.

62 Hynes, A.M., Shenton, M.J. and Badyal, J.P.S. (1996) *Macromolecules*, **29**, 18.

63 Hynes, A.M., Shenton, M.J. and Badyal, J.P.S. (1996) *Macromolecules*, **29**, 4220.

64 Butoi, C.I., Mackie, N.M., Gamble, L.J., Castner, D.G., Barnd, J., Miller, A.M. and Fisher, E.R. (2000) *Chem. Mater.*, **12**, 2014.

65 Hynes, A. and Badyal, J.P.S. (1998) *Chem. Mater.*, **10**, 2177.

66 Milella, A., Palumbo, F., Favia, P., Cicala, G. and d'Agostino, R. (2005) *Pure Appl. Chem.*, **77**, 399.

67 Cicala, G., Milella, A., Palumbo, F., Favia, P. and d'Agostino, R. (2003) *Diamond Relat. Mater.*, **12**, 2020.

68 Cicala, G., Milella, A., Palumbo, F., Rossini, P., Favia, P. and d'Agostino, R. (2002) *Macromolecules*, **35**, 8920.

69 Favia, P., Cicala, G., Milella, A., Palumbo, F., Rossigni, P. and d'Agostino, R. (2003) *Surf. Coat. Technol.*, **169–170**, 609.

70 Milella, A. (2002) PhD thesis, University of Bari.

71 Chen, R. and Silverstein, M.S. (1996) *J. Polym. Sci. A: Polym. Chem.*, **34**, 207.

72 Chen, R., Gorelik, V. and Silverstein, M.S. (1995) *J. Appl. Polym. Sci.*, **56**, 615.

73 Hashimoto, T., Murakami, Y. and Kawai, H. (1975) *J. Polym. Sci., Polym. Phys. Ed.*, **13**, 1613.

74 Chanzy, H.D., Smith, P. and Revol, J.F. (1986) *J. Polym. Sci., Polym. Lett. Ed.*, **24**, 557.

75 Bunn, C.W. and Howells, E.R. (1954) *Nature*, **174**, 549.

76 Qui, H. (2001) PhD thesis, University of Texas at Arlington.

10
Plasma CVD Processes for Thin Film Silicon Solar Cells
A. Matsuda

Growth processes of hydrogenated amorphous silicon (a-Si:H) and microcrystalline silicon (μc-Si:H) from SiH_4 and H_2/SiH_4-glow discharge plasmas are reviewed. Differences and similarities between a-Si:H and μc-Si:H growth reactions in the plasma and on the film-growing surface are discussed, and nucleus-formation processes followed by epitaxial-like crystal growth processes are explained as unique processes to μc-Si:H. Determination reaction of dangling bond defect density in the resulting a-Si:H and μc-Si:H films is emphasized in order to obtain a clue to improve optoelectronic properties of these materials for device applications especially in thin-film silicon-based solar cells. Material issues to realize low-cost and high-efficiency solar cells are described, and finally recent progresses in these issues are introduced.

10.1
Introduction

Hydrogenated amorphous silicon (a-Si:H) and microcrystalline silicon (μc-Si:H), thin-film silicon in general terms, are expected as promising materials for optoelectronic device applications such as solar cells, color sensors, thin-film transistors, etc. [1]. Among a variety of growth methods for a-Si:H and μc-Si:H, plasma-enhanced chemical vapor deposition (PECVD) is popularly used due to its potential to prepare high-quality thin-film silicon uniformly on a large-area substrate.

In this review, details of growth processes of a-Si:H and μc-Si:H from reactive plasmas are explained, and determination reaction of dangling bond defect density, one of the most important structural properties determining device performance, in the resulting films is discussed in order to obtain a clue to control the optoelectronic properties of those materials for device applications especially for solar cell applications. Finally, recent progresses are introduced in the material issues to realize low-cost/high-efficiency thin-film silicon solar cells.

Advanced Plasma Technology. Edited by Riccardo d'Agostino, Pietro Favia, Yoshinobu Kawai, Hideo Ikegami, Noriyoshi Sato, and Farzaneh Arefi-Khonsari
Copyright © 2008 WILEY-VCH Verlag GmbH & Co. KGaA, Weinheim
ISBN: 978-3-527-40591-6

10.2
Dissociation Reaction Processes in SiH₄ and SiH₄/H₂ Plasmas

The initial event in the growth process of a-Si:H and μc-Si:H is electron impact dissociation of source gas materials in monosilane (SiH_4) and monosilane/hydrogen (SiH_4/H_2) glow-discharge plasmas. Figure 10.1 shows schematically the concept of dissociation pathway of SiH_4 and H_2 molecules into a variety of reactive species through electronic excited states of these molecules by inelastic collisions with high-energy electrons in the plasma. Since the energy of electrons in the plasma usually obeys Maxwell distribution taking a wide variety from zero to several tens of electron volts (eV), ground-state electrons of source gas molecules are excited into their electronic excited states almost simultaneously in the plasma. Electronic excited states of so-called complicated molecules like SiH_4 are usually dissociating states, from which dissociation occurs spontaneously to SiH_3, SiH_2, SiH, Si, H_2, and H, as shown in Fig. 10.1 depending on the stereo-chemical structure of each electronic excited state. Hydrogen molecule is also decomposed to atomic hydrogen. Excitation of ground-state electron to vacuum state gives rise to ionization events, producing new electrons and ions to maintain the plasma.

Reactive neutral and ionic species produced in the plasma experience secondary reactions mostly with parent SiH_4 and H_2 molecules forming a steady state. Reaction rate constants for each reaction are summarized in the literature [2]. Steady-state

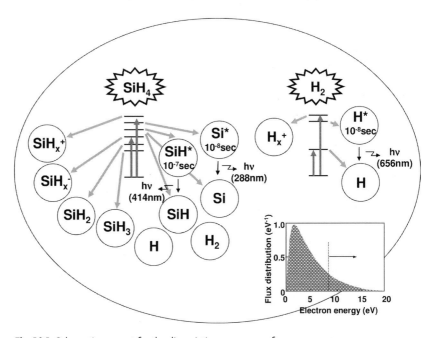

Fig. 10.1 Schematic concept for the dissociation processes of SiH_4 and H_2 molecules to a variety of chemical species in the plasma through their electronic excited states.

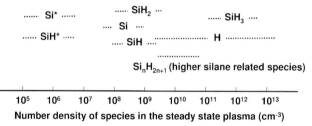

Fig. 10.2 Number density of chemical species in the realistic steady-state plasmas measured or predicted by various diagnostic techniques.

densities of reactive species are determined by the balance between their generation rate and their annihilation rate. Therefore, highly reactive species such as SiH_2, SiH, and Si (short-lifetime species) have much smaller values than SiH_3 showing low reactivity with SiH_4 and H_2 (long-lifetime species) in the steady-state plasma, although the generation rates of short-lifetime species are not so different from that of SiH_3.

Steady-state densities of reactive species have been measured using various gas-phase diagnostic techniques such as optical emission spectroscopy (OES), laser-induced fluorescence (LIF), infrared laser absorption spectroscopy (IRLAS), and ultraviolet light absorption spectroscopy (UVLAS). Figure 10.2 shows the steady-state number densities of neutral chemical species including emissive species in SiH_4 and SiH_4/H_2 realistic plasmas used for preparing device-grade a-Si:H and μc-Si:H [3]. It is concluded from the results that SiH_3 radical is dominant chemical species both for the growth of a-Si:H and μc-Si:H, although the density ratios of short-lifetime species and higher silane-related species (HSRS) such as Si_4H_9 to SiH_3 are varied when changing the plasma production condition and the optoelectronic properties of resulting films are very much influenced by the contribution of those short-lifetime species and HSRS to the film growth.

Steady-state density of atomic hydrogen (H) varies widely in the plasma as is also shown in Fig. 10.2. This is mainly due to the change in the hydrogen dilution ratio R (H_2/SiH_4) in the starting source gas materials, i.e. the density of atomic hydrogen increases with increasing R. When thinking of the fact that μc-Si:H is formed with increasing R at constant electron density in the plasma and constant substrate (film-growing surface) temperature, it is indicated that atomic hydrogen plays an important role in the case of μc-Si:H growth.

10.3
Film-Growth Processes on the Surface

10.3.1
Growth of a-Si:H

SiH_3 radicals reaching the film-growing surface start diffusing on the surface. During surface diffusion, SiH_3 abstracts surface-covering bonded hydrogen

(Si–H), forming SiH_4 and leaving Si dangling bonds (Si–) on the surface (growth-site formation). Toward the dangling bond site on the surface, another SiH_3 diffuses to find the site and to make Si–Si bond (film growth). This surface reaction scheme for the film growth has been proposed on the basis of two experimental results: substrate temperature dependences of radical (SiH_3) reflection probability measured by the step-coverage method and radical sticking probability estimated from film-growth rate [4]. Substrate temperature-independent reflection probability has suggested that almost all of surface sites are covered with bonded hydrogen, and temperature-dependent sticking probability above 350°C has indicated that radicals are diffusing on the film-growing surface and mobile SiH_3 can find a few dangling bond sites appearing thermally at higher temperature above 350°C.

10.3.2
Growth of μc-Si:H

As mentioned before, atomic hydrogen reaching the film-growing surface plays an important role for the growth of μc-Si:H [5]. This has been confirmed in the μc-Si:H formation map drawing on the space between radiofrequency (RF) electric power density applied to the plasma and hydrogen dilution ratio R in the source gas materials at constant substrate temperature during film growth (Fig. 10.3). As is seen in Fig. 10.3, μc-Si:H with larger crystallite size is prepared under higher hydrogen dilution and lower RF power density conditions, indicating the importance of atomic H and the negative effect of ionic species for crystal growth in μc-Si:H, since flux density of ionic species to the film-growing surface is increased linearly with increasing RF power density although the deposition rate tends to saturate. Figure 10.4 shows crystalline volume fraction in the resulting μc-Si:H as a

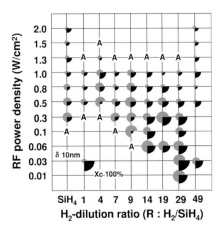

Fig. 10.3 Crystal size (three-quarter circle: δ) and volume % (quarter circle: X_c) of microcrystallites in the resulting films mapped out on the RF power density/hydrogen dilution ratio plane.

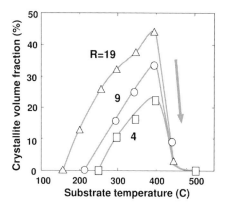

Fig. 10.4 Volume fraction (X_c) of microcrystallites in the resulting films plotted against the substrate temperature during film growth.

function of substrate temperature during film growth for three different hydrogen dilution ratios. Crystalline volume fraction increases with increasing substrate temperature, taking a maximum at around 350°C, and tends to zero sharply toward 500°C, suggesting the importance of enhanced surface diffusion of SiH_3 and the importance of surface hydrogen coverage. A large atomic H flux from the plasma realizes a full surface coverage with bonded hydrogen and also produces local heating through hydrogen exchanging reactions on the film-growing surface. These two actions enhance the surface diffusion of film precursors (SiH_3). Recently, more microscopic observations have been reported using *in situ* diagnostic techniques and a detailed mechanism underlying the formation process of μc-Si:H has been proposed.

10.3.2.1 Nucleus Formation Process

Figure 10.5 shows surface roughness evolution during film growth obtained by spectroscopic ellipsometry (SE) for three hydrogen dilution ratios R of 0, 10, and 20 [6]. As can be seen, after the formation of islands, enforced coalescence of islands takes place exhibiting smooth (flat) surface under μc-Si:H growth condition ($R = 20$). After exhibiting smooth surface (nucleus formation is confirmed at this moment), surface roughness is enhanced due to an orientation-dependent crystal growth rate difference. At the moment when the smooth surface appears, a specific surface absorption band has been observed in the infrared absorption spectrum measured using the *in situ* attenuated total reflection technique (ATR) during film growth [7]. Figure 10.6 shows the surface infrared absorption spectrum demonstrating the appearance of specific bands at 1897 and 1937 cm^{-1} together with usual Si-Hx surface and bulk absorption bands (between 2000 and 2150 cm^{-1}). This new absorption band is assigned to the SiH_2(Sid) complex as is also sketched in Fig. 10.6. It is noted that the number density (absorption intensity) of SiH_2(Sid) complex is found to be proportional to the magnitude of internal stress embedded in the film just before the appearance of those complexes.

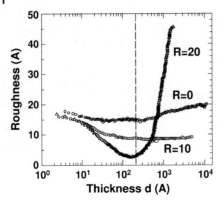

Fig. 10.5 Surface roughness evolution characteristics measured using spectroscopic ellipsometry during film growth for three different R values.

A nucleation model has been proposed from the experimental facts mentioned above. Enforced island coalescence due to an enhanced surface diffusion of SiH_3 gives rise to an internal stress involving many strained Si–Si bonds in the island coalescent regions. Atomic hydrogen attacks the strained Si–Si bond forming specific SiH_2(Sid) complex on the film-growing surface (or in the subsurface). These complexes give structural flexibility, which enables structure ordering by successive Si–SiH_3 bond formation on these sites, i.e. the SiH_2(Sid) complex acts as a pre-nucleation site on the film-growing surface.

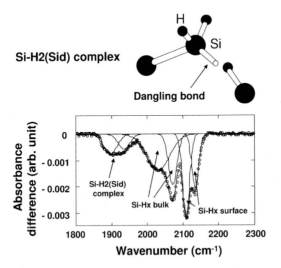

Fig. 10.6 Surface infrared absorption spectrum for the film just before nucleus formation showing the appearance of the Si-H$_2$(Sid) complex whose structure is also shown.

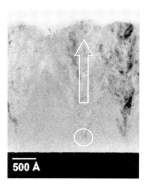

Fig. 10.7 Cross-sectional transmission electron microscopy image for typical μc-Si:H deposited on glass substrate.

10.3.2.2 Epitaxial-Like Crystal Growth

Figure 10.7 shows a cross-sectional transmission electron microscopy (TEM) image of typical μc-Si:H films deposited on a glass substrate [8]. As is clearly seen (indicated by the arrow), epitaxial-like crystal growth is observed from the nucleus. It is a well-known fact that epitaxial crystal growth occurs only when the surface diffusion length of film precursors is sufficiently long.

For both the nucleus formation step and epitaxial-like crystal growth step, enhanced surface diffusion of SiH_3 is a key factor. Enhanced surface diffusion of SiH_3 is given by atomic H flux through full coverage of surface reaction sites and local heating with many H exchanging reactions (abstraction reaction of surface-covering H with atomic H followed by saturation reaction of abstracted site with another atomic H, emitting heat by those two exothermal reactions). This is considered as one of the most crucial roles of atomic H in the course of μc-Si:H growth [9].

10.4
Defect Density Determination Process in a-Si:H and μc-Si:H

One of the most important structural properties in a-Si:H and μc-Si:H for device applications, especially for solar cell applications, is dangling bond defect density in these materials because dangling bonds make a deeply localized electronic state in their band gap and act as recombination centers for photoexcited electrons and holes.

10.4.1
Growth of a-Si:H and μc-Si:H with SiH_3 (H) Radicals

Figure 10.8 shows dangling bond defect density in the resulting a-Si:H and μc-Si:H as a function of substrate temperature under the conditions where SiH_3 (H) is well selected as film precursor. This substrate temperature-dependent dangling bond density (U-shaped curve) has been explained by taking into account the steady-state dangling bond density on the film-growing surface [10,11]. The steady-state dangling

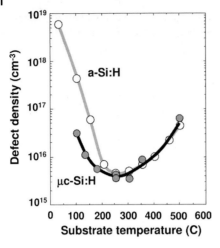

Fig. 10.8 Dangling bond defect density in a-Si:H and μc-Si:H films as a function of substrate temperature during film growth.

bond density on the film-growing surface for both a-Si:H and μc-Si:H growths is determined by the balance between the temperature-independent generation rate of dangling bonds on the surface by the abstraction reaction of bonded H with SiH₃ and the temperature-dependent annihilation rate of dangling bonds with surface-diffusing SiH₃. It should be noted here that a marked decrease of dangling bond defect density in the resulting μc-Si:H as compared to that in a-Si:H in the substrate temperature range below 250°C is caused by an enhanced surface diffusion of SiH₃ (enhanced annihilation rate of surface dangling bonds) in the case of μc-Si:H growth.

On the basis of understanding of the defect density determination reaction during film growth, several trials have been performed to control the defect density both in a-Si:H and μc-Si:H. For example, the steady-state defect density on the film-growing surface can be reduced when the growth rate is much faster than the thermal H removal rate in the substrate temperature range above 350°C where the generation rate of surface dangling bonds is dominated by the thermal H removal process. Actually a defect density of $10^{14}\,\mathrm{cm}^{-3}$ has been demonstrated in a-Si:H grown at 400°C when increasing the growth rate to $1\,\mathrm{nm\,s}^{-1}$[12].

10.4.2
Contribution of Short-Lifetime Species

Dangling bond defect density in the resulting film is basically determined on the film-growing surface as mentioned above, when SiH₃ (H) radicals contribute to the film growth. However, short-lifetime reactive species such as SiH₂, SiH, and Si contribute to the film growth when the secondary reactions of those short-lifetime species with parent molecules in the plasma (SiH₄ and H₂) are not enough due to a depletion of molecules owing to an application of high power density to the plasma (high electron density N_e) to try to obtain high rate growths of a-Si:H and μc-Si:H for the mass production of devices. In this case, the steady-state dangling bond density on the film-growing surface is increased, leading to an increase of dangling bond density in the resulting films, because dangling bonds are created by the contribution of short-lifetime species on the film-growing due to their strong reactivity into Si–H bonds on the surface (SiH and Si insertion reactions into Si–H bonds make new dangling bonds at their reaction sites) and they do not contribute to the annihilation reaction of dangling bonds due to their non-diffusion on the surface. This gives rise to the deterioration of optoelectronic properties in the resulting a-Si:H and μc-Si:H even when the substrate temperature is kept the same.

The contribution ratio of short-lifetime species to the growth of a-Si:H and μc-Si:H is expressed by the ratio of steady-state density of short-lifetime species $[SiH_x]$ to that of film precursors $[SiH_3]$ in the plasma when a conventional diode-type (capacitively coupled) plasma reactor is used. $[SiH_x]$ is given by the following rate equation:

$$\frac{d[SiH_x]}{dt} = N_{e2}\sigma_2 v_e[SiH_4] - k_2[SiH_x][SiH_4] - k_1[SiH_x][H_2] = 0 \tag{10.1}$$

Here, N_{e2}, σ_2, v_e, k_2, and k_1 are electron density, reaction cross-section responsible for the decomposition of SiH_4 into SiH_x, thermal velocity of electrons, representative reaction rate constant of SiH_x with SiH_4, and that of SiH_x with H_2, respectively. Since k_2 is much larger than $k_1[2]$, $[SiH_x]$ is given by

$$[SiH_x] = \frac{N_{e2}\sigma_2 v_e[SiH_4]}{k_2[SiH_4]} = \frac{N_{e2}\sigma_2 v_e}{k_2} \tag{10.2}$$

$[SiH_3]$ is given by the following rate equation:

$$\frac{d[SiH_3]}{dt} = N_{e3}\sigma_3 v_e[SiH_4] - \frac{[SiH_3]}{\tau_3} = 0 \tag{10.3}$$

Therefore,

$$[SiH_3] = N_{e3}\sigma_3 v_e \tau_3[SiH_4] \tag{10.4}$$

Parameters N_{e3}, σ_3, and τ_3 are electron density, reaction cross-section responsible for the decomposition of SiH_4 into SiH_3, and characteristic lifetime of SiH_3, respectively.

Therefore, the contribution ratio of short-lifetime species to the film growth is given by the ratio of Eq. 10.2 to Eq. 10.4:

$$\frac{[SiH_x]}{[SiH_3]} = \frac{N_{e2}\sigma_2 v_e}{k_2 N_{e3}\sigma_3 v_e \tau_3[SiH_4]} \propto \frac{N_{e2}}{N_{e3}\tau_3[SiH_4]} \tag{10.5}$$

Steady-state density of SiH_4 is also expressed as

$$\frac{d[SiH_4]}{dt} = FR - N_{et}\sigma_t v_e[SiH_4] - \frac{[SiH_4]}{\tau_4} = 0 \tag{10.6}$$

then

$$[SiH_4] = \frac{FR}{N_{et}\sigma_t v_e + (1/\tau_4)} \tag{10.7}$$

$[SiH_4]$ is strongly influenced by the plasma conditions. $[SiH_4]$ is rather constant when a sufficient amount of SiH_4 is fed into the reaction space as shown in Eq. (10.8), whereas $[SiH_4]$ is flow rate (FR) dependent when a high power is applied to the plasma (high total electron density N_{et}) with a limited supply of SiH_4 (SiH_4 depletion takes place) as expressed in Eq. (10.9):

$$[SiH_4] = \frac{FR}{1/\tau_4} = const. \quad (N_e \text{ limited}) \tag{10.8}$$

and

$$[SiH_4] = \frac{FR}{N_{et}\sigma_t v_e} \quad (FR \text{ limited}) \tag{10.9}$$

Therefore, the contribution ratio of short-lifetime species is strongly affected by the supply of parent molecule SiH_4 and plasma conditions as follows:

$$\frac{[SiH_x]}{[SiH_3]} \propto \frac{N_{e2}}{N_{e3}\tau_3} \propto f(T_e) \quad (N_e \text{ limited}) \tag{10.10}$$

and

$$\frac{[SiH_x]}{[SiH_3]} \propto \frac{N_{e2}N_{et}}{N_{e3}\tau_3 FR} \propto N_{et} f(T_e) \quad (FR \text{ limited}) \tag{10.11}$$

Namely, the contribution ratio of short-lifetime species to the film growth, an important defect density determination factor at constant substrate temperature, is given by the electron density ratio (N_{e2} to N_{e3}, corresponding to the electron temperature in the plasma) in a high flow rate and low power density regime (high-quality a-Si:H growth condition) as is seen in Eq. (10.10), whereas it depends on the total electron density (N_{et}), electron temperature, and FR in a limited-FR and high power density (high N_{et}) regime that is used for the high-rate growth of μc-Si:H.

10.5
Solar Cell Applications

Thin-film Si-based solar cells have been expected as low-cost photovoltaic systems. Actually, consumer-use a-Si:H-based solar cells are widely installed in pocket calculators, and large-area solar cells for electric power generation are now under development. A major advantage of a-Si:H for solar cell applications is its larger optical absorption coefficient in the visible wavelength range in contrast to single-crystalline and polycrystalline silicon counterparts showing indirect optical transition properties; hence less than $1\,\mu m$ in thickness is enough to absorb sufficient sunlight when using a-Si:H-based solar cells. Low-temperature processing using PECVD is also advantageous for cost reduction in a-Si:H-based solar cells.

However, there exists a well-known phenomenon in a-Si:H, i.e. photoinduced degradation, where initial conversion efficiency of a-Si:H-based solar cells of $\sim$10% is degraded to $\sim$7% after prolonged light soaking. A tandem-type stacked solar cell structure consisting of a top cell with thin a-Si:H layer and a bottom cell with narrow-gap materials such as a-SiGe:H and μc-Si:H has been proposed as a promising way to overcome photoinduced degradation and to achieve high conversion efficiency in a-Si:H-based solar cells. Initially, a-SiGe:H has been adopted for bottom cell materials; however, this material also shows severe photoinduced degradation. Recently, μc-Si:H has been proposed as a promising candidate for bottom cell material, because this material does not exhibit any photoinduced degradation. In this proposal, a high rate growth of this material is crucial for low-cost fabrication of tandem-type solar cells, since μc-Si:H possesses are basically of indirect optical transition nature.

Therefore, urgent material issues for realization of low-cost/high-efficiency thin-film silicon-based solar cell are to improve photoinduced stability in high-quality (low defect density) a-Si:H and to achieve high rate of growth of high-quality (low defect density) μc-Si:H.

10.6
Recent Progress in Material Issues for Thin-Film Silicon Solar Cells

10.6.1
Control of Photoinduced Degradation in a-Si:H

The relationship between the degree of photoinduced degradation and dihydride bonding (Si-H_2) density has been reported in a-Si:H prepared under a variety of deposition conditions [13]. Figure 10.9 shows the degree of photoinduced degradation defined as the difference of fill factor before and after light soaking in photo-I–V characteristics of an Ni-a-Si:H Schottky diode plotted against Si-H_2 density in a-Si:H film measured using infrared absorption spectroscopy. Furthermore, it has been suggested from the results of mass spectrometry during film growth that Si-H_2 density in the resulting a-Si:H is strongly increased by the contribution of HSRS such as Si_4H_9 when keeping the substrate temperature constant even when a high-quality (low defect density) a-Si:H is fabricated under proper substrate temperature conditions with low contribution ratio of short-lifetime species, as mentioned in Section 10.4. On the basis of understanding of responsible network structure (Si-H_2 bonding configuration) for the photoinduced degradation in a-Si:H and responsible

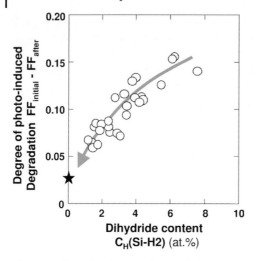

Fig. 10.9 Relationship between degree of photoinduced degradation and dihydride (Si-H$_2$) content in a-Si:H prepared under various deposition conditions. Star symbol represents a-Si:H prepared under the reduced contribution ratio condition of HSRS to SiH$_3$ at substrate temperature of 250°C.

relevant chemical species (HSRS) during film growth, a guiding principle for obtaining highly stabilized a-Si:H has been proposed [14].

Following the guiding principle, a-Si:H containing Si-H$_2$ density of almost 0% has successfully been prepared using a triode reactor (negatively biased mesh electrode is inserted between cathode and anode) at substrate temperature as low as 250°C by making good use of the difference in the gas-phase diffusion coefficient between SiH$_3$ (light) and HSRS (heavy) together with reaction of HSRS with parent SiH$_4$ during diffusion of SiH$_3$ and HSRS from the plasma to the substrate placed on the anode. Film containing almost 0% Si-H$_2$ density shows much more stable behavior against light soaking as shown in Fig. 10.9 by the star symbol. Stabilized conversion efficiency (after light soaking) of 9.4% has also been demonstrated using this triode method for actual p–i–n type a-Si:H solar cells [15].

10.6.2
High-Rate Growth of Device-Grade μc-Si:H

High hydrogen dilution method at relatively low working gas pressure in the range of several tens to several hundred millitorr has been conventionally used to obtain device-grade μc-Si:H. Recently, a simple concept for preparing device-grade μc-Si:H at high growth rate has been proposed: the narrow-gap/high-pressure (NG/HP) method. The flux density of SiH$_3$ is determined both by the generation rate of SiH$_3$ in the plasma and by the distance of the film-growing surface (substrate) from

SiH_3-generating region (plasma). To increase the generation rate of SiH_3 in the plasma, high power density is usually used (high electron density N_{et} in the plasma) under SiH_4 depletion conditions, being important to avoid scavenging reactions of atomic H with SiH_4 molecules for μc-Si:H growth. However, as is suggested by Eq. (10.11), defect density in the resulting μc-Si:H is increased through an increase of the contribution ratio of short-lifetime species to film growth under high electron density conditions. Therefore, to realize high flux density of SiH_3 to the film-growing surface, narrowing the electrode gap is more effective for the high-rate growth of high-quality (low defect density) μc-Si:H, since the flux density of film precursor (SiH_3) is increased super-linearly with decreasing distance between the radical-generating space (plasma) and the film-growing surface. For the stable production of plasma in the narrow-gap region, high total gas pressure is necessary to satisfy Paschen's law. As high total pressure condition is also beneficial to decrease the electron temperature in the plasma during film growth as is also expected from Eq. (10.11). The NG/HP method has recently been used for the high-rate growth of device-grade μc-Si:H [16]. The validity of the NG/HP method has been demonstrated in the fabrication process of μc-Si:H-based solar cells exhibiting high conversion efficiency of 9.1% at growth rate of 2.3 nm s^{-1}[17].

However, the NG/HP condition needs a reduced electrode distance (several millimeters under total pressure of 10 torr), causing non-uniform film growth by non-uniform plasma production when applying this method to large-area (e.g. 1 m^2) solar cell fabrication system. To overcome this problem, a new cathode with interconnected multi-hollows (Fig. 10.10) has been proposed to realize uniform plasma production even in a large-area parallel-plate electrode configuration under high-pressure conditions due to its rather independent discharge ability of Paschen's law. Using this newly designed cathode, quite a high growth rate of more than 8 nm s^{-1} has been obtained for the growth of high-quality μc-Si:H with defect density as low as 10^{15} cm^{-3}[18].

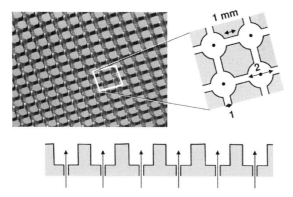

Fig. 10.10 Image of cathode surface with interconnected multi-hollows to produce high-density/uniform plasmas for large-area/high-rate film growth.

10.7
Summary

The growth process of a-Si:H as well as μc-Si:H from SiH_4 and SiH_4/H_2 plasmas has been reviewed. The temperature-dependent defect density determination reaction on the film-growing surface has been interpreted and the plasma condition-dependent defect density determination reaction is also explained using the concept of the contribution ratio of short-lifetime species to film growth in order to obtain a clue for improving the optoelectronic properties in these materials. Finally, recent progresses in material issues for solar cells applications have been introduced.

References

1 Spear, W.E. and LeComber, P.G. (1975) *Solid State Commun.*, **17**, 1193.

2 Perrin, J., Leroy, O. and Bordage, M.C. (1996) *Contrib. Plasma Phys.*, **36**, 3.

3 Matsuda, A. and Goto, T. (1990) *Mater. Res. Soc. Proc.*, **164**, 3.

4 Matsuda, A., Nomoto, K., Takeuchi, Y., Suzuki, A., Yuuki, A. and Perrin, J. (1990) *Surf. Sci.*, **227**, 50.

5 Matsuda, A. (1983) *J. Non-Cryst. Solids*, **59/60**, 767.

6 Koh, J., Lee, Y., Fujiwara, H., Wronski, C.R. and Collins, R.W. (1998) *Appl. Phys. Lett.*, **73**, 1526.

7 Fujiwara, H., Kondo, M. and Matsuda, A. (2002) *Surf. Sci.*, **497**, 333.

8 Fujiwara, H., Kondo, M. and Matsuda, A. (2001) *Phys. Rev., B*, **63**, 115306.

9 Suzuki, S., Kondo, M. and Matsuda, A. (2002) *J. Non-Cryst. Solids*, **299–302**, 93.

10 Ganguly, G. and Matsuda, A. (1993) *Phys. Rev., B*, **47**, 3361.

11 Nasuno, Y., Kondo, M. and Matsuda, A. (2001) *Tech. Digest of PVSEC-12, Jeju, Korea*, 791.

12 Ganguly, G. and Matsuda, A. (1992) *Jpn. J. Appl. Phys.*, **31**, L1269.

13 Nishimoto, T., Takai, M., Miyahara, H., Kondo, M. and Matsuda, A. (2002) *J. Non-Cryst. Solids*, **299–302**, 1116.

14 Takai, M., Nishimoto, T., Takagi, T., Kondo, M. and Matsuda, A. (2000) *J. Non-Cryst. Solids*, **266–269**, 90.

15 Shimizu, S., Kondo, M. and Matsuda, A. (2004) *Tech. Digest of PVSEC-14, Bangkok, Thailand*, 22.

16 Guo, L., Kondo, M., Fukawa, M., Saito, K. and Matsuda, A. (1998) *Jpn. J. Appl. Phys.*, **37**, L1116.

17 Matsui, T., Kondo, M. and Matsuda, A. (2004) *Tech. Digest of PVSEC-14, Bangkok, Thailand*, 33.

18 Niikura, C., Kondo, M. and Matsuda, A. (2003) *Proc. WCPEC-3, Osaka, Japan*, p. 5p-D4-03.

11
VHF Plasma Production for Solar Cells
Y. Kawai, Y. Takeuchi, H. Mashima, Y. Yamauchi, and H. Takatsuka

11.1
Introduction

There is great interest in the fabrication of large-area amorphous silicon films with high deposition rate to reduce the cost of amorphous silicon solar cells. Hydrogenated amorphous silicon (a-Si:H) is usually prepared by radio frequency (RF) discharge plasma chemical vapor deposition (CVD) where the parallel plate-type electrodes of frequency of 13.56 MHz are conventionally used. In this case, the increase in the deposition rate is achieved by increasing the power of the RF source and the deposition rate is at most $0.2–0.3\,\mathrm{nm\,s^{-1}}$. The method for increasing the deposition rate by applying very high frequency (VHF) to the plasma CVD has drawn attention because it can produce relatively high-quality films at a high speed [1–5]. In fact, Curtins *et al.* [1] achieved $2\,\mathrm{nm\,s^{-1}}$ at 70 MHz. These results were confirmed by other researchers [4,5].

Recently, a further improvement toward faster, uniform deposition over a larger area ($>1\,\mathrm{m^2}$) has been demanded from industry. In the VHF range, however, it is hard to obtain uniform films over a large area ($>1\,\mathrm{m^2}$) using a conventional parallel plate reactor because of nonuniform RF plasma potential, resulting in nonuniform power dissipation and consequently nonuniform deposition [6–8]. This nonuniform deposition comes from the voltage distribution due to standing wave effects. Some attempts were made together with simulation [9,10] in order to improve the conventional parallel plate reactor. Sansonnens *et al.* [7] pointed out numerically that the VHF plasma uniformity depends on how to feed VHF powers in the case of the parallel plate-type electrodes. In addition, Schmidt *et al.* [10] carried out experiments using a lens-shaped circular electrode to measure the correction of plasma nonuniformity due to the standing wave effect in a large area VHF plasma reactor. This work is the experimental verification of the theoretical reactor design in cylindrical geometry recently presented by Sansonnens and Schmitt [11]. They found that the lens-shaped electrode effectively compensates the standing wave effects by creating a uniform RF vertical electric field in the plasma volume. We have

Advanced Plasma Technology. Edited by Riccardo d'Agostino, Pietro Favia, Yoshinobu Kawai, Hideo Ikegami, Noriyoshi Sato, and Farzaneh Arefi-Khonsari
Copyright © 2008 WILEY-VCH Verlag GmbH & Co. KGaA, Weinheim
ISBN: 978-3-527-40591-6

started experiments on the production of VHF plasma with larger area using a ladder-shaped electrode of 1200 mm × 141 mm [12,13].

Microcrystalline silicon (μc-Si:H) is an attractive low band gap absorber material for integration in thin-film solar cells and has been widely investigated [14–16]. As is well known, to reduce production costs of solar cells, high deposition rates of μc-Si:H have to be achieved. Now the development of a high rate deposition process for microcrystalline silicon is indispensable in order to reduce manufacturing costs for microcrystalline silicon solar cells. Usually VHF plasma has been adopted to prepare μc-Si:H to obtain high deposition rates. Very recently, it was found [14–16] that higher deposition rate of μc-Si:H is achieved by a narrow-gap discharge at high pressure, maintaining the high quality. Paschen's law suggests that the higher the pressure becomes, the shorter the distance between the electrodes for discharge should be. As a result, discharge voltages will be greater because of larger losses of plasma to the electrodes. On the other hand, since a high quality of μc-Si:H can be obtained at high pressure, the wall potential is believed to be very low. The mechanism for higher deposition rates at high pressure has not been understood although it was discussed from the point of view of gas flow and silane concentration. Thus, it is one of the most important subjects in the production of microcrystalline silicon solar cells to clarify the mechanism of higher deposition rates at high pressure.

11.2
Characteristics of VHF H$_2$ Plasma

We proposed a VHF plasma reactor using a ladder-shaped electrode and succeeded [17–19] in preparing a high-speed and high-quality a-Si:H film on a substrate of 500 mm × 400 mm rectangle in size. A schematic diagram of the experimental apparatus is shown in Fig. 11.1. The system consists of a vacuum chamber, a ladder-shaped antenna and a RF power source. The vacuum chamber, 600 mm wide, 600 mm high and 400 mm deep, was electrically grounded. As shown in Fig. 11.2, the ladder-shaped electrode consists of a stainless steel ladder-shaped antenna, 422 mm long and 422 mm wide (17 rods of 6 mm in diameter, spacing of 20 mm), which was positioned within the vacuum chamber. The distance between the ladder-shaped antenna and the substrate is 40 mm. The RF power source used consists of an oscillator and a power amplifier. The RF power was applied to four loading points on the ladder-shaped antenna via a matching box and was varied between 30 and 150 W in the frequency range 13.56–200 MHz. The gas used was pure H$_2$ at a pressure of 20–300 mtorr and a flow rate of 50–200 sccm. The plasma parameters were measured with a movable Langmuir probe inserted in front of the substrate (Corning 7059 glass).

In order to examine whether the ladder-shaped antenna is useful for VHF plasma production or not, the plasma parameters were measured as a function of pressure for driving frequency with a movable Langmuir probe. It is well known that in the presence of RF fields the V–I curve of the Langmuir probe is deformed and does not provide correct plasma parameters. Here a filter was used to obtain a correct V–I curve. Furthermore, the electron density n_e was estimated from the ion saturation

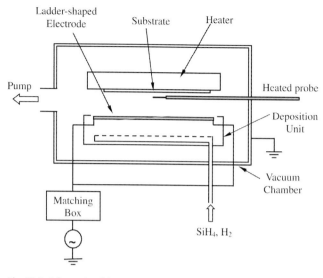

Fig. 11.1 Schematic of the experimental apparatus.

current because of less influence of RF fields on the V–I curve. The experimental results are shown in Fig. 11.3, where the gas flow rate was 50 sccm. Figure 11.3 indicates that when the RF driving frequency is increased, n_e increases, amounting to $n_e = 8 \times 10^{10}$ cm^{-3} at 120 MHz which is four times higher than that at 13.56 MHz. Thus, high-speed deposition of amorphous silicon films is expected by the VHF plasma produced with a ladder-shaped electrode. On the other hand, on increasing the RF driving frequency, the electron temperature T_e decreased. The decrease in the electron temperature means the reduction of ion bombardment and as a result

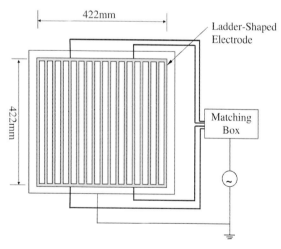

Fig. 11.2 Schematic of the ladder-shaped electrode of 422 × 422 mm.

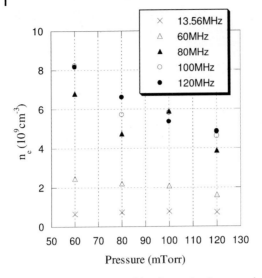

Fig. 11.3 The dependence of the plasma density, n_e, on the pressure for different VHF powers.

high-quality films is deposited on a substrate, that is, the VHF plasma is suitable for a high-speed deposition of amorphous silicon films with high quality. When the RF power was increased, n_e increased while the electron temperature was constant. Thus, it is concluded that the ladder-shaped electrode is useful for generation of a VHF plasma having advantage of high electron density.

11.3
Characteristics of VHF SiH$_4$ Plasma

A VHF-excited SiH$_4$ plasma up to 100 MHz was generated using the ladder-shaped electrode. Electric powers of discharge frequencies from 13.56 to 100 MHz were supplied to four power feeding points on the ladder-shaped electrode through an impedance matching transformer to generate the plasma between the ladder-shaped electrode and the heater (grounding potential electrically) for retaining and heating the substrate. The gas used was SiH$_4$ with a flow rate of 50 sccm, and the experiments were carried out in the relatively low pressure range from 24 to 40 mtorr in order to avoid disturbances of the plasma generation condition due to occurrence of powder formation in the gas phase. A heated Langmuir probe was used during parameter measurements of the SiH$_4$ plasma in order to solve the problem of measurements that is impeded by deposition of Si films onto the probe tip during the measurement. The heated probe used consisted of a tungsten wire 0.2 mm in diameter connected to a heating circuit to prevent film depositions by passing an electric current and heating the probe tip. Before measuring SiH$_4$ plasma parameters, the current range not disturbing plasma genera-tion due to heating was confirmed and calibration for the surface area of the heated probe was performed by measuring argon and hydrogen plasmas using an ordinary Langmuir probe. In addition, the electron density was estimated from the ion saturation current, which shows a smaller effect of the RF fields on I–V curves. The probe measurement point was located in the center of the ladder-shaped electrode of 422 mm × 422 mm, positioned at the distance of 1 cm from the substrate surface with the glass substrate (Corning 7059, 300 mm × 300 mm × 1.1 mm^t), which was placed on the heater in order to simulate the same condition of the film deposition. The dependence of the plasma characteristics on the frequency of VHF electric powers supplied to the ladder-shaped electrode were first examined to confirm that the ladder-shaped electrode could also be useful in the case of a VHF-excited SiH₄ plasma. The experimental results are shown in Fig. 11.4, where the pressure and the feeding power were 24 mtorr and 150 W, respectively. From Fig. 11.4(a), it is seen that with increasing

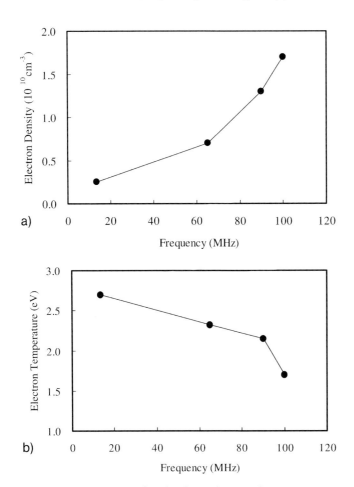

Fig. 11.4 The dependence of (a) the plasma density and (b) electron temperature on the VHF frequency; the VHF power and the pressure are 150 W and 24 mtorr, respectively.

the frequency of the supplied electric powers the electron density increases and reaches $1.7 \times 10^{10}\,\text{cm}^{-3}$ at 100 MHz, which is about 7 times as large as that at 13.56 MHz. Because the measurement point was near the substrate surface as described above, the reason for the faster deposition rate of a-Si:H films with the VHF-excited SiH_4 plasma using the ladder-shaped electrode is considered to a higher gas decomposition efficiency due to the high electron density compared with the conventional 13.56 MHz. The electron density of the SiH_4 plasma at 100 MHz is higher than that of the H_2 plasma described in Section 11.2, because the masses of ions such as SiH^+, SiH_2^+, SiH_3^+ in the SiH_4 plasma are greater than that in the H_2 plasma. Figure 11.4(b) clearly indicates that the electron temperature is 2.7 eV at 13.56 MHz whereas it decreases to 1.6 eV at 100 MHz, that is, the electron temperature decreases with increasing discharge frequency. It should be also noted that the electron temperature is considerably lower than that of H_2 plasma, as expected from the difference in the ionization potential. The drop in the electron temperature indicates a decrease of the plasma potential and as a result increased frequency reduces the ion impact on the film surface during deposition. The results of Fig. 11.4 thus show that a high-quality a-Si:H film with less damage during film deposition can be deposited at high speed using a VHF-excited SiH_4 plasma. These results are in good agreement with our film deposition experimental results as described earlier [20].

The pressure dependence of the electron density and the electron temperature of the VHF SiH_4 plasma at a frequency of 100 MHz is shown in Fig. 11.5, where the supplied power was 150 W. The electron density monotonically decreases with increasing pressure, declining by roughly half from $1.7 \times 10^{10}\,\text{cm}^{-3}$ at 25 mtorr to $0.8 \times 10^{10}\,\text{cm}^{-3}$ at 40 mtorr. In contrast, as shown in Fig. 11.5(b), the electron temperature tends to increase with increasing pressure, rising by a factor of about 1.3 from 1.6 eV at 24 mtorr to 2.2 eV at 40 mtorr. As the measurement accuracy of the electron temperature in this experiment can be estimated to be below ± 0.1 eV, based on the calibration test of measured values with the heated probe using hydrogen plasma and other gas plasmas, the increase of the electron temperature with increasing pressure as described above is considered to be a significant change. This differs from the pressure dependence of the weakly ionized plasma at the frequency of 13.56 MHz normally used in RF glow discharge [17]; i.e. the phenomenon wherein the electron temperature decreases with increasing pressure due to the increase of electron-to-atom or electron-to-molecule collisions.

Observed pressure dependence of the electron temperature is probably a cause of the substantial contribution to plasma confinement by electron trapping in the VHF plasma [21]. That is, when the variation time of oscillatory swarm motions of electrons between the two electrodes is larger than the oscillation period of the external VHF electric field, electrons may become spatially trapped between the two electrodes. As a result, electron loss at the surface of the electrodes is eliminated and the electron density increases. With increasing pressures, however, there are greater numbers of collisions between electrons and neutral particles, leading to a decrease in the number of trapped electrons, and this could lead to a higher electron temperature in order to make up the decrease of the electron density and maintain the discharge.

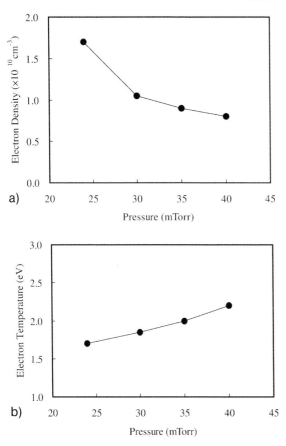

Fig. 11.5 The dependence of (a) the plasma density and
(b) electron temperature on the pressure; the VHF frequency
is 100 MHz at 150 W.

Figure 11.6 shows the VHF power dependence of the electron temperature at a frequency of 100 MHz with a pressure of 30 mtorr. It was confirmed that the electron density increased with increasing VHF electrical powers, although, as indicated in Fig. 11.6, the electron temperature decreased with increasing VHF powers. This phenomenon also differs from the supplied power dependence of the electron temperature in conventional RF glow discharges at 13.56 MHz, and is thought to be related to the effect of electron trapping in the formation of VHF Plasma, as pointed out in Fig. 11.5. Thus, the electron temperature does not increase substantially even with increasing supplied powers while the electron density increased. Given that plasma potential is maintained at a low level, it is reasonable to expect the reduction in ion flux energy onto the film surface during the deposition if VHF plasma were applied to film formation. This is considered to be a factor of simultaneous achievement of high-speed and high-quality deposition with the use of VHF-excited plasma, unlike the case of conventional plasma at 13.56 MHz.

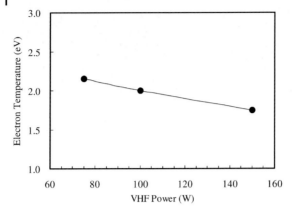

Fig. 11.6 The dependence of the electron temperature on the VHF power, where the VHF frequency and pressure are 100 MHz and 30 mtorr, respectively.

In order to investigate the effect of H_2 dilution on SiH_4/H_2 VHF plasma characteristics, we measured the parameters of the SiH_4/H_2 VHF plasma produced with the ladder-shaped electrode at a discharge frequency of 80 MHz. The dependence of the electron temperature on the RF power for different H_2 dilution rates is shown in Fig. 11.7, indicating that the electron temperature tends to decrease for high H_2 dilution rates. Here, the H_2 dilution rate D is defined as follows:

$$D = H_2 \text{ flow rate}/(SiH_4 \text{ flow rate} + H_2 \text{ flow rate})$$

This tendency is desirable for plasma CVD for the following reason. The decrease in the electron temperature means a decrease in the wall potential (the plasma potential), leading to a reduction of ion bombardment and, as a result, high-quality films will be deposited on a substrate. Figure 11.7 also shows that the dependence of the electron temperature on the RF power changes with H_2 dilution rates. Although the electron temperature increases with increases in the RF power for the H_2 plasma, it decreases with increases in the RF power for the SiH_4/H_2 plasma. It is considered that the RF power is dissipated mainly to produce the plasma, not to heat electrons for the H_2 plasma. On the other hand, as to the SiH_4 plasma, it is understood as follows. On increasing the RF power, SiH_4 gas becomes low enough and particles (Si_nH_m) occur in the plasma. The electron temperature increases as a result of these particles. Therefore, although the electron temperature at 100 W is lowest at $D = 91\%$, the electron temperature at 200 W is lowest at $D = 100\%$. At the condition that there is enough SiH_4 gas, about $D = 91\%$ is considered to be the optimum value of H_2 dilution rate for low electron temperature. This result suggests that there is an optimum H_2 dilution rate for the electron temperature becoming low. The dilution rate $D = 91\%$ is near the deposition condition of hydrogenated microcrystalline silicon films. The deposition condition of hydrogenated microcrystalline silicon films may be concerned with low electron temperature. More detailed measurement will

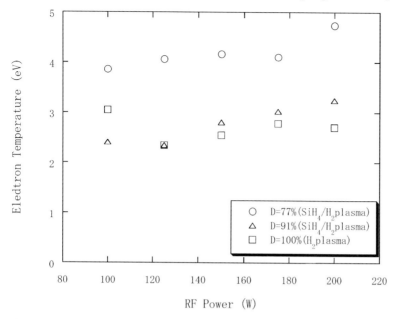

Fig. 11.7 The dependence of the electron temperature on the VHF power for different dilution rate, *D*, where the VHF frequency and the pressure are 80 MHz and 70 mtorr, respectively.

be necessary for a fuller explanation. Figure 11.7 also shows that the electron temperature at $D = 77\%$ is high at about 1.5 eV compared with that at other dilution rates, suggesting that there are negative ions in the SiH_4/H_2 plasma.

11.4
Characteristics of Large-Area VHF H₂ Plasma

Recently we have started experiments on the production of VHF plasma with larger area using a ladder-shaped electrode of 1200 mm × 141 mm. Figure 11.8 shows a schematic of the experimental apparatus. The system consists of a stainless steel vacuum vessel, a ladder-shaped electrode of 1200 mm × 141 mm and a VHF (RF) power supply. VHF powers of discharge frequencies up to 100 MHz were supplied to the feeding point on the ladder-shaped electrode shown in Fig. 11.9. The gas used was H_2 and the experiments were carried out in the pressure range from 30 to 200 mtorr, keeping a gas flow rate of 100 sccm. The plasma parameters and the profile of the ion saturation current were measured with a movable Langmuir probe.

At first we examined the spatial profile along the ladder (z-axis). Figure 11.10 shows the spatial profiles of the ion saturation current density, I_{is}, for different VHF powers, where the power and the pressure are 150 W and 30 mtorr, respectively. As the frequency of the VHF power source is increased, the plasma density increases while the spatial profile of the ion saturation current behaves less uniformly. Here, we

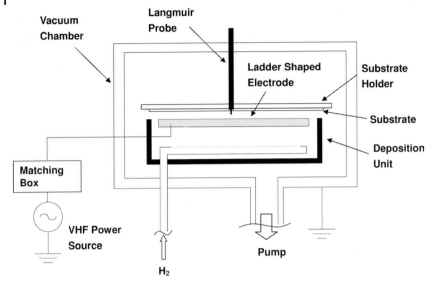

Fig. 11.8 Schematic of the experimental apparatus.

focused on the VHF discharge frequency of 60 MHz because the spatial profile of the ion saturation current for 60 MHz was relatively uniform.

As described in the previous section, a VHF plasma has an advantage that the plasma density becomes high at low pressure because electrons are trapped in the VHF electric fields, leading to improvement of particle confinement, that is, the plasma loss is reduced. Figure 11.11 indicates that the lower the pressure is, the higher the ion saturation current becomes. As shown in Fig. 11.10, the ion saturation current is relatively uniform at 60 MHz and 30 mtorr. When the VHF power was

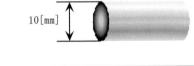

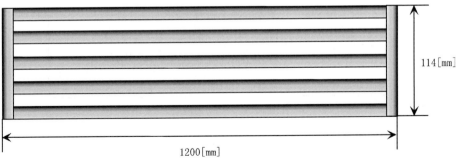

Fig. 11.9 Schematic of the ladder-shaped electrode.

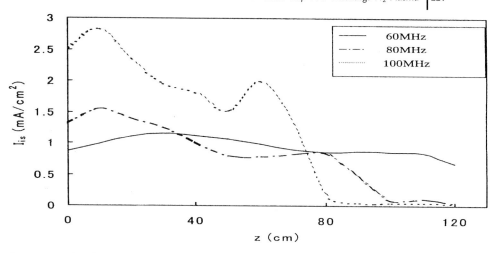

Fig. 11.10 Axial profile (z-axis) of the ion saturation current density, I_{is}, for different VHF frequencies, where the VHF power and pressure are 150 W and 30 mtorr, respectively.

increased, the ion saturation current increased. Here we measured the electron temperature, T_e, for different positions on the z-axis under the same conditions. Figure 11.12 indicates that the electron temperature is around 2.5 eV and is uniform. Thus, these results suggest that the VHF plasma produced with the ladder-shaped electrode is useful for the fabrication of amorphous silicon films with large area.

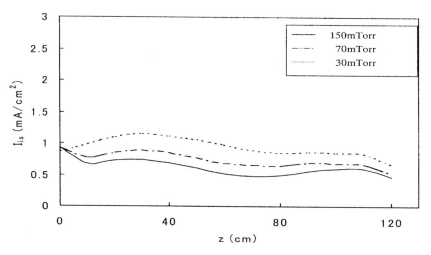

Fig. 11.11 Axial profile (z-axis) of the ion saturation current density, I_{is}, for different pressures, where the VHF frequency is 60 MHz at 150 W.

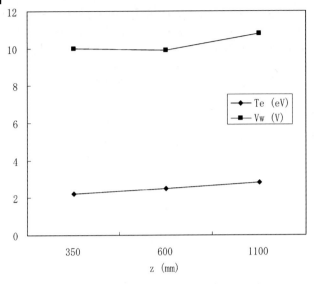

Fig. 11.12 Axial profile (z-axis) of the electron temperature, T_e, and wall potential, V_w, where the VHF frequency and power are 60 MHz and 150 W, and the pressure is 30 mtorr.

11.5
Short-Gap VHF Discharge H$_2$ Plasma

Since microcrystalline silicon is prepared by introducing small amount of silane gas (SiH$_4$/H$_2$ <10%) into hydrogen gas plasma with a large hydrogen gas flow rate, the parameters of SiH$_4$/H$_2$ gas mixture plasma are not so much different from those of hydrogen plasma. In fact, we have observed such a tendency in VHF plasma, so far, although it was measured for a pressure lower than the present case [17–19]. Thus, we have investigated the characteristics of VHF hydrogen plasma with a Langmuir probe [18] in order to clarify the mechanism for higher deposition rates at high pressure. A VHF plasma was produced using a ladder-shaped electrode of 1200 mm × 114 mm, which was positioned within a stainless steel rectangular vacuum chamber. As shown in Fig. 11.9, the ladder-shaped electrode consists of five stainless steel thin rods. The VHF power with frequency of 60 MHz up to 450 W was supplied to the ladder-shaped electrode. The distance between the ladder-shaped electrode and the substrate was 5 mm. The forward and reflected RF power was measured with a power meter. The gas flow rate of hydrogen gas was 50 sccm. The pressure ranged from 30 mtorr to 4 torr. The plasma parameters were measured with a very small cylindrical Langmuir probe (diameter 0.2 mm, length 1.2 mm) which was inserted into the plasma at a distance of 3 mm from the ladder-shaped electrode. The plasma parameters estimated from the Langmuir probe characteristics contain some errors which mainly arise from the evaluation of plasma sheath. Although same errors exist here, the plasma parameters were obtained assuming that such errors are of the same order.

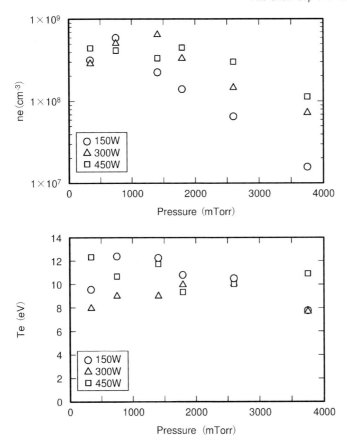

Fig. 11.13 The dependence of the plasma parameters on the pressure for different VHF powers: (a) the plasma density and (b) the electron temperature.

The plasma parameters were examined as a function of pressure for different VHF powers. Figure 11.13 shows that when the pressure is increased, the plasma density decreases, while the electron temperature, T_e, is around 10 eV at high pressure. In this experiment, the plasma density was estimated from the ion saturation current of the V–I curve of the Langmuir probe. As shown in Fig. 11.13(a), the plasma density is below $10^9\,cm^{-3}$, which is due to the narrow gap of the discharge electrodes. When the gap was 34 mm, the plasma density amounted to $1.5 \times 10^9\,cm^{-3}$ for hydrogen gas pressure of 30 mtorr and VHF power of 150 W. Furthermore, at higher pressure the plasma density becomes lower, which will be improved by using a narrower gap electrode. Figure 11.13(a) also shows that when the VHF power is increased, the plasma density increases. In order to further increase the deposition rate of μc-Si:H, it is necessary to raise the plasma density at high pressure by increasing the VHF power.

In this experiment, the fact that the electron temperature became as high as 10 eV is not surprising, because the distance between the electrodes for VHF discharge is very short. Generally, the electron temperature is proportional to the electric field between the electrodes for discharge. When the gap of 34 mm was used, the electron temperature decreased to about 3 eV at 30 mtorr, which is not shown in Fig. 11.13.

The higher the electron temperature becomes, the larger the probability of the electron attachment is in hydrogen plasma; that is, negative ions are produced in hydrogen plasma [22]. According to the total cross-section for production of negative ions [22,23], it takes a peak value around the electron energy of 10 and 14 eV. It is well known [24,25] that when negative ions exist in plasma, the electron saturation current of the V–I curve of the Langmuir probe anomalously decreases. A typical V–I curve is illustrated in Fig. 11.14, where the pressure and VHF power are 2000 mtorr and 300 W, respectively. Figure 11.14 indicates that the electron saturation I_{es} at high pressure is anomalously low compared with the ion saturation current I_{is} because, according to the probe theory, the ratio $I_{es}/I_{is} \sim 28$ for H_2 plasma. That is, in the high-pressure region, the electron density, which is proportional to the deposition rate, did not increase for higher VHF powers and VHF power was dissipated to produce negative ions. Saturation of the deposition rate was reported [16] in the high-pressure region, which is understood from the above results.

The wall potential, which is defined as the difference between the plasma potential and the floating potential, is a key value in plasma CVD because when the wall potential is high, the ion bombardment increases and as a result the film quality

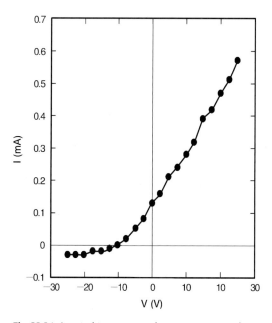

Fig. 11.14 A typical Langmuir probe curve at 2 torr, where VHF power is 300 W.

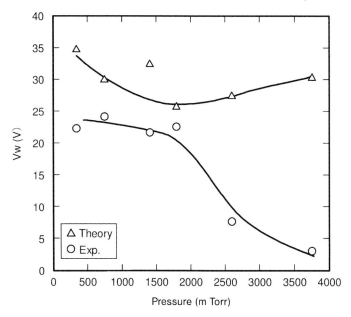

Fig. 11.15 The dependence of the wall potential on the pressure, where the VHF power is 450 W.

obtained becomes poor. According to the probe theory, the wall potential V_w is given by [24,26]

$$V_w = \frac{\kappa T_e}{q} \ln\left[\frac{4}{\sqrt{e}}\left(\frac{\pi m_e}{8 m_i}\right)^{1/2}\right] \tag{11.1}$$

Here, κ and q are the Boltzmann constant and electron charge, respectively, and m_e and m_i are the mass of electron and ion, respectively. As seen from Eq. (11.1), since the electron temperature is almost constant at high pressure in this experiment, the wall potential should be constant at high pressure. The wall potential was measured as a function of pressure, where the theoretical curve calculated from Eq. (11.2) is plotted in Fig. 11.15, indicating that observed wall potentials are anomalously lower than the theoretical values above 2 torr. This is due to the existence of negative ions. Although the electron temperature in hydrogen plasma is high at high pressure, the wall potential is quite low due to the production of negative ions. Thus, the VHF plasma at high pressure provides high-quality films. However, as already mentioned, the present electron density is not enough for obtaining a high deposition rate of μc-Si:H. Furthermore, the small amount of addition of SiH_4 gas was attempted and the plasma parameters were measured with a heated Langmuir probe. The results showed almost the same tendency as those of hydrogen gas plasma. As seen in Figs. 11.14 and 11.15, it was found that negative ions exist in the hydrogen plasma produced by the narrow-gap VHF discharge. Identification of the negative ions should be necessary for understanding such a plasma, which is work for the future.

References

1 Curtins, H., Wyrsch, N., Favre, M. and Shah, A.V. (1987) *Plasma Chem. Plasma Process,* **7**, 267.

2 Oda, S., Noda, J. and Matsumura, M. (1990) *Jpn. J. Appl. Phys.,* **29**, 1889.

3 Howling, A.A., Dorier, J.-L., Hollenstein, Ch., Kroll, U. and Finger, F. (1992) *J. Vac. Sci. Technol.,* A **10**, 1080.

4 Heintze, M., Zedlitz, R. and Bauer, G.H. (1993) *J. Phys. D: Appl. Phys.,* **26**, 1781.

5 Heintze, M. and Zedlitz, R. (1996) *J. Non-Cryst. Solids,* **198–200**, 1038.

6 Schwarzenbach, W., Howling, A.A., Fivaz, M., Brunner, S. and Hollenstein, Ch. (1996) *J. Vac. Sci. Technol.,* A **14**, 132.

7 Sansonnens, L., Pletzer, A., Magni, D., Howling, A.A., Hollenstein, Ch. and Schmitt, J.P.M. (1997) *Plasma Sources Sci. Technol.,* **6**, 170.

8 Kuske, J., Stephan, U., Steinke, O. and Rohlecke, S. (1995) *Mater. Res. Soc. Symp. Proc.,* **27**, 377.

9 Lieberman, M.A., Booth, J.P., Chabert, P., Rax, J.M. and Turner, M.M. (2002) *Plasma Sources Technol.,* **11**, 283.

10 Schmidt, H., Sansonnens, L., Howling, A.A., Hollenstein, Ch., Elyaakoubi, M. and Schmitt, J.P.M. (2004) *J. Appl. Phys.,* **95**, 4559.

11 Sansonnens, L. and Schmitt, J. (2003) *App. Phys. Lett.,* **82**, 182.

12 Takatsuka, H., Yamauchi, Y., Takeuchi, Y., Urabe, S. and Kawai, Y. (2004) *Proc. XV Int. Conf. on Gas Discharges and Their Application, Toulouse, Sept.,* Vol. 2, 645.

13 Takatsuka, H., Yamauchi, Y., Takeuchi, Y., Mashima, H., Yamashita, H. and Kawai, Y. (2005) *Jpn. J. Appl. Phys.,* **44**, L38.

14 Kondo, M., Fukawa, M., Guo, L. and Matsuda, A. (2000) *J. Non-Cryst. Solids,* **266–269**, 84.

15 Isomura, M., Kondo, M. and Matsuda, A. (2002) *Jpn. J. Appl. Phys.,* **41**, 1947.

16 Graf, U., Meier, J., Kroll, U., Bailat, J., Droz, C., Vallat-Sauvain, E. and Shah, A. (2003) *Thin Solid Films,* **427**, 37.

17 Murata, M., Mashima, H., Yoshioka, M., Nishida, S., Morita, S. and Kawai, Y. (1997) *Jpn. J. Appl. Phys.,* **36**, 4563.

18 Takeuchi, Y., Murata, M., Uchino, S. and Kawai, Y. (2001) *Jpn. J. Appl. Phys.,* **40**, 3405.

19 Mashima, H., Takeuchi, Y., Noda, N., Murata, M., Naitou, H., Kawasakio, I. and Kawai, Y. (2003) *Surf. Coat. Technol.,* **171**, 167.

20 Takeuchi, Y., Nawata, Y., Ogawa, K., Serizawa, A., Yamauchi, Y. and Murata, M. (2001) *Thin Solid Films,* **386**, 133.

21 Makabe, T. (1998) 45th Spring Meeting, Japan Society of Applied Physics and Related Societies, Ext. Abstr., 29p-ZR-4 (in Japanese).

22 Schulz, G.J. (1959) *Phys. Rev.,* **113**, 816.

23 Rapp, D., Sharp, T.E. and Griglia, D.D. (1965) *Phys. Rev. Lett.,* **14**, 533.

24 Amemiya, H. (1990) *J. Phys.,* D **23**, 999.

25 Bacal, M. (2000) *Rev. Sci. Instrum.,* **71**, 3981.

26 St. Brithwaite, N. and Allen, J.E. (1988) *J. Phys.,* D **21**, 1733.

12

Growth Control of Clusters in Reactive Plasmas and Application to High-Stability a-Si:H Film Deposition

Y. Watanabe, M. Shiratani, and K. Koga

12.1
Introduction

It is well known that particles are generated due to homogeneous or inhomogeneous processes in reactive plasmas. For the former process, nucleation and subsequent growth of particles are due to gas-phase reactions in the plasmas, and they usually tend to have a blueberry-like structure [1]. In contrast, for the latter process, particles originate from molecular or cluster-like species which are produced due to sputtering of electrodes or peeling off of films deposited on walls (mainly electrodes) in the reactor, and they usually tend to have a cauliflower-like structure [2].

Until now, the growth of particles in reactive plasmas has been studied mainly using silane (SiH_4) high-frequency capacitively coupled plasmas (HFCCP), because such kind of plasmas are indispensable to prepare electronic devices such as solar cells and thin-film transistors. The existence of particles in SiH_4 HFCCP was firstly reported by Roth *et al.* in 1985 [3]. They found using the laser light scattering method that the particles were mainly observed around the plasma/sheath (P/S) boundary.

The clue to clarifying the growth processes of particles and controlling their growth in SiH_4 HFCCP is given by Watanabe *et al.* [4,5]. They found that growth of particles in SiH_4 HFCCP was suppressed by modulating periodically the amplitude of the discharge voltage. This discharge modulation has provided not only a method for suppressing the growth of particles, but also for reproducing complicated processes of particle growth after discharge initiation with good accuracy. After this work, they proceeded to clarify the growth kinetics of particles in a range of sizes below a few tens of nanometers by developing various methods for observation of their growth [6–10].

The groups of Bouchoule *et al.*, Orleans University, France, and Hollenstein *et al.*, Ecole Polytechnique Federale Lausanne, Switzerland, also have paid great attention to clarifying the growth kinetics of particles and controlling their growth in SiH_4 HFCCP. The former group has carried out interesting studies regarding the effects of gas temperature and flow on growth of particles by developing a sophisticated reactor for studying the growth of particles in SiH_4 HFCCP [11–13]. The latter group has observed the growth of negatively charged species using a mass spectrometer which

Advanced Plasma Technology. Edited by Riccardo d'Agostino, Pietro Favia, Yoshinobu Kawai, Hideo Ikegami, Noriyoshi Sato, and Farzaneh Arefi-Khonsari
Copyright © 2008 WILEY-VCH Verlag GmbH & Co. KGaA, Weinheim
ISBN: 978-3-527-40591-6

can detect them in a wide mass range of 1–500 amu (or 1 to about 1300 amu in some cases). Their results have given the basis of a model that the particles grow through a series of reactions of negative ions which begin with SiH_3^- [14–18].

The growth processes of particles in SiH_4 HFCCP observed by these researchers can be classified into two phases: the initial growth phase in which the size is in the range below about 10 nm (such small particles are referred to as "clusters" hereafter); and the subsequent growth phase in which their size is in the range above about 10 nm. The former cluster growth phase will be further classified into two phases: the phase before nucleation in which the sizes of clusters are below about 0.5 nm almost corresponding to Si_nH_x ($n = 4$) (hereafter, such small clusters, Si_nH_x ($n = 2, 3, 4$), will be referred to as higher-order silanes (HOSs)); and the subsequent period after nucleation during which clusters continue to grow in the presence of SiH_x ($x = 2, 3$) and HOSs (sometimes, such nucleated clusters will be referred to as large clusters).

In this chapter, after reviewing the main studies carried out until now regarding the growth of clusters in Section 12.2, we explain such results using the simple model which takes into account the production and transport loss processes of clusters in Section 12.3. In Section 12.4, we show how to control the growth of clusters based on the knowledge of growth kinetics described in Sections 12.2 and 12.3, and then present both the relationship between the amount of a-Si clusters and the quality of a-Si:H films and realization of high-stability films for solar cells by suppressing their growth in Section 12.5.

12.2
Review of Cluster Growth Observation in SiH₄ HFCCP

Until now, we have studied the issues as described below regarding the growth processes of clusters by developing various new sophisticated methods for *in situ* observation of the time evolution of their size and density.

12.2.1
Precursor for Cluster Growth Initiation

To identify the particle species which contributes to cluster formation as precursor, two kinds of experiments have been carried out: (i) observation of the region where clusters begin to grow, between the powered (HF) and grounded (GND) electrodes; and (ii) observation of the spatial profiles of cluster amount, and density and generation rate of short-lifetime silane radicals between the HF and GND electrodes.

Concerning item (i), the spatial density profiles of Si_nH_x ($n < 10$) and Si_nH_x ($n < 200$) were observed at times after discharge initiation $T_{on} = 5, 10, 20,$ and 30 ms using the photodetachment method which utilized the dependence of electron affinity of clusters on their size and employed the second and third harmonics of a YAG laser (wavelengths of 532 and 355 nm) as light sources. The result, Fig. 12.1, shows that clusters begin to grow in the radical generation region around the P/S boundary near the HF electrode [19].

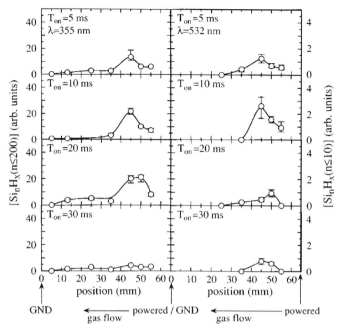

Fig. 12.1 Growth of clusters, Si_nH_x ($n < 10$) and Si_nH_x ($n < 200$), between HF and GND electrodes observed using dependence of cluster electron affinity on cluster size and photodetachment method with a YAG laser (wavelengths of 355, 532 nm). Experimental conditions: 3% SiH_4 + He, 80 Pa, 10 sccm, 6.5 MHz, 60 W.

Concerning item (ii), the spatial profiles of interest were measured with optical absorption spectroscopy using a visible laser and UV light (the former for cluster amount and SiH_2 density and the latter for Si density), optical emission spectroscopy of SiH and Si (radical generation rate), and the Langmuir probe method (positive- and negative-ion densities). The result, Fig. 12.2, shows the spatial profiles of cluster amount, SiH and Si emission intensities, and SiH_2 and Si densities between the HF and GND electrodes at 0.25, 0.5, and 0.75 s after discharge initiation [20]. The clusters tend to grow localizing around the P/S boundary near the HF electrode, and their spatial profiles are similar to those of short-lifetime radical (SiH_2 and Si) densities and also similar to Si_nH_x ($n < 10$) and Si_nH_x ($n < 200$) densities shown in Fig. 12.1. The density of SiH_2 is of the order of 10^9 cm^{-3} and quite low compared to that ($\sim 10^{11}$ cm^{-3}) of SiH_3 in spite of the fact that both species generate at almost the same rates due to dissociation of SiH_4, because the former short-lifetime radicals react rapidly with the parent molecules. Further, from our experimental results (not shown here), the spatial distribution of cluster amount is different from those of densities of negative ions, positive ions and SiH_3 densities, which are rather flat between the electrodes [20–22].

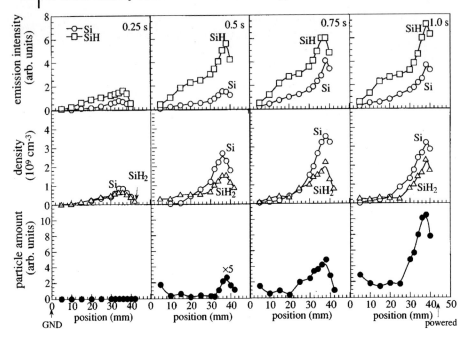

Fig. 12.2 Spatial profiles of radical generation rate (emission intensities of SiH, Si), radical densities (SiH$_2$, Si), and cluster amount at 0.25, 0.5, 0.75, and 1.0 s after discharge initiation. Experimental conditions: 10% SiH$_4$ + Ar, 13 Pa, 20 sccm, 6.5 MHz, 80 W.

From a series of such experimental results, we have concluded that, under our experimental conditions, the radical species mainly contributing to the initiation of cluster growth are SiH$_2$, which generate at quite a large rate compared to the other short-lifetime radicals.

12.2.2
Cluster Nucleation Phase

In our experiments carried out before, we found that particles of about 10 nm in size and with a density of about 10^{10} cm^{-3} existed at an early time of about 1 s after discharge initiation, which subsequently coagulated to grow to larger particles [23,7]. In order to clarify the process by which such 10 nm particles were produced, we developed a double pulse discharge (DPD) method for SiH$_4$ HFCCPs [10]. This method uses two pulse discharges: the main discharge of a period T_{on} to grow the clusters of interest, and the supplemental discharge of a very short period t_{on}, generated at t_{off} after the main discharge, which does not cause the generation of unnecessary clusters. The time evolution of size and density of clusters which is given as a function of T_{on} is deduced by observing the decay of electrons, produced by the

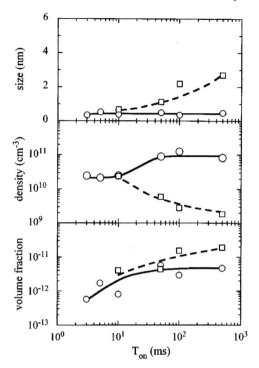

Fig. 12.3 Dependence of size, density, and volume fraction of clusters on discharge period T_{on}. Experimental conditions: 100% SiH$_4$, 5 sccm, 13.3 Pa, 13.56 MHz, 10 W, $T_g = $ RT.

supplemental pulse discharge, due to their attachment to the clusters as a parameter of t_{off}. The result, Fig. 12.3, shows that clusters nucleate at a size of about 0.5 nm, almost corresponding to Si$_n$H$_x$ ($n = 4$), and a HOS density of about 3×10^{10} cm^{-3}. Also, the HOS density reaches about 10^{11} cm^{-3} after nucleation, which is extremely high compared to that of positive ions, about 10^9 cm^{-3} [10]. Figure 12.3 also shows that, after nucleation, the size of nucleated clusters continues to grow, decreasing their density under the existence of HOSs and radicals SiH$_x$ ($x = 0$–3). Taking into account these results in addition to those shown in Fig. 12.2, it can be concluded that the SiH$_2$ insertion reactions are the most promising for production of HOSs under our experimental conditions. Further, we have experimentally shown that such nucleated clusters continue to grow toward particles of 10 nm in size.

12.2.3
Effects of Gas Flow on Cluster Growth

To study the effects of gas flow on cluster growth, we prepared a glass-tube reactor which had HF and GND electrodes made of stainless mesh and a gas flow in the

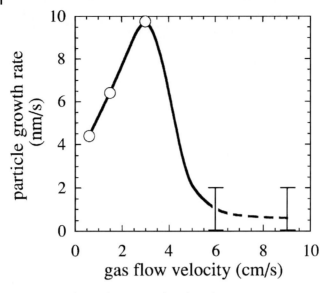

Fig. 12.4 Particle growth rate around P/S boundary near HF electrode as function of gas flow velocity. Experimental conditions: 5% SiH_4 + He, 2–30 sccm, 80 Pa, 6.5 MHz, 80 W.

direction from the HF electrode to the GND, and measured the particle growth rate, around the P/S boundary at which its rate was the highest between the electrodes, as a function of gas flow velocity using the laser light scattering (LLS) method. Figure 12.4 shows the result [24]. The result that the growth rate becomes maximum at a gas flow velocity $V_g = 3 \, \mathrm{cm \, s^{-1}}$ can be explained by taking into account the production and transport loss of clusters in the radical generation (RG) region near the HF electrode.

12.2.4
Effects of Gas Temperature Gradient on Cluster Growth

When depositing a-Si:H, the substrate is heated to around 250 °C to obtain a good-quality films. This heating usually causes a gradient of gas temperature in the direction from the HF electrode to the GND equipped with the substrate, inducing a force (thermophoretic force) driving some fraction of clusters in the direction against the gradient. Figure 12.5 shows spatial profiles of the particle amount as a parameter of GND electrode temperature T_g in the case of HF electrode temperature $T_H = $ room temperature (RT) [25]. The particles which localize around the P/S boundary near the HF electrode for $T_g = $ RT are driven toward the HF electrode with increasing T_g. For $T_g = 300$ °C, neutral particles are driven out of the discharge space and only negatively charged particles exist around the P/S boundary near the HF electrode. Based on the result in Fig. 12.5, we can estimate a size above which particles are driven toward the HF electrode by taking into account a diffusive force against the thermophoretic

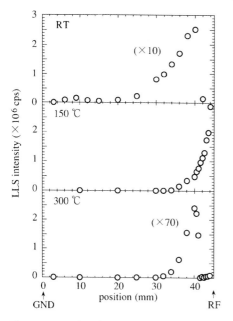

Fig. 12.5 Spatial profiles of particle amount (LLS intensity) at GND electrode temperature T_g = RT, 150 °C and 300 °C. Experimental conditions: HF electrode temperature = RT, 100% SiH$_4$, 5 sccm, 13 Pa, 8 W (0.1 W cm^{-2}), T_{on} = 0.8 s.

one. As shown in Fig. 12.5, the density gradient of particles which leads to the diffusive force toward the grounded electrode has a characteristic length of about 1 cm in our case. For a temperature gradient of 50 K cm^{-1} between the electrodes, which is a plausible value in the case of depositing device-quality a-Si:H films, the particle size of interest is estimated to be about 2 nm [25].

12.2.5
Effects of H$_2$ Dilution on Cluster Growth

Diluting SiH$_4$ gas with H$_2$ has often been employed to prepare high-quality a-Si:H films. We have studied the effect on cluster growth as a parameter of H$_2$ partial pressure. Figure 12.6 shows the spatial profiles of the cluster amount (LLS intensity) and generation rates of hydrogen and silane radicals between the HF and GND electrodes at 0.2, 0.4, and 0.8 s after discharge initiation for 100% SiH$_4$ and 20% SiH$_4$ + H$_2$ [25]. For the 100% SiH$_4$ case, all the spatial profiles have the maximum values at around the P/S boundary near the HF electrode. On the other hand, for the 20% SiH$_4$ + H$_2$ case, the maximum growth of clusters is observed around the P/S boundary near the GND electrode, while the generation rates of H atoms and silane radicals are maximum around the P/S boundary near the HF electrode. Further, from our experimental result (not shown here), the size growth rate of clusters has been

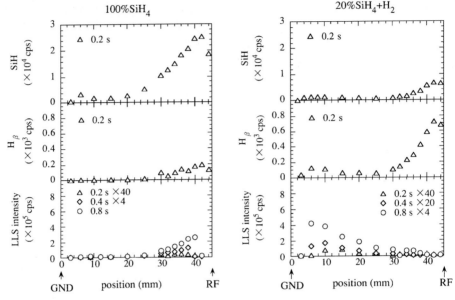

Fig. 12.6 Spatial profiles of cluster amount (LLS intensity) and generation rates of silane radicals and hydrogen (SiH and H emission-intensities) between the HF and GND electrodes at 0.2, 0.4 and 0.8 s after discharge initiation for 100% SiH$_4$ and 20% SiH$_4$ + 80% H$_2$. Experimental conditions: flow rate 5 sccm, pressure 13 Pa, temperature of both electrodes RT for 100% SiH$_4$; flow rate 25 sccm, total pressure 65 Pa, temperature of both electrodes RT for 20% SiH$_4$.

found to decrease with the decrease in the concentration ratio of SiH$_4$ to H$_2$ gas, while their density is almost independent of its decrease [26]. These results suggest that hydrogen species, may be H atoms, are effective for suppressing the size growth of clusters.

12.2.6
Effects of Discharge Modulation on Cluster Growth

We found that discharging on and off periodically (discharge modulation) is quite effective for suppressing particle growth [4]. To clarify this mechanism, we developed a method for observing the time evolution of size and density of clusters which utilized both dependence of their diffusion on their size and dependence of electron attachment to them on their density. Figure 12.7 shows the cluster density as a function of power-off period t_{off} in one modulation cycle and as a function of power-on period t_{on} in one modulation cycle which is related to the size (number of Si atoms n) of Si$_n$H$_x$ clusters grown during t_{on} [9]. In the figure, the smaller clusters (corresponding to the shorter t_{on}) begin to decrease at a shorter t_{off} because of their faster diffusion. From this result, we have found the important fact that a t_{off} value at which the density of clusters begin to decrease with increasing t_{off} under a constant t_{on} (e.g. $t_{on} = 1$ ms) is almost equal to the time within which they diffuse out of the RG

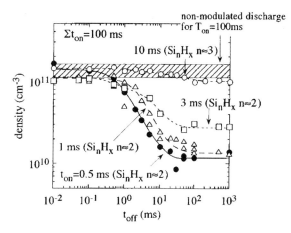

Fig. 12.7 Dependence of cluster density on t_{off} as a function of t_{on}. Experimental conditions: 100% SiH_4, 5 sccm, 13 Pa, 14 MHz, 40 W, total discharge time T_{on} (summation of t_{on} values) = 100 ms.

region near the HF electrode. Thus we conclude that growth suppression by the modulation is brought about by the fact that the clusters grown in the RG region during t_{on} diffuse out of the region during t_{off}.

12.3
Cluster Growth Kinetics in SiH₄ HFCCP

Up to now, two reaction processes have been proposed for cluster growth: one are the neutral SiH_2 insertion reactions, and the other are the negative ion reactions initiated by SiH_3^-. As described in Section 12.2, the results obtained under our experimental conditions carried out until now support that the former SiH_2 insertion reactions participate in cluster formation. In addition, we have obtained experimental results that the density ratio of clusters to positive ions N_i decreases significantly when increasing the excitation frequency from RF to VHF. The decease in electron temperature T_e due to the excitation frequency should bring about the increase in electron attachment rate to silane radicals, resulting in the increase in SiH_3^- density. If SiH_3^- is a precursor of clusters, the density ratio of clusters to N_i should increase on increasing the excitation frequency [27,28].

It has been pointed out that while SiH_2 reacts rapidly with SiH_4, the subsequent reactions of SiH_2 with Si_2H_6, which are considered to be induced by three-body collisions, should not be so fast under such low-pressure conditions as in our experiments. However, for the experiment carried out using SiH_4 HFCCP in a microwave cavity by Nomura *et al.*, many Si_nH_x ($n > 2$) clusters, the density of which is above 10^{11} cm^{-3}, are observed in the case of pressure below 0.1 torr [29]. This result suggests that, even in our case, the clusters are possible to grow through the insertion reactions, while their growth is also affected by their transport loss due to diffusion and gas flow. From the result shown in Fig. 12.3, such insertion reactions are considered to be needed for producing HOSs which induce nucleation.

Perrin and Hollenstein calculated the effective electron attachment rate on Si clusters by taking into account the electron attachment cross-section, electron affinity, and ionization potential of clusters as functions of their size and T_e [30]. As shown in Fig. 2.16 in [30], the effective attachment rate at $T_e = 2.5$ eV, typical for SiH$_4$ HFCCPs, increases drastically around Si$_n$H$_x$ ($n = 4$) with cluster growth. This cluster size of interest almost coincides with that of nucleation shown in Fig. 12.3, suggesting that the rapid increase in density of Si$_n$H$_x$ ($n = 4$) due to their trapping caused by such enhancement of electron attachment triggers cluster nucleation.

In order to study the effects of transport loss of clusters on their growth, we consider rate equations of SiH$_3$, related to film deposition, and Si$_n$H$_x$ ($n < 20$) clusters in the plasma with the gas flow from the HF mesh electrode to the GND mesh as shown in Fig. 12.8(a), where a cluster size of $n = 20$ corresponds to almost 0.7 nm. While effects of charging and additional growth reactions besides the SiH$_2$ insertion ones should be taken into account in a size range of $n > 4$ as described above, we here neglect these effects for simplicity. In this simple model, the steady-state rate equations of SiH$_3$ and Si$_n$H$_x$ ($n = 2, 4, 20$) are expressed as follows:

$$k_d n_e [\text{SiH}_4] - k_r [\text{SiH}_3]^2 - (D_1/L^2)[\text{SiH}_3] - [\text{SiH}_3]/T_{\text{res}} = 0 \text{ for SiH}_3$$

$$k_{n1}[\text{SiH}_2][\text{Si}_{n-1}\text{H}_{x-2}] - (D_n/L^2)[\text{Si}_n\text{H}_x] - [\text{Si}_n\text{H}_x]/T_{\text{res}} = 0 \text{ for Si}_n\text{H}_x (n = 2, 4, 20)$$

where [ ] means the density of species in the brackets, k_d is the SiH$_3$ production rate due to gas dissociation by electron collisions, k_r the SiH$_3$ recombination rate, k_{n1} the reaction rate of Si$_n$H$_x$ with SiH$_2$ (SiH$_2$ insertion reactions), D_1 and D_n the diffusion

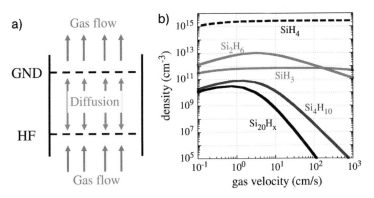

Fig. 12.8 Effects of gas flow on growth of clusters. (a) SiH$_4$ HFCCP reactor with gas flow used in simplified model. (b) Densities of SiH$_4$, SiH$_3$, Si$_2$H$_6$, Si$_4$H$_{10}$, and Si$_{20}$H$_x$ as function of gas velocity.

coefficients of SiH_3 and Si_nH_x ($n = 2, 4, 20$), respectively, L the characteristic diffusion length, and T_{res} the gas residence time. The calculation was carried out for 100% SiH_4, pressure 13.3 Pa, gas temperature 150 °C, gap between the electrodes 3 cm, and electrode diameter 10 cm. In the calculation, we gave values of $[SiH_3]$, $[Si_2H_6]$, $[Si_4H_{10}]$ and $[Si_{20}H_x]$ at $V_g = 3\,cm\,s^{-1}$ which were measured with a quadrupole mass spectrometer and by the DPD method in our experiments. Figure 12.8(b) shows the densities of SiH_4, SiH_3, Si_2H_6, Si_4H_{10}, and $Si_{20}H_x$ as a function of gas flow velocity V_g. In the range of V_g below a few centimeters per second, that is, $T_{res} >$ about 1 s, the loss of clusters is dominated by their diffusion and the Si_nH_x density increases with V_g. This increase is due to the increase in the amount of gas supplied to the plasma. On the other hand, in the range of V_g more than a few centimeters per second, that is, $T_{res} <$ about 1 s, their loss is affected by the gas velocity and their densities decrease with increasing V_g. Since most of our experiments have been carried out in a range of V_g below a few centimeters per second, the loss of clusters is due to their diffusion. Further, as can be seen in Fig. 12.3, the time for clusters to grow to $Si_{20}H_x$ of about 0.7 nm in size is about 20 ms. This means that the characteristic cluster growth time T_c is quite short compared to T_{res} in such a low V_g range. We also have found that the clusters grown under such conditions have an amorphous structure.

For the experiments carried out by the Orleans group, the following values are obtained from their paper: gap length 3.3 cm, electrode diameter 13 cm, 3% $SiH_4 + Ar$, 0.117 torr [11]. In this case, V_g is estimated to be about $30\,cm\,s^{-1}$ ($T_{res} \sim 0.1\,s$) and SiH_4 gas is considered to be highly dissociated. Under such condition that the transport loss of clusters is dominated by the gas flow, the densities of clusters which grow through the insertion reactions are considered to be too low to bring about their nucleation. Hence, the charging of clusters becomes important for their growth, while it is not clear yet regarding which of $SiH_3{}^-$ and SiH_2 are precursor for them. In their case, they report that the time for clusters to grow to 1 nm is of the order of 0.1 s, being close to T_{res} [31], and the clusters of a few nanometers in size have a crystalline structure. For the experiments carried out by the Lausanne group, V_g is estimated to be over $100\,cm\,s^{-1}$. Thus, while 100% SiH_4 is used in their case, the situation is similar to the Orleans case and negative ion reactions are inevitable for cluster growth [16,32].

12.4
Growth Control of Clusters

As discussed, the growth of clusters can be controlled by changing various parameters such as supplied HF discharge power, gas residence time, gas temperature, concentration ratio of material to dilution gas, and dissociation degree of material gas. The structure of clusters has been found to change with the concentration ratio of SiH_4 to dilution gas such as Ar and H_2. When the ratio is not so small, they tend to have an amorphous structure; when it is quite small, they tend to have a crystalline structure. Such structural differences may be caused by difference of cluster temperature values for both cases.

In this section, our discussion is focused on suppressing the growth of a-Si clusters in SiH_4 HFCCP in the case of $T_c < T_{res}$, because the reactors employed for deposition of a-Si:H films in industry are usually large and hence such a condition is easily satisfied.

12.4.1
Control of Production Rate of Precursor Radicals

The best way for suppressing the growth of a-Si clusters is to decrease the production rate of precursor radicals SiH_2 by decreasing the electron temperature T_e. The decrease in T_e leads to the decrease in density ratio of SiH_2, contributing to HOS production, to SiH_3, contributing to good-quality a-Si:H deposition. It is well known that the increase of excitation frequency from RF to VHF brings about not only a decrease in T_e but also an increase in plasma density. Thus, recently VHF discharges have been favorably compared to the RF ones for the preparation of a-Si:H films of higher quality at a high deposition rate.

12.4.2
Control of Growth Reactions and Transport Loss of Clusters

The flow and temperature of the gas are also important parameters to control the growth of clusters, which are related to controlling the rates of growth reaction and transport loss of clusters. The size and even structure can be controlled by changing the relationship between T_{res} and T_c as described above. Gas heating has been pointed out to affect the reaction rates for cluster growth [11,31], and the thermophoretic force due to gas temperature gradient induced in the discharge space is useful to limit the upper size of clusters [24].

The periodical discharge modulation affects the production of radical species as precursors of clusters and their loss due to diffusion and/or gas flow, resulting in controlling the cluster growth. The combination of discharge modulation and gas temperature gradient is quite useful for suppressing cluster growth [24,32]. While its mechanism is not clear yet, the dilution of SiH_4 with H_2 is also effective in suppressing the growth of cluster size.

12.5
Application of Cluster Growth Control to High-Stability a-Si:H Film Deposition

As described above, the structure of clusters depends on both the concentration ratio of SiH_4 to dilution gas such as H_2 and Ar and the gas residence time in the reactor. Rocca *et al.* have fabricated a-Si:H solar cells under the conditions that clusters of crystalline structure are generated, and found that those cells do not have the light-induced degradation peculiar to such types of solar cell [35].

Recently, Matsuda and co-workers have reported that, under the experimental conditions for which a-Si clusters are grown, the decrease in hydrogen content,

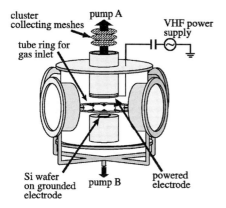

Fig. 12.9 Cluster-suppressed plasma CVD reactor developed for deposition of high-quality a-Si:H.

$C_{H(SiH2)}$, associated with Si–H$_2$ bonds in the a-Si:H films results in the decrease in degree of the light-induced degradation, and $C_{H(SiH_2)}$ is closely related to the amount of HOSs Si$_n$H$_x$ ($n = 2$, 3, 4) [36,37]. As shown in Fig. 12.3, large a-Si clusters of size about 0.5 nm coexist with the HOSs. This suggests that such large clusters may be closely related to $C_{H(SiH_2)}$. Thus, we have studied the relationship between the amount of clusters and quality of deposited a-Si:H films, and tried to prepare high-stability a-Si:H films for solar cells by suppressing the growth of a-Si clusters.

For this purpose, we have developed the reactor, as shown in Fig. 12.9, in which the amount of a-Si clusters is suppressed by (a) utilizing the gas flow across the RG region near the HF electrode, (b) inducing the thermophoretic forces from the GND electrode to the HF, (c) removing stagnation regions of gas flow in the reactor, and (d) employing VHF discharges (60 MHz) [38]. The material gas passed through the RG region was evacuated through the HF electrode which is made of stainless mesh. The stagnation regions were removed by evacuating through four additional pumping ports on the side wall of the reactor. In some experiments, such a reactor of diode-discharge type was operated as triode-discharge type by employing the GND electrode made of stainless mesh and placing the substrate on the other side of plasma. The amount of clusters was observed using the downstream cluster collection (DCC) and the photon-counting LLS methods developed by us. Especially, the former method has a high sensitivity and can detect clusters of size down to about 1 nm. $C_{H(SiH_2)}$ was deduced from profiles of Si–H$_2$ and Si–H absorbance bands centered at wavenumbers of 2090 and 2000 cm^{-1} which are obtained using Fourier transform infrared (FTIR) spectroscopy. The results are shown in Fig. 12.10. $C_{H(SiH_2)}$ is found to decrease almost linearly with decreasing volume fraction of clusters incorporated into the films, which is given by the volume of clusters of interest [39], and its value decreases from about 1 at.%, which is the value of device-quality films employed for conventional solar cells, to about 0.05 at.% [39]. Further, we fabricated

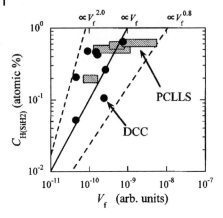

Fig. 12.10 Dependence of hydrogen content $C_{H(SiH_2)}$ associated with Si–H$_2$ bonds in a-Si:H films on volume fraction V_f of clusters incorporated into films. Solid circles and hatched rectangles indicate results obtained using DCC and PCLLS (photon-counting LLS) methods respectively. Experimental conditions: 100% SiH$_4$, 30 sccm, 9.3 Pa, $T_g = 250\,°C$, 60 MHz, 3.5–10 W.

Shottkey cells using the a-Si:H films deposited with the reactors in order to evaluate their fill factors before light soaking and after 7.5 hours light soaking (FF$_{in}$ and FF$_{st}$ respectively). Table 12.1 gives FF$_{in}$, FF$_{st}$, and degradation ratio (FF$_{in}$ − FF$_{st}$)/ FF$_{in}$ obtained using the cluster-suppressed diode-type and triode-type reactors together with those for conventional diode-type plasma CVD. The amount of large clusters for the triode-type has been found to be much less than that for the diode-type. FF$_{in}$ is much improved from 0.53 for conventional diode-type to 0.60 for triode-type by decreasing the amount of large clusters. The former FF$_{in}$ = 0.53 corresponds to a PIN single solar cell efficiency of 12%. The degradation ratio is also improved with the decrease in cluster amount. The degradation ratio for the triode-type reactor is a small value of 6.7%, while this value is 17% for conventional diode-type reactors [40].

However, further studies are necessary regarding which large clusters and HOSs are the main origin of Si–H$_2$ bonds of interest, because the latter may decreases with decreasing amount of large clusters.

Tab. 12.1 FF$_{in}$, FF$_{st}$, degradation ratio for cluster-suppressed triode-type, cluster-suppressed diode-type, and conventional plasma CVD reactors.

Deposition method	Initial FF	Stabilized FF	Degradation rate (%)
Cluster-suppressed triode plasma CVD	0.60	0.56	6.7
Cluster-suppressed diode plasma CVD	0.56	0.49	12.5
Conventional diode plasma CVD	0.53	0.44	17.0

12.6
Conclusions

The growth processes of clusters (particles in the size range below about 10 nm) in SiH$_4$ HFCCP have been understood fairly well by taking into account their production and transport loss processes. Based on this knowledge, it has been discussed that the growth control of clusters is possible as regards size, density, and structure.

The growth of a-Si clusters can be suppressed by utilizing the gas-drag and thermophoretic forces together with removing the stagnation regions of gas flow. Growth suppression is quite effective in improving quality of a-Si:H films for solar calls. Both $C_{H(SiH_2)}$, associated with Si–H$_2$ in deposited a-Si:H films, and the initial and stabilized fill factors, FF$_{in}$ and FF$_{st}$, of a Shottky cell fabricated using the films have been much improved by using the newly developed cluster-suppressed reactors with triode- and diode-type discharges. $C_{H(SiH_2)}$ decreases down to 0.05 at.% in contrast to about 1 at.% for the diode-type reactors which are conventionally employed to prepare device-quality a-Si:H films. Regarding the fill factors, FF$_{in}$ increases from the highest value of 0.53 for conventional diode-type reactors to 0.60 for the triode-type, and the degradation ratio decreases from the highest value of 17% for the conventional reactor to 6.7% for the triode-type.

These results indicate that the stability of a-Si:H films for solar cells will be further improved by decreasing the amount of clusters.

References

1 Shiratani, M., Kawasaki, H., Fukuzawa, T., Tsuruoka, H., Yoshioka, T. and Watanabe, Y. (1994) *Appl. Phys. Lett.*, **65**, 1900.

2 Garscadden, A., Ganguly, B.N., Haaland, P.D. and Williams, J. (1994) *Plasma Sources Sci. Technol.*, **3**, 239.

3 Roth, R.M., Spears, K.G., Stein, G.D. and Wong, G. (1985) *Appl. Phys. Lett.*, **46**, 253.

4 Watanabe, Y., Shiratani, M., Kubo, K., Ogawa, I. and Ogi, S. (1988) *Appl. Phys. Lett.*, **53**, 1263.

5 Watanabe, Y., Shiratani, M. and Makino, S. (1989) *Proc. 9th Int. Symp. on Plasma Chemistry*, Pugnochiuso, Italy, **2**, 1329.

6 Watanabe, Y., Shiratani, M. and Yamashita, M. (1992) *Appl. Phys. Lett.*, **61**, 1510.

7 Shiratani, M., Kawasaki, H., Fukuzawa, T., Yoshioka, T., Ueda, Y., Singh, S. and Watanabe, Y. (1996) *J. Appl. Phys.*, **79**, 104.

8 Shiratani, M. and Watanabe, Y. (1998) *Rev. Laser Eng.*, **26**, 449.

9 Fukuzawa, T., Kushima, S., Matsuoka, Y., Shiratani, M. and Watanabe, Y. (1999) *J. Appl. Phys.*, **86**, 3543.

10 Koga, K., Matsuoka, Y., Tanaka, K., Shiratani, M. and Watanabe, Y. (2000) *Appl. Phys. Lett.*, **77**, 196.

11 Bouchoule, A., Plain, A., Boufendi, L., Blondeau, J.Ph. and Laure, C. (1991) *J. Appl. Phys.*, **70**, 1991.

12 Fridman, A.A., Boufendi, L., Hbid, T., Potapkin, B.V. and Bouchoule, A. (1996) *J. Appl. Phys.*, **79**, 1303.

13 Boufendi, L. and Bouchoule, A. (1996) *J. Vac. Sci. Technol.*, **A14**, 572.

14 Howling, A.A., Dorier, J.-L. and Hollenstein, Ch. (1993) *Appl. Phys. Lett.*, **62**, 1341.

15 Howling, A.A., Sansonnens, L., Dorier, J.-L. and Hollenstein, Ch.

(1993) *J. Phys. D: Appl. Phys.*, **26**, 1003.

16 Howling, A.A., Sansonnens, L., Dorier, J.-L. and Hollenstein, Ch. (1994) *J. Appl. Phys.*, **75**, 1340.

17 Courteille, C., Dorier, J.-L., Hollenstein, Ch., Sansonnens, L. and Howling, A.A. (1996) *Plasma Sources Sci. Technol.*, **5**, 210.

18 Hollenstein, Ch., Scwarzenbach, W., Howling, A.A., Courteille, C., Dorier, J.-L. and Sansonnens, L. (1996) *J. Vac. Sci. Technol.*, **A14**, 535.

19 Fukuzawa, T., Obata, K., Kawasaki, H., Shiratani, M. and Watanabe, Y. (1996) *J. Appl. Phys.*, **80**, 3202.

20 Watanabe, Y., Shiratani, M., Fukuzawa, T., Kawasaki, H., Ueda, Y., Singh, S. and Ohkura, H. (1996) *J. Vac. Sci. Technol.*, **A14**, 995.

21 Fukuzawa, T., Shiratani, M. and Watanabe, Y. (1994) *Appl. Phys. Lett.*, **64**, 3098.

22 Kawasaki, H., Ohkura, H., Fukuawa, T., Shiratani, M., Watanabe, Y., Yamamoto, Y., Suganuma, S., Hori, M. and Goto, T. (1997) *Jpn. J. Appl. Phys.*, **36**, 4985.

23 Shiratani, M., Kawasaki, H., Fukuzawa, T., Tsuruoka, H., Yoshioka, T. and Watanabe, Y. (1994) *Appl. Phys. Lett.*, **65**, 1900.

24 Matsuoka, Y., Shiratani, M., Fukuzawa, T., Watanabe, Y. and Kim, K. (1999) *Jpn. J. Appl. Phys.*, **38**, 4556.

25 Shiratani, M., Sakamoto, K., Maeda, S., Koga, K. and Watanabe, Y. (2000) *Jpn. J. Appl. Phys.*, **39**, 287.

26 Watanabe, Y., Shiratani, M. and Koga, K. (2001) *Phys. Scr.*, **T89**, 29.

27 Watanabe, Y., Harikai, A., Koga, K. and Shiratani, M. (2002) *Proc. 5th Int. Conf. on Reactive Plasmas/16th European Conf. on Atomic and Molecular Physics of Ionized Gases*, **2**, 329.

28 Shiratani, M., Koga, K., Harikai, A., Ogata, T. and Watanabe, Y. (2003) *Matr. Res. Soc. Symp. Proc.*, **762**, A9.5.1.

29 Nomura, H., Kono, A. and Goto, T. (1996) *Jpn. J. Appl. Phys.*, **35**, 3603.

30 Perrin, J. and Hollenstein, Ch. (1999) in *Dusty Plasmas* (ed. A. Bouchoule), Wiley, Chichester, **77**.

31 Fridman, A.A., Boufendi, L., Hbid, T., Potapkin, B.V. and Bouchoul, A. (1996) *J. Appl. Phys.*, **79**, 1303.

32 Bhandarkar, U.V., Swihart, M.T., Girshick, S.L. and Kortshagen, U.R. (2000) *J. Phys. D: Appl. Phys.*, **33**, 2731.

33 Bhandarkar, U., Kortshagen, U. and Girshick, S.L. (2003) *J. Phys. D: Appl. Phys.*, **36**, 1399.

34 Watanabe, Y., Shiratani, M., Fukuzawa, T. and Koga, K. (2000) *J. Tech. Phys.*, **41**, 505.

35 Roca, P., Cabarrocas, I., Hamma, S., Sharma, S., Viera, G., Bertran, E. and Costa, J. (1998) *J. Non-Cryst. Solids*, **227–230**, 871.

36 Nishimoto, N., Takai, M., Miyahara, H., Kondo, M. and Matsuda, a. (2002) *J. Non-Cryst. Solids*, **299**, 1116.

37 Shimizu, S., Miyahara, H., Shimosawa, M., Kondo, M. and Matsuda, A. (2003) *Proc. 3rd World Conf. on Photovoltanic Energy Conversion*, 5P-A9-03.

38 Koga, K., Kai, M., Shiratani, M., Watanabe, Y. and Shikatani, N. (2002) *Jpn. J. Appl. Phys.*, **41**, L168.

39 Koga, K., Kaguchi, N., Shiratani, M. and Watanabe, Y. (2004) *J. Vac. Sci. Technol.*, **A22**, 1536.

40 Shiratani, M., Koga, K., Kaguchi, N., Bando, K. and Watanabe, Y. (2004) *Proc. 7th Asian Pacific Conf. on Plasma Science and 17th Symp. on Plasma Science for Materials*, I-A10.

13

Micro- and Nanostructuring in Plasma Processes for Biomaterials: Micro- and Nano-features as Powerful Tools to Address Selective Biological Responses

E. Sardella, R. Gristina, R. d'Agostino, and P. Favia

An in-depth investigation of *in vitro* and *in vivo* biological responses to biomaterials has clarified that materials properties such as surface charge, surface energy, material stiffness, and porosity can regulate different biological behaviors that range from cell proliferation, migration, and differentiation to the precise control of protein adsorption. By integrating these factors into the design of substrate for cell/protein patterning, we can engineer a micro- and nano-environment that can have potential applications in several biomedical fields. Plasma processes can in fact be considered one of the most successful methods to texture material surfaces. They can produce complex patterning in a single step, when complex mechanisms of reactions are active, or can be part of a more elaborate patterning procedure. The main advantage of plasma processes in this field is the possibility to change chemistry and/or topography in an independent way in order to study the effect of topography disentangled from chemistry on cell adhesion or protein adsorption. In this way by recognizing the types of topography/chemistry that are favorable for preserving desired protein adsorption and cell behavior it is possible to manufacture functional patterns for crucial biomedical applications.

13.1
Introduction: Micro and Nano, a Good Point of View in Biomedicine

Cell-based artificial organs, cellular biosensors, "smart" biomaterials, and tissue engineering are nowadays intensely pursued in many disciplines to improve the quality of human life. New interesting insights emerging from a close cooperation of material science and biology show that, by carefully controlling surface chemistry and micro-/nano-topography of material surfaces, one can design devices that enhance a selective cell population to grow (or specific biomolecules to adsorb) in specific regions of a device [1,2]. Over the past two decades these results have attracted interest from governments, private enterprises, and academic researchers. The remarkable recognition capabilities of cells and biomolecules can lead to novel tissue substitutes, biological electronics like biosensors, diagnostic systems,

Advanced Plasma Technology. Edited by Riccardo d'Agostino, Pietro Favia, Yoshinobu Kawai, Hideo Ikegami, Noriyoshi Sato, and Farzaneh Arefi-Khonsari
Copyright © 2008 WILEY-VCH Verlag GmbH & Co. KGaA, Weinheim
ISBN: 978-3-527-40591-6

and drug delivery systems. Although micro- and nano-biotechnology is a recent concept this field benefits from the advances made in the semiconductor industry in the last half century for the production of microprocessors, mostly due to plasma processes.

Before describing the technologies of pattering and their application in biomedical field, it is quite interesting to introduce some definitions about irregular surfaces. Hansen and von Recum [9] defined surface irregularities as deviations from a geometrically ideal (flat) surface. From these words it is possible to gather that all material surfaces can be considered irregular at microscopic level. One of the possible surface irregularities, beside flatness, waviness, and roundness, is defined as *roughness*, i.e. the surface topography on which the space between two hills is 5 to 100 times larger than the depth [1]. The roughness can be periodic or random. When surface roughness is periodic it is defined as *surface texture*, which is represented by spatially, arranged micro- and nanofeatures. Often surface texture is used as synonymous of *surface pattern*, including also surfaces characterized by chemically different domains regularly distributed on the substrate without changing surface morphology. Another surface irregularity is *porosity*. It can be a surface property or can be considered as a bulk property when pores cross through the bulk of the substrate, like in membranes or in three-dimensional (3D) scaffolds for cell seeding. Beside bulk porosity, another example of bulk irregularity is the dispersion of a material in another chemically different from the first one (composite materials). Often such kinds of dispersions are embedded in the substrate as nanofeatures; they can be released in contact with a fluid [3], or can contribute to increase the hardness of a material [4–8]. Rough, porous, and patterned materials are widely used in biomaterial engineering.

Due to the wide range of applications of micro- and nano-patterned materials, and to the large variety of technologies used to obtain micro- and nanofeatures, the authors suggest classifying patterned surfaces into three categories, according to their final application.

1. *Supports for cell seeding*, whose main application is represented by tissue engineering. Basically, when a surface is left in contact with a physiological fluid (e.g. cell culture medium, blood) a protein layer is adsorbed, whose chemical composition and conformational state depend on the original surface properties. When cells contact the surface, they interact with the proteins adsorbed by means of cell membranes, that is, by means of membrane receptors [9,10]. Both protein–surface and cell–surface interactions are influenced by surface micro- and nanotopography. A remarkable number of surfaces have been engineered to show the presence of micro- and nanofeatures capable to address controlled cell growth and attachment and migration along predefined directions [11] and to study the organization of the proper cytoarchitecture in specific tissues. The ability to position cells on a substrate with control over their size and spatial arrangement has facilitated fundamental

studies in cellular research. Micropatterned cell cultures are ideal to address fundamental issues like cell–cell interactions and cell–substrate interactions.

Although the regeneration of complex organs is far from being achieved, 2D and 3D patterned materials have been used for producing tissues for medical purposes, like skin [12] and cartilage [13], or as a versatile tool to study *in vitro* the complex phenomenon of cell–material interactions.

Besides tissue regeneration, live cellular arrays appear promising as platforms for pharmaceutical drug development. In these devices cells are manipulated by means of biophysical "handles", that have been classified as passive or active [14]. Passive handles include chemistry and topography while active handles include some forms of energy to move cells as physical objects such as polarizable or bend light bodies [15]. A combination of passive and active handles can be used to manufacture, for example, a platform to manipulate cells in drug development and functional genomic.

2. *Diagnostic devices*, i.e. surfaces containing chemical and physical cues at micrometric and nanometric levels. Molecular biosensors utilize biomolecules such as enzymes, antibodies, nucleic acids, receptors, etc. The basic feature of the biosensor is the immobilization of biomolecules onto a conductive or semiconductive support, and the electronic transduction of the biological functions associated with the biological matrices. The integration with micropatterning allows the simultaneous determination of a large variety of parameters from a minute amount of sample within a single experiment. This approach is advantageous for cost-effective disease screening, to efficiently monitor the effectiveness of patient therapies, or to recognize, in real time, biological molecules indicative of specific physiological activities [16].

Cell-based biosensors, for example, represent one of the possible devices, achievable by patterning cells on 2D textured materials [17]. They incorporate cells as a sensing element that convert an environmental stimulus to signals conducive for processing. Cells express an array of potential molecular sensors (such as receptors, channels, and enzymes) that are maintained in a physiological relevant manner by native cellular machinery. The powerful effect of such devices would be due to their high sensitivity and to the specific responses of cells to specific stimuli (e.g. olfactory neurons respond to single odorant molecules, retinal neurons are triggered by single photons). A cell-based biosensor has a distinct advantage over molecular sensor; in a cell-based biosensor, the sensing element is maintained in the native state, whereas in a molecular sensor, there is always the problem of protein degradation which will affect the affinity and accuracy of the sensor [18].

3. *Drug delivery systems* can make use of nanostructured and nanocomposite materials to control the amount of drug released in the body. Nanocomposites are characterized by biological molecules or inorganic compounds organized as nanodomains or microdomains randomly dispersed inside an organic/inorganic matrix. Such compounds can be released in a physiological milieu through three different primary time-dependent mechanisms: diffusion, degradation, and swelling followed by diffusion [19,20]. In case of diffusion mechanism, it is driven by a release of the drug either by passing through the pores or between polymer chains, and these are the processes that control the release rate. Organic coatings/materials containing silver clusters can be successfully used as antibacterial systems [21]. Silver ions exhibit antimicrobial activity against a broad range of micro-organisms. As a consequence, silver is included in many commercially available healthcare products, e.g. silver-containing dressings for wound treatment.

Plasma processes are today one of the most successful technology for texturing materials, because of their ability to modify surfaces in a very cost effective way. Ratner stated in a commentary [22] that "micro- and nanotechnologies are not so new as many perceive it to be, because plasma scientists have been working with modifications of 'small things' even before Langmuir coined the term *plasma* . . .".

Micropatterning procedures have been performed with plasma processes in microelectronics since the 1970s to produce integrated circuits; they have been transferred and adapted to process material for biomedical purposes in the following years. Many examples exist in literature, demonstrating that plasma processes are successfully used during one or more stages of multi-step texturing procedures [4].

This review shows some examples of micro- and nanopatterning procedures of utility in biomedical applications, where different plasma processes are involved. Biological responses to such surfaces are also discussed.

13.2
Micro- and Nanofeatures Modulate Biointeractions *In Vivo* and *In Vitro*

Before starting any discussion about the techniques used for micro-and nanopatterning materials, it is necessary to focus our attention on the elements included in all complex levels of organization of the living matter. Biomolecules and cells are at the base of a more complex organization. The dimensions of biomolecules are up to 20 nm, while cells are up to 100 μm wide. Recent advances in microscopic analytical techniques enable researchers to observe well-organized 3D structures on which cells are microassembled to form tissues *in vivo*. The basement membrane is an example of a natural 3D scaffold [23]; it is composed by extracellular matrix (ECM) components, whose rigidity and chemical composition exert a strong influence on cell behavior. Topography of the basement membrane is varied: pores, fibers, ridges, bumps, all with dimensions in the 2–200 nm range [24]. More recently, the issue of whether the form of the tissue can feed back to regulate patterns of proliferation was addressed by Nelson *et al.* [25]. This group demonstrated that the shape of the cellular island has a prominent effect on the pattern of proliferative process by using microfabricated ECM protein islands of different geometric shape to control the organization of bovine endothelial cells. The regions of concentrated growth corresponded to sites with high tractional stress generated within the cell sheet. Their results demonstrate the existence of patterns of mechanical forces that originate from the contraction of cells, emerge from their multicellular organization, and result in patterns of growth.

In vitro investigations show that although the importance of topographic cues may vary for different cell types, its relevance is unquestionable [26]. Moreover, considering that *in vivo* cells and proteins interact with nanofeatures, investigating cell–protein substrate interactions at the submicrometer scale is essential to control biological responses. Micro- and nanofeatures are probably able to elicit cell adhesion and/or migration along favored directions, inducing the correct organization of cells in a cytoarchitecture, leading eventually to a tissue with specific functions. Topographical irregularities of surface used as support for cell adhesion could be reflected on the cytoskeleton of the cell, on the amount of ECM produced, and also on more complex multicellular behaviors such as angiogenesis [27]. *In vitro* and developmental biology studies strongly suggest that particular cellular morphologies are closely linked, and may be required to activate cell proliferation and/or differentiation.

For example, Dalby *et al.* [28] have demonstrated that hTERT-BJ1 fibroblasts reorganize their cytoskeleton and adhesion depending on the initial reactions with nano-bumped surfaces: the morphology of filopodia depends on the distance between cells and nano-islands and this affects protein expression. Lee *et al.* [29] have studied human ligament fibroblasts (HLF) on aligned polyurethane nanofibers and have observed that cells grown along aligned fibers produce much more collagen with respect to the same cells cultured on randomly distributed fibers. In the latter case the material was considered promising for ligament engineering. Cells can indeed respond to nanometric cues *in vitro* [30]; many studies exist showing that, when fibroblasts are seeded on nanostructured materials, they probe around the surface of the substrate by means of filopodia. When a nanostructure is detected by filopodia, generally it is recognized as a site suitable for adhesion, and variations in the cytoskeleton of the cells are stimulated [31]. The dimension of the surface features as well as their shape besides influencing ECM organization can induce some mechanical stresses on cells that can affect, for example, the long-term cell differentiation. A recent focus in this field is based on the immobilization and patterning of bacteria on surfaces which appear to provide new opportunities for sensing and detecting biomolecules using whole cells and for studying cell–cell interactions and interactions between cells and their surroundings [32].

Although a vast amount of literature is available on the effect of topography on cell behavior, there is relatively little research on protein conformation and functionality on a nanostructured substrate. This may be due to the difficulty in fabricating nanostructures which are similar in size to the protein and the complexity involved in studying the conformation and functionality of protein attached to such substrates [33].

Recent studies focus their attention on the involvement of bio-macromolecules networks, like ECM, in promoting specific cell responses. It appears that besides chemistry and conformation mechanical constraints of bio-macromolecule networks can also affect cell responses. As an example we can consider the role of supramolecular organization of type I collagen, a major ECM protein of most mammalian tissues, into muscularized arteries. This protein complex is synthesized by cells as hydroxylated subunits that form long triple helical domains 2 nm wide and 300 nm long [34]. Once transported outside the cell, such polypeptide complexes assemble into large supramolecular fibrils, a structural network also devoted to stimulate the physiological response of smooth muscle cells (SMCs) within adult muscularized arteries when a vascular injury occurs. The conformation of type I collagen is modified (denatured) by ECM-degrading proteases; in this case denatured collagen I is able to promote SMC proliferation for wound healing [35]. Within healthy tissue type I collagen does not alter SMC proliferation, the SMCs remaining in a mostly non proliferative state. Although it appears that mechanical stresses induced by collagen fibrils on cells represent the key parameter of further SMCs proliferation, how mechanical constraints of collagen I influence morphology and proliferation of cells is not clear yet. Toward a deep understanding of this phenomenon scientists try to synthesize reference materials on which the conformation and chemistry of ECM is controlled by tuning material surface characteristics. Elliott *et al.* [36] show that by synthesizing ECM-mimics, surfaces in which thin films of collagen type I are adsorbed, in different conformations, onto alkanethiol monolayers it is possible to verify that cells grown onto thin films of very small fibrils of collagen are totally different from cells grown onto thin films with large fibrils and native collagen *in vitro*. Assuming that chemical composition of topographically different collagen films is the same, the differences in cell behavior have been correlated with other features such as mechanical forces. In fact, such factors play a vital role in cell differentiation, proliferation [37], motility [38], cell adhesion, and morphology [39].

Denis and co-workers [40] showed that a conformationally different collagen can be adsorbed by tuning surface topography of a material on which the biomolecule is adsorbed. They studied the collagen on smooth surfaces and on nanopatterned ones, with dots 15 ± 5 nm high and 60 ± 15 nm wide. They found that although similar amounts of collagen were adsorbed on both surfaces, elongated aggregates were formed on smooth substrates but not on rough ones. This effect was attributed to the differences in protein mobility on the substrate since the collagen molecule is relatively free to move on smooth surface, but nanopillars on the rough substrate hindered the collagen mobility.

Such observations suggest that very complex cascade reactions often start from simple substrate features, on which cells grow directly or through adsorbed biomolecules. On the other hand, we can conclude that, since the dimension of a protein is in the nanometer range, a surface with nano-topographical features similar or smaller to the size of the protein may be sensed by proteins affecting the conformation and a substrate with features larger than the size of the protein may appear smooth to the protein and have minimal effect on protein adsorption kinetics and adsorbed amount as confirmed by Han *et al.* [41] for nanoporous surfaces. The key role of material surfaces, to avoid undesirable biological interactions, is to retain protein functionality without altering their correct conformation.

The results discussed above emphasize the need to design model substrates exhibiting controlled variations of topography and/or chemistry, in order to drive different biointeractions that occur on the material, e.g. after its implantation in a viable organism. On the other hand it is often strictly required that the bulk properties of the material are retained to fulfill its functions. A large amount of literature shows that it is possible to spatially control the biointeractions by producing chemically different domains spatially arranged on a material surface, or textured surfaces, without altering the surface composition of the material. The use of patterning techniques based on wet chemistry procedures hardly modify in an independent way surface chemistry and topography. Therefore, the growing interest in plasma processes is justified, due to their versatility in tuning independently surface chemistry and topographical characteristics of substrates. Hence, we believe that valuable insights can be drawn from these studies.

13.3
Micro- and Nano-fabrication Technologies

Different kinds of technologies can be used to promote patterning of materials. A control of dimension and spatial distribution of the micro- and nanofeatures is desirable for a better correlation between the experimental observations and material properties. For this reason very elaborate processes are developed. This aspect has been described in detail in many interesting reviews [42,43]. Some of the most used patterning procedures applied on material surfaces are listed in Table 13.1. The following paragraphs are intended as a brief overview of the most common texturing procedures.

13.3.1
Photolithography: The Role of Photolithographic Masks

Photolithography [44,45] is the most consolidated patterning procedure, by which three thousand transistors per second are produced in the USA. Such a technique,

Tab. 13.1 Examples of Techniques for Micro- and Nanopatterning Surfaces.

Technique	Acronym	Physical tool	Dimensions of the features	Advantages	Disadvantages	Ref.
Photolithography	—	UV radiation	>100 nm (width)	Consolidated background	Expensive	[54,110]
			>70 nm (height)	Fast replication	Diffraction problems	
Extreme UV lithography	EUVL	Extreme UV radiation ($\lambda = 0.2$–100 nm)	~10 nm	Low diffraction	Expensive	[111–112]
X-ray lithography	XRL	X rays ($\lambda = 0.2$–40 nm)		High resolution	Mask damaging	
Lithography by particles	FIBL (fast ion beam lithography) EBL (electron beam lithography)	Ions, electrons, neutrals; metastable species	0.1–50 nm	Lowest diffraction ($\lambda_B < 0.1$ nm) High resolution	Expensive Low rate of replication	[113]
Scanning probe lithography	SPL	Probe	Atomic resolution	High resolution Ability to realize different geometries of patterns Ability to structure nonplanar surfaces	Expensive Low rate of replication	[114]
Soft lithography	REM (replica molding) µCP (microcontact printing)	Mould used as template	10–100 nm	Cheap High rate of replication	Low resolution Difficulties in dimensional control of the features	[114–120]

	SAMIM (solvent-assisted micromolding) NFPSP (near-field phase shifting photolithography) MIMIC (micromolding in capillary) μTM (microtransfer molding)		Absence of diffraction phenomenon	Defects of the pattern
Simple patterning procedures	CL (colloidal lithography)	Colloidal separation	>10 nm	Cheap	Low control of pattern dimensions Low replication	[121,122]
	MPSBC (microphase separation of block copolymers)	Copolymerization			

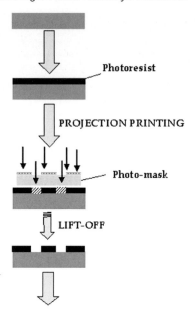

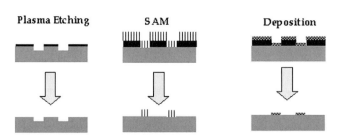

Fig. 13.1 Photolithography: after lift-off, different processes can be used to transfer a pattern to the material surface. *Plasma etching*: chemical ablation of the substrate due to volatile species produced in a plasma; coating with a SAM for example of alkanethiols on gold used as substrate; *deposition* of a thin layer for example by means of plasma processes.

widely used to produce integrated circuits on silicon, has been refined to realize structures smaller than 100 nm, also for several other applications. The photolithographic process can be schematized into two main steps, as represented in Fig. 13.1: (1) producing of a master; and (2) replica of the pattern by means of the master, with the first step slower and much more expensive than the second one.

During the first step a substrate, generally silicon, is coated with a thin layer of a light-sensitive polymer, the photoresist. A photolithographic mask composed of a patterned metallic film, generally with a chromium opaque portion, supported by a transparent material (e.g. quartz) is used to expose only defined zones of the photoresists to UV light. The zones of the photoresist exposed to the light undergo chemical photoinduced reactions that can result in two possible modifications (projection printing): the photoresist zones illuminated by the light (a) degrade, so they become more soluble in a developing solution with respect to the other zones (*positive photoresist*) or (b) are crosslinked, resulting in insoluble domains for the developing solution (*negative photoresist*). The positive/negative of the mask pattern can be transferred to the exposed photoresist layer by rinsing the substrate into a developer solution. The resulting pattern can be used as lift-off mask in the following steps:

- transferring the pattern of the photoresist to the underlying substrate (i.e. silicon);
- patterning the thin film of another polymer, or a thin layer of a biomolecule, on the substrate.

13.3.1.1 Role of Plasma Processes in Photolithography

Plasma processes can be successfully involved in one or more steps of photo-lithography since their versatility makes such processes much more attractive than wet chemistry. Industrial manufacturing of integrated circuits often includes dry etching of the substrate uncoated by the photoresist stencil mask (not to be confused with the "photo-mask" in photolithography), to produce topographical features on the substrate. Further lift-off procedures are applied, afterwards, to remove the photoresist. Patterned surfaces for biomedical applications can be obtained by means of the photolithographic steps described before. Chemically different domains with specific biological activities are required for manufacturing cell-based sensors, DNA chips, supports for cell seeding and other surfaces of interest for biomedical applications. They can be spatially arranged at the surface of a substrate by means of deposition processes after the lift-off of the photoresist. The deposition can be realized by means of a plasma process, e.g. by depositing a cell-repulsive layer (protein repellent) such as polyethylene oxide (PEO)-like coating [46]. In this case cell-repulsive (protein repellent) domains can be alternated with cell-adhesive ones (uncoated substrate), aimed for example to control adhesion and movement of cells along predefined directions, or to promote protein adsorption only in certain zones of the substrate. The role of PEO-like domains consists in controlling the undesired nonspecific protein adsorption outside defined zones of the pattern.

Besides cell confinement, PEO-like micropatterned surfaces have also been used to guide the transport of certain proteins along predefined directions. Cells regulate their active transport of intracellular cargo by means of a conversion of chemical energy stored in ATP into mechanical work produced by motor proteins (e.g. kinesin) which move along protein filaments (e.g. microtubules). Such motor proteins can be used to power the movement and assembly of synthetic materials and to synthesize molecular motors for manufacturing hybrid nanodevices [47]. Spatial and temporal control of a transport system in general requires that motion is confined along tracks. Clemmens *et al.* [48] report a study on microtubules used as shuttle units which are guided along microtracks of kinesin surrounded by PEO-like coatings. A schematic that represents the movement of motor proteins on kinesin tracks used to study

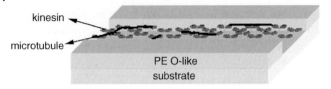

Fig. 13.2 Microtubules move along chemical tracks labeled with kinesin surrounded by kinesin-free domains coated by means of a PEO-like coating. The latter discourages aspecific adsorption both of kinesin and microtubules. This approach represents a good strategy to study guiding of molecular motors at track boundaries: microtubules collisions with the boundaries are analyzed by measuring the angle of approach and the outcome of the collision a key parameter to determine the total travel distance on the track.

guiding of the protein at track boundaries is shown in Fig. 13.2. The substrates can be obtained by means of photolithography performed on glass, a PEO-like coating (20 nm thick) was deposited soon after the projection-printing step and the photoresist development, as sketched in Fig. 13.1. A further selective adsorption of kinesin in the uncovered glass tracks was obtained. To study the effect of topography disentangled from that one of chemistry on microtubules guidance, differently patterned surfaces have been compared; in conclusion the authors derived physical models to predict an efficient guidance of motor-protein shuttles. The kinesin–microtubule system is implicated in several serious diseases, which is one reason why medical researchers are particularly interested in understanding the fine details of how kinesins and microtubules interact.

Besides physisorption of biomolecules inside uncovered tracks, biomolecules can be covalently bonded onto a material through photoinduced reactions [49–52]. In this last procedure, biomolecules like hyaluronan (Hyal) or its sulfated derivative (HyalS) are previously conjugated with a photoreactive unit (4-azidoaniline hydrochloride). The solution of the photoreactive polysaccharide ($HyalN_3$ or $HyalSN_3$) is then dispersed onto a substrate which displays, for example, NH_2 functionalities on the surface (i.e. aminosilanised glasses), and dried at room temperature. This step can be performed with a spin coater to speed up the procedure. After that, a conventional photolithographic step is performed with a photomask in close contact with the coated material. The irradiation induces covalent bonding of the $HyalN_3$ ($HyalSN_3$) to the NH_2 groups only on the domains exposed to the UV source. After a lift-off procedure, with distilled water, a pattern of Hyal (or HyalS) domains in correspondence with the holes of the photomask is obtained. Plasma processes can extend the possibility of photoimmobilization also to materials that do not exhibit NH_2 functionalities at their surface (e.g. polyethylene terephthalate, polystyrene). As an example, the immobilization of $HyalN_3$ was successfully obtained on NH_3 plasma-treated polyethylene terephthalate (NH_2-PET) [53]. Cell culture experiments performed on micropatterned surfaces containing 25 and 5 μm wide Hyal stripes (~35 nm thick) have shown that cell adhesion and alignment was promoted along NH_2-PET track direction. Furthermore, cell proliferation and metabolic activity of chondrocytes was found to be dramatically influenced by the presence of surface features; patterned surfaces have been found to play a vital role in maintaining original chondrocyte phenotype, an important requirement for cartilage tissue regeneration.

13.3.1.2 Limits of Photolithography

The lateral dimension of the smallest feature that can be generated by means of photolithography is of 100 nm, a limit that derives from optical diffraction in the projection printing step [54]. Interesting improvements of photolithography derive from UV, X-ray, ion, and electron beams used instead of UV radiation, aimed to reduce the size of the smallest feature achievable. Although these last methods are able to generate very small features (<10 nm wide), often they are very expensive, therefore hardly affordable in academic laboratories. For these economical reasons new technologies have been developed and they will be described in the following sections.

Another drawback of photolithography consists in the lift-off procedure: the photoresist is generally lifted off by means of sonication of the material in acetone, a harsh procedure for most polymers. Recent advances in this field are water-soluble photoresists and the replacement of acetone with water [55].

13.3.2
Soft Lithography

13.3.2.1 Description of the Technique

An improvement of photolithography consist in using silicon substrates, patterned by means of wet or dry etching, as master for molding an elastomer, usually polydimethylsiloxane (PDMS), poly(methyl methacrylate) (PMMA), polyurethanes (PUs). Whitesides pioneered these "soft lithography" techniques, which provide simple and cost-effective procedures for texturing materials [56,57]. With the term "soft lithography" a group of different processes are described (e.g. replica molding, REM; microcontact printing, μCP; micromolding in capillaries, MIMIC; solvent-assisted micromolding, SAMIM). These techniques appear to be very effective in producing submicrometer features in polymers with high fidelity due to the ability of the elastomer to function as a reversible seal on smooth surfaces [58].

13.3.2.2 Role of Plasma Processes in Soft Lithography

In soft lithography plasma processes can be involved in the modification of the replica, or of the substrate to be patterned. One of the possible processes in which soft lithography can be developed is represented by microcontact printing (μCP) [59], where the mold is used as a stamp for transferring a pattern on material surfaces. The mold is rinsed in a solution of an "ink" (e.g. alkanethiols [24] and alkylsiloxanes [60], catalytic precursors [61], lipids [62], proteins [63], etc.). The stamp is then positioned in close contact with the material to be patterned. In this way, the raised parts of the stamp can transfer spatially micro-arranged self-assembled monolayers (SAMs) of the ink to the substrate [56]. Several examples are known of applications of plasma processes in μCP, successfully involved in modifying substrate to be patterned later by means of deposition of a protein-repellent PEO-like layer [64]. In this case the ability of PEO-like

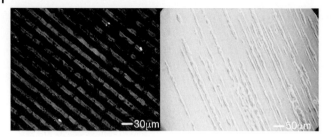

Fig. 13.3 Patterned surfaces obtained by means of μCP procedure. After a photolithographic process aimed to obtain the stamp (see Fig. 13.1), a solution of a PDMS prepolymer is streamed on the stamp to obtain the replica (mold or master). The replica inked with an appropriate solution of protein is contacted with PEO-like coated substrates in dry conditions, to transfer the protein pattern to the substrate. After a reaction with a fluorescent antibody it is possible to see the fibronectin stripes of the pattern (a). 3T3 human fibroblasts align along pre-defined directions marked by the fibronectin tracks (b).

coatings to adsorb proteins in dry conditions during the transferring of the pattern is exploited. Tanaka *et al.* [65] demonstrated that micrometer-sized tracks of adsorbed laminin can be successfully transferred on PEO-like coated substrates by means of μCP. Similar patterned substrates have been produced as supports for cell seeding. As shown in Fig. 13.3, cell culture experiments performed on substrates obtained in our laboratory with a similar technique show that it is possible to induce 3T3 fibroblasts alignment along 15 μm wide fibronectin stripes totally avoiding cell attachment in the surrounding plasma deposited PEO-like domains.

13.3.2.3 Limits of Soft Lithography

Soft lithography is often proposed as valid alternative to photolithography due to its simplicity, convenience, and ability to transfer patterns also on curved surfaces. On the other hand the elastomeric character of PDMS is a clear limit for the technique: due to its flexibility and, therefore, to its deformability, the stamp can in fact collapse under mechanical and physical stresses [65]. This represents a technical problem in particular when complex patterns have to be realized. By using new materials and new patterning strategies it should be possible to overcome the most common problems of μCP [66].

13.3.3
Plasma-Assisted Micropatterning: The Role of Physical Masks

Most part of the methods described so far are expensive, and require several steps which increase the risk of contamination, for example from solvents used during the procedures. This effect generally compromises applications in biomedical field. These limitations have prompted the development of many alternative complementary methods of patterning that differ in the way the features are generated.

Plasma-assisted micropatterning by means of physical masks is a valid alternative, because it capitalizes on the many advantages of plasmas: cost-effective [67]; clean and efficient dry processes; conformal, sterile and nontoxic coatings; broad range of different chemistries; reduced contaminations; quick, one-step processes [68]. A physical mask (not to be confused with a "photomask" in photolithography) is characterized by micro-/nano-arranged holes. It is generally used as a "stencil" to transfer a pattern to an underlying layer.

13.3.3.1 Micropatterning

A typical physical mask is represented by a polymer or metal sheet pierced by means of laser cut, wet or dry etching procedures as in the case of copper grids widely used for transmission electron microscopy (TEM). After positioning the pierced sheet onto the substrate, the texturing procedure can be performed out by filling through the openings sheet with thin film of another material, or by etching the uncovered zones of the substrate. By means of physical mask lithography it is possible to accomplish different patterns, aimed at:

- changing only the topography of the substrate, keeping its surface chemistry unaltered;
- changing the surface chemistry of the substrate without altering its surface topography;
- changing both chemistry and topography of the substrate.

Many researchers have shown that it is possible to induce a micro-arrangement of cells on a surface modified by means of plasma-assisted physical mask surface functionalization process [69]. In this case only chemical, and no topographical constraints are provided to cell attachment, as for μCP procedure (Section 13..2). The two-step process consists of a first plasma treatment (2.45 GHz discharge) using ammonia as gas feed, and a second plasma treatment fed by H_2 performed through a physical mask. During the H_2 treatment the functionalities grafted in the first process are selectively removed from the zones of the substrate uncovered by the mask. This procedure allow one to obtain cell-adhesive tracks, corresponding to N-functionalized tracks, that promote cell alignment along predefined directions, alternated with cell-repulsive domains, corresponding to the H_2-treated zones.

Wu *et al.* [70] proposed a method combining microwave plasma-enhanced CVD (MPECVD) and vacuum ultraviolet (VUV) light irradiation, to produce nanostructured materials exhibiting superhydrophobic vs. superhydrophilic character: MPECVD of trimethylmethoxysilane superhydrophobic coating [71] is followed by a physical mask modification with UV light irradiation or with a laser (157 nm) to obtain regularly arranged micro-areas with different chemical/physical properties. After light irradiation the exposed areas of the sample became superhydrophilic (WCA $< 10°$). SEM characterizations of the substrate shown that both superhydrophobic and superhydrophilic regions of the patterns have similar rough morphology ($R_{rms} = 34 \pm 6$ nm) but Kelvin force microscopy (KFM) in tapping mode shown that the surface potential in the superhydrophilic regions is higher than of superhydrophobic ones. Chemical characterization of such surfaces demonstrates that oxygen content in superhydrophilic areas is higher with respect to superhydrophobic ones.

Cell culture results on such materials shown that cells prefer superhydrophilic areas probably due to the presence of COOH and OH groups in the VUV-irradiated surface.

Micro-lithographic technique can be used to obtain patterns in which a simultaneous presence of chemical and topographical cues produces different effects on cell adhesion. A combination of plasma deposition processes can be considered an effective method to produce cell-adhesive tracks surrounded by cell-repulsive domains; cells align in tracks due to the presence of cell-adhesive chemistry, and to the walls of the tracks. In our laboratory, a plasma-deposited acrylic acid coating (pdAA) followed by a plasma deposition of PEO-like coatings through TEM copper grids with different patterns has been performed on PS substrates [72]. Human fibroblasts (hTERT) seeded on such patterned substrates aligned along pdAA tracks (cell-adhesive composition) [73,74]. The topographical effect on cell morphology has been clearly observed on cells adhering on pdAA domains of different dimensions: they were randomly distributed on wide pdAA zones, while they stay aligned along the edge of the boundary close to the PEO-like step, and inside the tracks. The surface of cells adhering along PEO-like steps appeared to be ruffled, as shown in Fig. 13.4. Moreover, cells migrated along the pdAA tracks, an important hint of cell viability, fundamental for potential applications in tissue engineering and cell sorting.

In Ref. [73] it is described how it is possible to tune cell adhesion by light changes in chemical composition of the plasma deposited films used in the pattern: patterns have been obtained, for example, with combinations of plasma depositions of different PEO-like coatings, obtained by changing process parameters, which produce different degree of cell-adhesive character to the surface. A plasma deposition of highly crosslinked PEO-like coating (i.e. deposited at high input power) has been performed before a deposition of a "non-fouling" PEO-like coating, produced in low fragmentation regime (i.e. low power, high retention of monomer structure) through a physical mask. In this case chemically different PEO-like domains (cell repulsive/adhesive) are alternated in the same topographic pattern; therefore it is possible to study the effect of chemical composition of different domains on cell behavior: cell align along highly crosslinked PEO-like tracks totally avoiding PEO-like wide unfouling domains.

Plasma deposition allows a number of surface chemistries with a wide selection of functionalities. These chemistries enable surface functionality tailoring for specific molecular site binding, which may not otherwise be achievable [75]. These peculiar characteristics of plasma processes can be used to micropattern biomolecules. Slocick *et al.* [76] deposited plasma-polymerized allylamine (ppAAm) through a TEM copper grid, on a substrate coated by SAM of (3-mercaptopropyl)trimethoxysilane (SH-SAM). Quantum dots functionalized with a cysteine conjugate have been immobilized on SH-SAM domains while an immobilization of a green fluorescent protein has been linked to ppAAm micro-areas through its carboxylic groups. Such kinds of substrates are very attractive for biosensors, microreactors, and microfluidic devices, because it is possible to immobilize different kinds of biomolecules with excellent spatial resolution.

Plasma aided physical mask micropatterning could be also used to texture materials with pH-sensitive surface areas [77]. Spatially resolved alternated micro-

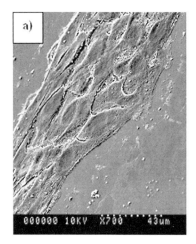

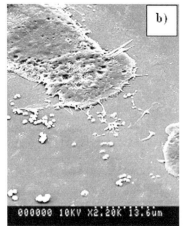

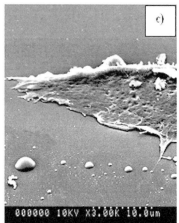

Fig. 13.4 Scanning electron microscopy images of hTERT BJ1 fibroblasts spatially microarranged on 40 μm wide pdAA tracks surrounded by wider PEO-like domains. Cells are in close contact in a confluent layer into the pdAA tracks (a) continuously probing the underlying environment by pseudopodes (b) along the direction of the tracks. When in close contact with PEO-like step their membrane appears ruffled (c) [3].

areas coated with plasma-deposited polyallylamine and plasma-deposited acrylic acid have shown a different behavior when exposed to solutions at different pH. These substrates are very attractive to design biosensors for target molecules that selectively adsorb/desorb onto defined regions of the substrates in a reversible way.

Recent results have shown the possibility to use also atmospheric pressure plasmas as an alternative to low-pressure ones [78] in patterning procedures. Wertheimer and co-workers [77] demonstrated that it is possible to produce an array of cell-adhesive islands surrounded by a cell-repulsive matrix using atmospheric pressure dielectric barrier discharges (DBDs) [79] through Kapton masks. Cell-adhesive islands (30 μm in diameter) are produced by means of a plasma deposition of nitrogen-rich polymerized polyethylene (ethylene/N_2 fed discharges) while cell-repulsive ones are represented by the untreated material (biaxially oriented polypropylene).

All described processes require a close contact between the physical mask and the substrate to avoid that, during the process, active species of the plasma phase can pass underneath the physical mask and modify in an undesirable way the chemistry of the covered domains. This undesired phenomenon dramatically limits the range of application of the patterning procedure described, so it is impossible to transfer patterns at a resolution lower than 5–10 μm. The features that are possible to obtain with this technique thus have lateral wideness comparable to cell dimension. Despite the broad range of materials and patterns available, emerging strategies are being pioneered to improve spatial resolution of the transferred features (i.e. ultrathin alumina mask patterning [80], colloidal lithography [81]), and to study the effects of nanofeatures on protein adsorption and cell adhesion.

13.3.3.2 Nanopatterning

Colloidal lithography (see Table 13.1) is an example of *bottom-up* texturing strategies [82–84]. The physical mask in these processes is a monolayer of metal or polymer micro- or nanospheres that are assembled into close-packed colloidal crystal lattices. Colloidal crystallization can be accomplished through both convective self-assembly and simple drop coating. In the first case a substrate is placed vertically into a colloidal suspension of the spheres until a complete evaporation of the liquid occurs [85]. The process may take several days before the deposition of a close packed layer of spheres is completed. Alternatively, a drop of the colloidal suspension can be dispersed on the substrate and quickly dried (spin coated) to produce a colloidal crystal with localized order [86]. As shown in Fig. 13.5, plasma processes can be applied during one or more steps of patterning procedure ensuring a broad range of combination of chemistries and textures.

The main advantages of colloidal lithography with respect to other techniques are the high spatial resolution and the close contact between the physical mask (layer of colloidal particles) and the substrate. A critical point of the patterning is the self assembly step of the nanospheres. When they are functionalized with polar groups the wettability of the substrate is strictly necessary. Plasma processes can be successfully used to extend the range of applicability of colloidal lithography in tailoring the surface chemistry of the substrate to address a correct self assembly of the beads [3].

By comparing topographically different substrates characterized by the same chemistry it is possible to disentangle the effect of topography on cell adhesion and protein adsorption from that of chemistry. It has been shown that by depositing a pdAA coating on flat and nanopitted substrates it is possible to see that the nanofeatures are effective in promoting cell adhesion and spreading with respect to flat pdAA surfaces (unpublished results of the authors' laboratory).

Recent studies [87] on substrates produced with colloidal lithography have shown that regularity and symmetry of the nanofeatures play a key role in affecting cell responses since a different adhesive behavior can be observed on ordered dots (pillars), opposed to that on random dots.

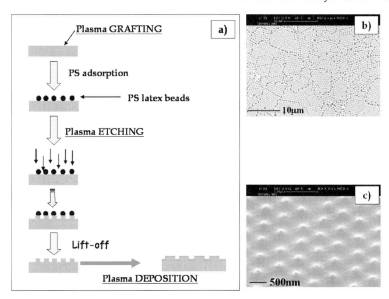

Fig. 13.5 (a) Schematic representation of colloidal lithography. Functionalized polystyrene (PS) latex beads are generally used for this purpose. The colloidal crystallization occurs through self-assembly of the nanoparticles on the substrate; the substrate must be wettable to promote their organization in a monolayer. To improve the wettability of the substrate plasma treatments can be performed. (b) Mosaic arrays of spherical particles can be formed with a spin coating process. The ordered hexagonal array can be used as physical mask for an etching/deposition plasma process to transfer the pattern to the substrate. (c) After a lift off it is possible to obtain a nanopitted substrate. Plasma processes can be finally used at the end of the nanopatterning procedure to tune conformally the chemistry of the substrate.

A combination of plasma deposition of pdAA and PEO-like coatings can be used during colloidal lithography to obtain nanodomes characterized by different chemistry and different attitude to interact with proteins. Valsesia *et al.* [88,89] proposed a nanocolloidal patterning consisting a multistep approach: plasma deposition of pdAA (protein adhesive) coating; surface masking with colloidal nanobeads; oxygen plasma etching; plasma polymerization of PEO-like (protein repulsive) coatings; ultrasound lift-off of the bead residuals. The authors demonstrate that the nanodomes (about 100 nm wide) of pdAA surrounded by PEO-like unfouling coatings are able to selectively bind bovine serum albumin protein through a N-hydroxysuccinimide (NHS) and (3-ethylaminopropyl)carbodiimide (EDC) coupling procedure [90]. These substrates are very attractive in biosensor technology.

Besides colloidal lithography, plasma processes can be involved also in the production of template-imprinted nanostructured surfaces for protein recognition [90]. Synthetic materials capable of selectively recognizing proteins were produced by a multistep procedure [94]: a protein was adsorbed on mica surfaces; a disaccharide layer was then adsorbed on the biomolecules to protect them from drying-induced denaturation, and to constitute recognition cavities for proteins; a fluoropolymer was deposited to coat everything and was successively properly mounted on a solid support to facilitate the detachment from mica. After protein lift-off, a negative pattern containing disaccharide cavities able to recognize proteins was produced, whose efficiency was ascertained. These surfaces can be potentially applied in separation systems for proteins, biosensor technology, and for new biomedical materials.

13.3.4
Novel Approaches in Plasma-Patterning Procedures

Recent advances in plasma processing have shown powerful alternatives to the patterning procedures described so far, in which plasma plays a leading role. In this section we will explore progresses made recently, with special emphasis on two possible plasma patterning roots applied without the use of physical masks: plasma deposition of thermoresponsive films and plasma deposition of nanostructured Teflon-like coatings.

13.3.4.1 Plasma Polymerization and Patterning of "Smart" Materials

Some of the main efforts in materials engineering are today devoted on designing materials that are sensitive to external stimuli like temperature, pH, light, and electric fields. Such stimuli generally promote specific responses in the material, which can be expressed in terms of a change of shape, surface characteristics, and solubility, and in a sol–gel transition. Such kinds of materials are defined as "smart", since they are able to communicate with the surrounding environment, via proper stimuli. Smart materials are widely used in delivery of therapeutics, tissue engineering, bioseparators, and sensors, as attested by many papers and patents in the last 20 years. Among others, poly(N-isopropylacrylamide) (PNIPAAm) represents one the most applied in several fields, including the biomedical one. It is a thermoresponsive polymer because it is soluble below $32\,°C$ (lower critical solution temperature, LCST) and precipitates above $32\,°C$ in water [91]. Below LCST hydrogen bonds between amide groups of the polymer and water are favored, thus leading to the dissolution of the polymer in aqueous media; above $32\,°C$, instead, hydrogen bonds are broken and water molecules are expelled from the polymer, resulting in its precipitation. Scanning probe microscopy studies have shown the coil-to-globule transition of the polymer chains [93]. By controlling the polymer composition and topology, the coil-to-globule transition could be kinetically and thermodynamically controlled [94]. As an example, copolymerization of NIPAAm with hydrophobic moieties results in a decreasing of the LSCT, while copolymerization with hydrophilic comonomers results in its increasing. Thermoresponsive NIPAAm is widely used also for cell manipulation. For example, decreasing the temperature few degrees above the LSCT, a NIPAAM substrate seeded with eukaryotic cells promotes a detachment of a sheet of living cells without enzymatic treatment [94]. This phenomenon is due to the transition of the polymer changing from hydrophobic to hydrophilic when the temperature decreases below $32\,°C$. Cheng et al. [95] proposed a study in which plasma-polymerized (pp)NIPAM is deposited on a glass substrate properly interfaced to a wire heating system [96]. This approach is particularly suited for microelectromechanical systems (MEMS) fabrication. When the device is exposed to proteins they adsorb exclusively on heated areas of the pattern, where the (pp)NIPAM is above the LCST, hydrophilic. Such materials can be also used to promote cell adhesion in defined zones of the substrate, or cell detachment when and where the substrate is cooled. In this case cell confinement is promoted by means of a programmable surface chemistry.

13.3.4.2 Deposition of Micro- and Nanostructured Coatings

Modulated radiofrequency (13.56 MHz) glow discharges fed with fluorocarbons allow one to obtain Teflon-like coatings with high retention of the chemical structure of the gas feed, remarkable surface roughness, and very low surface tension [97–99]. Fluorocarbon films deposited in continuous wave (CW) discharges are characterized by a variable fluorination degree and crosslinking character, tunable with the experimental conditions, and by low roughness. When deposited in modulated wave (MW) conditions (i.e. the discharge is pulsed on/off at defined time intervals) [100,101], particularly at very low duty cycle (short time on with respect to time off), highly fluorinated coatings result, which are characterized by crystalline polytetrafluoroethylene (PTFE) nanostructures randomly distributed on an amorphous fluorocarbon background. The roughness is due to the presence of nanostructured features all along the surface. To disentangle the effect of chemistry from that of morphology, morphologically different Teflon-like coatings have been coated with a conformal Teflon-like film deposited in CW plasma. It was thus possible to compare the effect of morphologically different, chemically similar coatings toward cell adhesion and growth [102–104]. It has been demonstrated that the presence of nanostructured features improves cell growth when compared with flat samples with the same surface chemistry. Rosso *et al.* [105] found that Teflon-like structured coatings stimulate the cytoskeleton organization; therefore, considering that the shape of cells and the size of their adhesion contacts govern the balance between cell proliferation and death, he suppose that these coatings could be successfully used as cell-seeding supports.

Thin-film networks of multiwalled carbon nanotubes [106], obtained by means of hot filament plasma-enhanced chemical vapor deposition (HF-PECVD), can be successfully used as 3D supports for cell seeding. Correa-Duarte *et al.* [107] produced networks of interconnected nanotubes by exerting chemically induced capillary forces upon nanotubes properly aligned on a substrate: a transformation from being a vertically aligned structure to an interlocking resistive network of interconnected nanotubes produces a 3D architecture of cavities. Isolated large stretched mouse L929 fibroblasts were found with elongated cytoplasm projections attaching to the walls of the cavities attesting for a good basis to stimulate robust tissue formation.

A plasma deposition of bi-layer films with different softness can be applied to produce functional films with defined morphology and wettability [108]. A deposition of a hard top layer on top of a soft film that adheres to a substrate could induce a ''buckling phenomenon'' as a result of compressive forces: equilibrium between the energy required to bend the stiff upper film and that one involved in the soft underlying substrate deformation produces wavy-like structures with wavelengths ranging from 100 to 10 000 nm. Wang and Grundmeier [109] synthesized patterned surfaces by means of a plasma deposition from heptadecafluoro-1-decene (HDFD) for the soft layer and hexamethyldisiloxane (HMDS) for the hard one. Very interesting wavy structures have been obtained and a correlation between surface topography and experimental parameters of plasma deposition has been discussed.

13.4
Conclusions

Carefully surface-engineered materials can dramatically affect biological responses of living cells in contact with the surface, a useful finding for biomedical applications. Proteins and cells on surfaces can be influenced by chemistry and/or topography of the substrate, and this aspect has being studied in depth to understand if it is possible to "switch on" a desirable cell behavior. Investigations of these phenomena include the production of artificial surfaces with well-defined chemistry and/or topography, to be used as models, where plasma treatment, deposition, and etching processes often play a crucial role. Many different plasma processes have been found effective in the last few years, and now compete with conventional surface modification techniques improving cell–surface interactions. Micro- and nanopatterning plasma-aided techniques are being implemented into experimental tissue and cell engineering protocols, in the fabrication of diagnostic devices for health care, or in drug delivery systems.

The plasma-assisted procedures described in this chapter are some examples of the significant advances that plasma processes can provide to biomedical field. The use of micro- and nanopatterned surfaces produced via plasma, and with highly defined arrays of chemically and/or topographically different features, allows one to control the behavior of eukaryotic and prokaryotic cells in terms of adhesion, spreading, and death probably by tuning chemical composition and/or conformation of ECM components adsorbed on material surfaces before cell attachment.

References

1 Sniadecki, N.J., Desai, R.A., Ruiz, S.A. and Chen, C. (2006) *Ann. Biomed. Eng.*, **34**, 59.

2 Liu, H. and Webster, T.J. (2007) *Biomaterials*, **28**, 354.

3 Sardella, E., Favia, P., Gristina, R., Nardulli, M. and d'Agostino, R. (2006) *Plasma Process. Polym.*, **3**, 456.

4 Biederman, H. and Osada, Y. (1992) in *Plasma Polymerization Processes* (eds H. Biederman and Y. Osada), Elsevier, Dordrecht, p. 5.

5 Biederman, H. and Martinu, L. (1990) in *Plasma Deposition, Treatments and Etching of Polymers, Plasma–Materials Interactions Series* (ed R. d'Agostino), Academic Press, New York.

6 d'Agostino, R., Martinu, L. and Pische, V. (1991) *Plasma Chem. Plasma Proc.*, **11**, 1.

7 d'Agostino, R., Fracassi, F., Lamendola, R. and Palumbo, F. (1993) *High Temp. Chem. Proc.*, **2**, 287.

8 Hauert, R., Gampp, R., Muller, U., Schroeder, A., Blum, J., Mayer, J., Birchler, F. and Wintermantel, E. (1997) *Polym. Prepr. (Am. Chem. Soc. Div. Polym. Chem.)*, **38**, 994.

9 Jansen, J.A. and von Recum, A.F. (2004) Textured and porous materials, in *Biomaterials Science, An Introduction to Materials in Medicine, Vol. 2*, (eds B.D. Ratner,A.S. Hoffmann,F.J. Schoen,J.E. Lemons), Elsevier Academic Press, UK, p. 218.

10 Nimeri, G., Fredriksson, C., Elwing, H., Liu, L., Rodahl, M. and Kasemo, B. (1998) *Colloids Surf. B: Biointerfaces*, **11**, 255.

11 Folch, A. and Toner, M. (2000) *Annu. Rev. Biomed. Eng.*, **2**, 227.

12 Morgan, J.R., Sheridan, R.L., Tompkins, R.G., Yarmush, M.L. and Burke, J.F. (2004) in *Burn dressing and skin substitutes, Biomaterials Science, An Introduction to Materials in Medicine, Vol. 7* (eds B.D. Ratner,A.S. Hoffmann,F.J. Schoen,J.E. Lemons), Elsevier Academic Press, UK, p. 602.

13 Cancedda, R., Dozin, B., Giannoni, P. and Quarto, R. (2003) *Matrix. Biol.*, **22**, 81.

14 Ozkan, M., Pisanic, T., Scheel, J., Barlow, C., Esener, S. and Bhatia, S.N. (2003) *Langmuir*, **19**, 1532.

15 Fuhr, G., Glasser, H., Mueller, T. and Schnelle, T. (1994) *Biochim. Biophys. Acta*, **1201**, 353.

16 Willner, I. and Katz, E. (2000) *Angew. Chem. Int. Ed.*, **39**, 1180.

17 Park, T.H. and Shuler, M.L. (2003) *Biotechnol. Prog.*, **19**, 243.

18 Pancrazio, J.J., Whelan, J.P., Borkholder, D.A., Ma, W. and Stenger, D.A. (1999) *Ann. Biomed. Eng.*, **27**, 697.

19 Heller, J. and Hoffman, A.S. (2004) in *Drug delivery systems, Biomaterials Sscience: An Introduction to Materials in Medicine*, 2nd edn (eds B.D. Ratner,A.S. Hoffman,F.J. Schoen and J.E. Lemons), Elsevier Academic Press, San Francisco, CA, p. 628.

20 Robinson, J.R., and Lee V.H.L. (eds) (1987) *Controlled Drug Delivery:*

Fundamentals and Applications, 2nd edn , Marcel Dekker, New York.

21 Furno, F., Morley, K.S., Wong, B., Sharp, B.L., Arnold, P.L., Howdle, S.M., Bayston, R., Brown, P.D., Winship, P.D. and Reid, H.J. (2004) *J. Antimicrobial Chemother.*, **54**, 1019.

22 Ratner, B.D. (2001) *Plasmas Polym.*, **6**, 189.

23 Flemming, R.G., Murphy, C.J., Abrams, G.A., Goodman, S.L. and Nealey, P.F. (1999) *Biomaterials*, **20**, 573.

24 Abrams, G., Goodman, S.L., Nealey, P.F., Franco, M. and Murphy, C.J. (1997) *Proc. Am. Coll. Vet. Ophthalmol.*, **28**, 50.

25 Nelson, C.M., Jean, R.P., Tan, J.L., Liu, W.F., Sniadecki, N.J., Spector, A.A. and Chen, C.S. (2005) *Proc. Natl Acad. Sci. USA*, **102**, 11594.

26 Britland, S., Clark, P., Connolly, P. and Moores, G. (1992) *Exp. Cell. Res.*, **198**, 124.

27 Yim, E.K.F. and Leong, K.W. (2005) *Nanomed. Nanotechnol. Biol. Med.*, **1**, 10.

28 Dalby, M.J., Giannaras, D., Riehle, M.O., Gadegaard, N., Affrossman, S. and Curtis, A.S.G. (2004) *Biomaterials*, **25**, 77.

29 Lee, C.H., Shin, H.J., Cho, I.H., Kang, Y.-M., Kim, I.A., Park, K.-D. and Shin, J.-W. (2005) *Biomaterials*, **26**, 1261.

30 Curtis, A.S.G. and Wilkinson, C.D.W. (2001) *Trends Biotechnol.*, **19**, 97 and references therein.

31 O'Connor, T.P., Duerr, J.S. and Bentley, D. (1990) *J. Neurosci.*, **10**, 3935.

32 Weibel, D.B., DiLuzio, W.R. and Whitesides, G.M. (2007) *Nature*, **5**, 209.

33 Yap, F.L. and Zhang, Y. (2007) *Biosens. Bioelectron.*, **22**, 775.

34 Kadler, K.E., Holmes, D.F., Trotter, J.A. and Chapman, J.A. (1996) *J. Biochem.*, **316**, 1.

35 Jones, P.L., Jones, F.S., Zhou, B. and Rabinovitch, M.J. (1999) *J. Cell Sci.*, **112**, 435.

36 Elliott, J.T., Tona, A., Woodward, J.T., Jones, P.L. and Plant, A.L. (2003) *Langmuir*, **19**, 1506.

37 Ingber, D.E. and Folkman, J. (1989) *J. Cell Biol.*, **109**, 317.

38 Verkhovsky, A.B., Svitkina, T.M. and Borisy, G.G. (1999) *Curr. Biol.*, **9**, 11.

39 Lo, C.M., Wang, H.B., Dembo, M. and Wang, Y.L. (2000) *Biophys. J.*, **79**, 144.

40 Denis, F.A., Hanarp, P., Sutherland, D.S., Gold, J., Mustin, C., Rouxhet, P.G. and Dufrene, Y.F. (2002) *Langmuir*, **18**, 819.

41 Han, M., Sethuraman, A., Kane, R.S. and Belfort, G. (2003) *Langmuir*, **19**, 9868.

42 Xia, Y., Rogers, J.A., Paul, K.E. and Whitesides, G.M. (1999) *Chem. Rev.*, **99**, 1823.

43 Xia Y. (ed.) (2004) *Adv. Mater., 16*, (special issue dedicated to George Whitesides).

44 Moreau, W.M. (1988) *Semiconductor Lythography: Principles and Materials*, Plenum, New York.

45 Brambley, D., Martin, B. and Prewett, P.D. (1994) *Adv. Mater. Opt. Electron.*, **4**, 55.

46 Pan, Y.V., Hanein, Y., Leach-Scampavia, D., Bohringer, K.F., Ratner, B.D. and Denton, D.D. (2001) *14th IEEE International Conference on MEMS;*435.

47 Clemmens, J., Hess, H., Howard, J. and Vogel, V. (2003) *Langmuir*, **19**, 1738.

48 Clemmens, J., Hess, H., Lipscomb, R., Hanein, Y., Böringer, K.F., Matzke, C.M., Bachand, G.D., Bunker, B.C. and Vogel, V. (2003) *Langmuir*, **19**, 10967.

49 Blawas, A.S. and Reichert, W.M. (1998) *Biomaterials*, **19**, 595.

50 Barbucci, R., Lamponi, S., Magnani, A. and Pasqui, D. (2002) *Biomol. Eng.*, **19**, 161.

51 Magnani, A., Priamo, A., Pasqui, D. and Barbucci, R. (2003) *Mater. Sci. Eng. C*, **23**, 315.

52 Dorman, G. and Prestwich, G.D. (2000) *TIBTECH*, **18**, 64.

53 Barbucci, R., Torricelli, P., Fini, M., Pasqui, D., Favia, P., Sardella, E., d'Agostino, R. and Giardino, R. (2005) *Biomaterials*, **26**, 7596.

54 Okazaki, S. (1991) *J. Vac. Sci. Technol. B*, **9**, 2829.

55 Li, N., Tourovskaia, A. and Folch, A. (2003) *Crit. Rev. Biomed. Eng.*, **31**, 423.

56 Xia, Y. and Whitesides, G. (1998) *Angew. Chem. Int. Ed.*, **37**, 550.

57 Kane, R.S., Takayama, S., Ostuni, E., Ingber, D.E. and Whitesides, G.M. (1999) *Biomaterials*, **20**, 2363.

58 Geissler, M. and Xia, Y. (2004) *Adv. Mater.*, **16**, 1249, and references therein.

59 Aizenberg, J., Black, A.J. and Whitesides, G.M. (1999) *Nature*, **398**, 495.

60 Ha, K., Lee, Y.-J., Jung, D.-Y., Lee, J.H. and Yoon, K.B. (2000) *Adv. Mater.*, **16**, 6968.

61 Kind, H., Geissler, M., Shmid, H., Michel, B., Kern, K. and Delamarche, E. (2000) *Langmuir*, **16**, 6367.

62 Hovis, J.S. and Boxer, S.G. (2001) *Langmuir*, **17**, 3400.

63 Tan, L., Tien, J. and Chen, C.S. (2002) *Langmuir*, **18**, 519.

64 Vickie Pan, Y., McDevitt, T.C., Kim, T.K., Leach-Scampavia, D., Stayton, P.S., Denton, D.D. and Ratner, B.D., (2002) *Plasmas Polym.*, **7**, 171.

65 Tanaka, T., Morigami, M. and Atoda, N. (1993) *Jpn. J. Appl. Phys.*, **32**, 6059.

66 Rogers, J.A., Paul, K. and Whitesides, G.M. (1998) *J. Vac. Sci. Technol. B*, **16**, 88.

67 Yasuda, H. and Matsuzawa, Y. (2005) *Plasma Process. Polym.*, **2**, 507.

68 Favia, P., Sardella, E., Gristina, R. and d'Agostino, R. (2003) *Surf. Coat. Technol.*, **169–170**, 707.

69 Schröder, K., Meyer-Plath, A., Keller, D. and Ohl, A. (2002) *Plasmas Polym.*, **7**, 103.

70 Wu, Y., Kouno, M., Saito, N., Nae, F.A., Inoue, Y. and Takai, O. (2007) *Thin Solid Films*, **515**, 4203.

71 Wu, Y., Sugimura, H., Inoue, Y. and Takai, O. (2002) *Chem. Vap. Depos.*, **8**, 47.

72 Sardella, E., Gristina, R., Senesi, G.S., d'Agostino, R. and Favia, P. (2004) *Plasma Process. Polym.*, **1**, 63.

73 Detomaso, L., Gristina, R., Senesi, G.S., d'Agostino, R. and Favia, P. (2005) *Biomaterials*, **26**, 3831.

74 France, R.M., Short, R.D., Duval, E., Jones, F.R., Dawson, R.A. and McNeil, S. (1998) *Chem. Mater.*, **20**, 1176.

75 Siow, K.S., Britcher, L., Kumar, S. and Griesser, H.J. (2006) *Plasma Process. Polym.*, **3**, 392.

76 Slocik, J.M., Beckel, E.R., Jiang, H., Enlow, J.O., Zabinski, J.S., Jr., Bunning, T.J. and Naik, R.R. (2006) *Adv. Mater.*, **18**, 2095.

77 Valsesia, A., Silvan, M.M., Ceccone, G., Gilliland, D., Colpo, P. and Rossi, F. (2005) *Plasma Process. Polym.*, **2**, 334.

78 Girard-Lauriault, P.-L., Mwale, F., Iordanova, M., Demers, C., Desjardins, P. and Wertheimer, M.R. (2005) *Plasma Process. Polym.*, **2**, 263.

79 Kogelschatz, U. (2003) *Plasma Chem. Plasma Process*, **23**, 1.

80 Lei, Y., Cai, W. and Wilde, G. (2007) *Prog. Mater. Sci.*, **52**, 465.

81 Krozer, A., Nordin, S.-A. and Kasemo, B. (1995) *J. Colloid Interf. Sci.*, **176**, 479.

82 Hanarp, P., Sutherland, D.S., Gold, J. and Kasemo, B. (2003) *Colloid Surf. A: Physicochem. Eng. Aspects*, **214**, 23.

83 Whitesides, G.M. and Grzybowski, B.A. (2002) *Science*, **295**, 2418.

84 Bretagnol, F., Valsesia, A., Ceccone, G., Colpo, P., Gilliland, D., Cerotti, L., Hasiwa, M. and Rossi, F. (2006) *Plasma Process. Polym.*, **3**, 443.

85 Jang, P., Bertone, J.F., Hwang, K.S. and Colvin, V.L. (1999) *Chem. Mater.*, **11**, 2132.

86 Tien, J., Terfort, A. and Whitesides, G.M. (1997) *Langmuir*, **13**, 5349.

87 Curtis, A.S.G., Casey, B., Gallagher, J.G., Pasqui, D., Wood, M.A. and Wilkinson, C.D.W. (2001) *Biophys. Chem.*, **94**, 275.

88 Valsesia, A., Colpo, P., Manso, M., Meziani, T., Ceccone, G. and Rossi, F. (2004) *Nanoletters*, **4**, 1047.

89 Valsesia, A., Colpo, P., Meziani, T., Bretagnol, F., Lejeune, M., Rossi, F., Bouma, A. and Garcia-Parajo, M. (2006) *Adv. Func. Mater.*, **16**, 1242.

90 Shi, H., Tsai, W.-B., Garrison, M.D., Ferrari, S. and Ratner, B. (1999) *Nature*, **398**, 593.

91 Schild, H.G. (1992) *Prog. Polym. Sci.*, **17**, 163.

92 Zareie, H.M., Bulmus, E.V., Gunning, A.P., Hoffman, A.S., Piskin, E. and Morris, V.J. (2000) *Polymer*, **41**, 6723.

93 Kujawa, P. and Winnik, F.M. (2001) *Macromolecules*, **43**, 4130.

94 Shimizu, T., Yamato, M., Kikuchi, A. and Okano, T. (2001) *Tissue Eng.*, **7**, 141.

95 Cheng, X., Wang, Y., Hanein, Y., Bohringer, K.F. and Ratner, B.D. (2004) *J. Biomed. Mater. Res. A*, **70**, 159.

96 Pan, Y.V., Wesley, R.A., Uginbuhl, R.L., Denton, D.D. and Ratner, B.D. (2001) *Biomacromolecules*, **2**, 32.

97 Favia, P., Cicala, G., Milella, A., Palumbo, F., Rossini, P. and d'Agostino, R. (2003) *Surf. Coat. Technol.*, **169–170**, 609.

98 Milella, A., Palumbo, F., Favia, P. and d'Agostino, R. (2005) *Pure Appl. Chem.*, **77**, 399.

99 Kay, E., Coburn, J.W. and Dilks, A. (1980) *Topics Curr. Chem.*, 94.

100 Han, L.C.M., Timmons, R.B. and Lee, W.W. (2000) *J. Vac. Sci. Technol. B*, **18**, 799.

101 Limb, S.J., Lau, K.K., Edell, D.J., Gleason, E.F. and Gleason, K.K. (1999) *Plasma Polym.*, **4**, 21.

102 Gristina, R., D'Aloia, E., Senesi, G.S., Sardella, E., d'Agostino, R. and Favia, P. (2004) *Eur. Cells Mater.*, **7**, 8.

103 D'Aloia, E., Senesi, G.S., Gristina, R., d'Agostino, R. and Favia, P. (2005) Proceedings of 17th International Symposium on Plasma Chemistry (ISPC-17), Toronto, Canada, 7–12, August.

104 Senesi, G.S., D'Aloia, E., Gristina, R., Favia, P. and d'Agostino, R. (2007) *Surf. Sci.*, **601**, 1019.

105 Rosso, F., Marino, G., Muscariello, L., Cafiero, G., Favia, P., D'Aloia, E., d'Agostino, R. and Barbarisi, A. (2006) *J. Cell Physiol.*, **207**, 636.

106 Li, W.Z., Wen, J.G., Tu, Y. and Ren, Z.F. (2001) *Appl. Phys. A*, **73**, 259.

107 Correa-Duarte, M.A., Wagner, N., Rojas-Chapana, J., Morsczeck, C., Thie, M. and Giersig, M. (2004) *Nano-letters*, **4**, 2233.

108 Bowden, N., Brittain, S., Evans, A.G., Hutchinson, J.W. and Whitesides, G.M. (1998) *Nature*, **393**, 146.

109 Wang, X. and Grundmeier, G. (2006) *Plasma Process. Polym.*, **3**, 39.

110 Jeong, H.J., Markle, D.A., Owen, G., Pease, F., Grenville, A., von, R. and Nau, B. (1994) *Solid State Technol.*, **37**, 39.

111 White, D.L., Bjorkholm, J.E., Bokor, J., Eichner, L., Freeman, R.R., Jewell, T.E., Mansfield, W.M., MacDowell, A.A., Szeto, L.H., Taylor, D.W., Tennant, D.M., Waskiewicz, W.K., Windt, D.L. and Wood, O.R. (1991) *Solid State Technol.*, 37.

112 Dunn, P.N. (1994) *Solid State Technol.*, 49.

113 Jones, R.G. and Tate, P.C.M. (1994) *Adv. Mater. Opt. Electron.*, **4**, 139.

114 Minne, S.C., Manalis, S.R., Atalar, A. and Quate, C.F. (1996) *Appl. Phys. Lett.*, **68**, 1427.

115 Xia, Y., Kim, E., Zhao, X.-M., Rogers, J.A., Prentiss, M. and Whitesides, G.M. (1996) *Science*, **273**, 347.

116 Kumar, A. and Whitesides, G.M. (1993) *Appl. Phys. Lett.*, **63**, 2002.

117 Kim, E., Xia, Y., Zhao, X.-M. and Whitesides, G.M. (1997) *Adv. Mater.*, **9**, 651.

118 Xia, Y. and Whitesides, G.M. (1998) *Angew. Chem., Int. Ed. Engl.*, **37**, 550.

119 Kim, E., Xia, Y. and Whitesides, G.M. (1995) *Nature*, **376**, 581.

120 Xia, Y. and Whitesides, G.M. (1998) *Annu. Rev. Mater. Sci.*, **28**, 153.

121 Allard, M., Sargent, E.H., Lewis, P.C. and Kumacheva, E. (2004) *Adv. Mater.*, **16**, 1360.

122 Jenekhe, S.A. and Chen, X.L. (1999) *Science*, **283**, 372.

14

Chemical Immobilization of Biomolecules on Plasma-Modified Substrates for Biomedical Applications

L.C. Lopez, R. Gristina, Riccardo d'Agostino, and Pietro Favia

The research of new synthetic biomaterials is nowadays an active research field which combines together several scientific disciplines such as medical sciences, biology, chemistry, physics, and engineering. The aim of this multidisciplinary approach is to achieve the synthesis of materials intended to interact with biological systems in order to perform, augment, or replace a living natural function. The big challenge in the biomaterial science is represented by the understanding on how the surface chemistry of a material can be used to control the biological activity of a cell interacting with its surface. The evidence that the biological response to a material is strictly related to its surface highlights the important role of surface modification techniques in achieving a "physiological" biological response. In order to achieve this goal, plasma modification processes have been largely used as a tool to produce surface functionalization of biomaterials.

A step forward in the control of cell–surface interactions of biomaterials has been reached by the immobilization of bioactive molecules on plasma-modified materials in order to provide specific targets for cell recognition and growth. Different plasma processes, both in deposition and grafting modes, have been used to produce surfaces characterized by functional groups that allow one to immobilize biomolecules either through chemical attachment (covalent bonding, ionic or hydrophilic interaction), or through physical entrapment. Surfaces functionalized with specific biomolecules can be applied to trigger a specific biological response and thus to drive the cell–protein interaction with the surface.

This chapter focuses on the role that different plasma modification processes (deposition, treatments, grafting polymerization) of polymeric materials represent in the immobilization of peptides, carbohydrates, enzymes, and other biomolecules. Biocompatibility assays aimed to validate the specific biological response to those modified materials will also be described.

Advanced Plasma Technology. Edited by Riccardo d'Agostino, Pietro Favia, Yoshinobu Kawai, Hideo Ikegami, Noriyoshi Sato, and Farzaneh Arefi-Khonsari
Copyright © 2008 WILEY-VCH Verlag GmbH & Co. KGaA, Weinheim
ISBN: 978-3-527-40591-6

14.1
Introduction

Biomaterials are synthetic, natural, or a combination of both materials, used, in medical devices, to treat or replace tissues, organs, or functions of the body, which intimately interact with biological environments [1]. For these aims, the research of new and improved biomaterials, capable at the same time of replacing a missing biological function and controlling the host response, is attracting a growing scientific interest due to the possibility of saving every year millions of human lives and/or improving the quality of life. Heart valves, dent implants, intraocular lenses, dialysis membranes, orthopedic devices, stents, and vascular grafts are only a few well-known examples of medical devices that are improved by the application of new type of biomaterials.

Biomaterials can be built with a combination of conventional materials: ceramics, metals, composites, and, above all, due to their excellent properties and to their widespread applicability, with polymeric materials. Polymers display indeed stability properties, good mechanical features, and low weight and therefore have been widely used in the fabrication of biomedical devices such as prostheses, catheters, and intraocular lenses, just to name a few significant examples. Despite those excellent "bulk" materials, polymers need to be properly surface modified in order to be used in the biotechnology field since they do not present an "attractive" surface suitable for interactions with biological environments. In the biomaterial science jargon it means that polymeric surfaces do not fulfill biocompatibility requirements, that is, "the ability of a material to perform with an appropriate host response in a specific application" [2]. The main requirements of a biomaterial rely on its ability in avoiding a chronic inflammatory response, in resisting bacterial colonization, and in promoting normal differentiation in the surrounding tissues. Moreover, as it is clear from the biocompatibility definition, specific requirements for biomaterials largely vary with the application and with the particular site of implantation. For example, for stable tissue integration, a surface modification which promotes specific cell adhesion is highly desirable; on the contrary, for fluid-contacting applications it might be necessary to avoid adhesive substrates. Therefore, the main challenges in biomaterials sciences are represented by the possibility to control adhesion and proliferation of cells on materials (e.g. to promote the reconstruction of a new organ), as well as to avoid proteins and platelet adsorption on sites were no cell adhesion is required. To reach these goals, the very recent advances in this field aim to an in-depth understanding of how the surface chemistry can be used on a material to control the biological activity of a cell interacting with a surface and, therefore, to precisely control its biological response [3,4].

In the efforts to guide cell material interactions *in vitro* and *in vivo*, biological events are strongly influenced by the presence of specific molecules on cell membrane, like integrins, that specifically interact with molecules found *in vivo*, on other cells or on the extracellular matrix (ECM), or with biomolecules adsorbed on synthetic materials *in vitro*. These types of interaction have to be strictly specific in order to drive cell adhesion and behavior. In order to obtain a specific cell response toward biomaterials many attempts have been exploited in modifying surfaces with appropriate signaling molecules, e.g. by immobilizing biomolecules that can be specifically recognized by cells. Different kinds of biomolecules have been immobilized onto different natural and synthetic substrates to drive a specific cell–surface interaction: whole proteins (vitronectin, fibronectin) as well as short peptides, enzymes, and even DNA fragments. A list of biological moieties that have been immobilized onto various substrates is reported in Table 14.1.

Biomolecule immobilization can be performed to obtain temporary or permanent bonding of the biomolecule to a material. For example, in the case of drug delivery systems, the immobilized drug has to be released in a predefined time after its introduction into the organism, while for specific tissue growth on a biomaterial, molecules have to be permanently bonded.

Several immobilization methods have been applied and can be summarized in two main categories: chemical attachments by means of covalent bonding, and physical entrapment and adsorption methods [5]. Physical adsorption methods mainly consist in a nonspecific molecule attachment to substrates due to van der Waals and electrostatic forces. On the contrary, covalent immobilization requires the presence of surface "anchor" functionalities; although many synthetic and natural polymers already possess pendant anchor groups, most others need to be modified to provide suitable surface anchor functionalities for chemical immobilization. Therefore, in order to perform chemical immobilization, a preliminary surface functionalizing step, which can be obtained by wet chemical or physical methods, is always required. In particular, material surfaces can be modified by:

- chemical methods (wet oxidations, acid treatments, etc.),
- physical methods (adsorption, Langmuir–Blodgett films, etc.), or
- physicochemical methods (laser treatments, plasma processes, etc.).

Among this wide variety of techniques, wet chemical procedures very often use harsh conditions and solvents which may remain entrapped in the bulk of materials, causing possible harm to the biological system in contact. Physicochemical methods,

Tab. 14.1 Biologically Active Molecules for Immobilization on Biomaterial Surfaces.

Proteins (collagen, vitronectin, laminin, fibronectin, etc.)
Enzymes
Antibodies
Short peptides (RGD, YIGSR, etc.)
Carbohydrates
Polysaccharides
Oligosaccharides
Lipids
Antibiotics
Antithrombogenic molecules
DNA

instead, such as plasma processes, represent an efficient, versatile, nondestructive, easy-to-use way to afford the first functionalizing step of polymers for biomedical applications. Plasma processes appear very profitable because it is possible to achieve surface chemical and physical modifications of the few topmost layers of materials while retaining their mechanical, physical, and chemical bulk properties unaltered [6,7]. Furthermore, different materials with different shapes can be plasma processed in a convenient way allowing an homogeneous surface modification, even conformal to the topography of the substrate. Plasma modification of surfaces can be considered less harmful than chemical methods and no chemical leaching occurs from stable plasma-modified surfaces. Moreover, the possibility to functionalize the surface of materials, even of thermolabile substrates, makes plasma techniques really appealing among other surface modification methods. Plasma modification processes have been already widely used in modifying materials for biomedical purposes [8,9], since they allow one to obtain surfaces with a tunable density of functional groups. By changing the feed gas, in the case of treatment, *"grafting"* processes (NH_3, O_2, H_2O, etc.), or by using non-polymerizable organic compounds, for plasma depositions, it is possible to obtain surfaces rich in a wide variety of functional groups such as amines, carboxyl, hydroxyl, imine, epoxy, isocyanate, etc. Functional groups obtained via plasma might *per se* confer adhesive and/or repulsive "biological properties" to the material, or can be used as anchor groups for biomolecule immobilization, and provide a more specific biological response.

It is well documented that both oxygen- [10] and nitrogen-containing groups promote cell adhesion and growth [11]. For example, glow discharges fed with vapors of acrylic acid (AA) that deposit thin functional organic films are characterized by –COOH and other O-bearing groups and have been extensively used as cell-adhesive surfaces [12]. Plasma-deposited poly(ethylene oxide) (PEO)-like coatings have been extensively used as unfouling surfaces for different bio-applications [13].

In the schemes of Fig. 14.1(a)–(c) it is shown how different types of plasma modification processes produce suitable surfaces for immobilization purposes: plasma treatments (*grafting*); plasma deposition by means of plasma-enhanced chemical vapor deposition (PE-CVD); and plasma-induced graft polymerization. Plasma treatments by means of non-polymerizable gases (H_2, O_2, NH_3, etc.) can lead to the synthesis of surfaces rich of N- and O-containing functional groups; also plasma deposition of monomers (allylamine, acroleyn, acrylic acid, etc.) by means of PE-CVD leads to plasma-deposited layers rich in N- or O-containing groups such as amino, carboxylic, and aldheyde. Surfaces displaying surface chemical functionalities can be also obtained by means of plasma-induced graft polymerization of organic monomers [9b], which formally is a plasma treatment process necessary to activate the surface, followed by a chemical interaction of the treated surface with an organic monomer.

After the functionalizing step the modified surfaces can be used in wet chemical reactions to perform the immobilization of various biomolecules (enzymes, peptides, carbohydrates, etc.) even by means of "tether" molecules that tie them at the surface with the correct active conformation in order to maintain their biological activity unaltered [14]. In order to avoid the use of solvents and harsh conditions the immobilization step is obtained in water-based media, using mild coupling procedures.

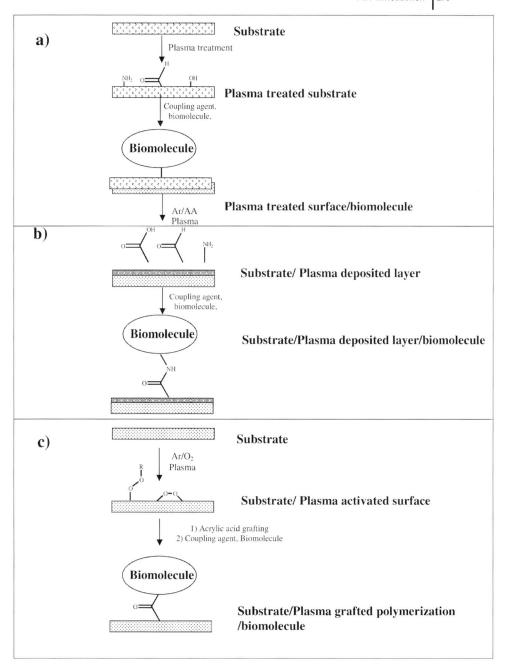

Fig. 14.1 Surface modification procedures: (a) plasma treatment, (b) plasma deposition of thin coatings, and (c) plasma-induced graft polymerization. The produced surfaces are all rich in chemical anchor groups (–COOH, –NH₂, –OH, etc.) that can be used as primer anchors for the immobilization of biomolecules.

In the following part of the chapter it will be shown how, in the last two decades, plasma processes have been used to produce functionalized surfaces suitable for the immobilization of various biomolecules. The results obtained using different plasma processes and different biomolecules, such as PEO chains, polysaccharides, proteins, short peptides, carbohydrates, and enzymes, will be investigated. Particular attention will be devoted to the various chemical coupling procedures applied and to the biological tests that prove the biocompatibility of the produced surfaces.

14.2
Immobilization of Biomolecules

14.2.1
Immobilization of PEO Chains (Unfouling Surfaces)

One of the major challenges in designing materials intended for biological environments is the production of surfaces that resist nonspecific protein adsorption. When a biomaterial is implanted *in vivo*, in fact, after a few minutes a layer of nonspecific adsorbed proteins is formed. Subsequently, many events which generate the so called "foreign body reaction", such as cytokine production or macrophage attack, will occur to develop the formation of a thin collageneous capsule which isolates the implant from the body [15]. The layer of adsorbed proteins plays therefore the crucial role of mediating the interaction of cells with the biomaterial surface. Understanding this mechanism is fundamental for engineering materials that are able to drive the correct biological response for the intended use. Proteins randomly adsorbed at the biomaterial surface will not specifically interact with receptors, proteins, or glycoproteins located at the cell membrane, in order to drive a cell-specific response. The only possibility to drive a specific cell–surface interaction (e.g. a specific function) is therefore to avoid a nonspecific protein adsorption which may occur immediately after the implantation.

To prevent protein adsorption poly(ethylene glycol) (PEG), also known as poly(ethylene oxide) (PEO), a synthetic, water-soluble polymer has been largely employed in several practical applications ranging from chemistry to biomedicals and cosmetics [16]. The large interest that PEO coatings generate in these applications is mainly due to its low degrees of protein adsorption [17], together with low cell and bacterial adhesion [18]. These important features make PEO surfaces extremely appealing for the development of implants, biosensors, and for the construction of diagnostic assays.

Physical adsorption [19], graft polymerization [20], plasma polymerization [21], covalent attachment [22], and direct PEO molecules attachment [23] have been all performed in order to graft PEO chains to various artificial surfaces. Nevertheless, it is worth noting that the ability of PEO coatings in preventing nonspecific protein adsorption may also depend not upon the immobilization method but rather upon the chain density [13].

In order to generate permanent nonfouling surfaces, PEO surfactants have been covalently immobilized on polyethylene (PE) surfaces dip-coated with a PEO surfactant layer and subsequently treated with a very low-power Ar plasma treatment [24]. Both *in vitro* platelet adhesion and fibrinogen adsorption tests showed the nonfouling character of the surface. Another approach to immobilize PEO chains on plasma-modified surfaces consisted in the use of carboxylic groups as anchor functionalities. In particular, PE surfaces have been exposed to an Ar plasma treatment and, subsequently, the surfaces have been reacted in a 50% acrylic acid aqueous solution obtaining a surface rich in carboxylic groups. The covalent attachment procedure of bis-amino PEO chains of different molecular weights has been accomplished with 1-cyclohexyl-3-(2-morpholinoethyl) carbodiimide (CMC) used as coupling agent [25].

Other interesting approaches report on the immobilization of amino-PEO chains onto fluorinated ethylene propylene copolymer (FEP) without any use of coupling agent, simply using an aldehyde-rich plasma-deposited layer [26] obtained from acroleyn- or acetaldehyde-fed discharges. The immobilization of amino-PEO chains has been obtained through the formation of a Schiff base with the aldehyde groups, followed by a reduction with cyanoborohydride. The successful immobilization of PEO chains has been confirmed by XPS, FTIR, SIMS, and WCA measurements. The resulting PEO surface has been tested for its repulsion to fibrinogen adsorption and compared with native FEP- and aldehyde-based surfaces. The fibrinogen adsorption on PEO surfaces was 28 and 43% less than those adsorbed on aldehyde and FEP, respectively.

14.2.2
Immobilization of Polysaccharides

In the efforts to design protein-resistant surfaces, polysaccharide coatings have also been found to behave as unfouling surfaces [27]. Nevertheless, although most polysaccharide-based coatings show unfouling features, in the literature there are some evidences of cell attachment onto coatings obtained from hyaluronic acid [28], and onto esterified hyaluronic acid [29] as well as examples of protein adsorption [30]. This peculiar dual behavior seems to be due to the surface architecture of polysaccharide molecules that, in some cases, drives cell adhesion while in others generates nonfouling layers. The following examples will show this unique dual behavior of polysaccharide coatings: generating nonfouling coatings or improving cell adhesion.

Polysaccharides and in particular alginate and hyaluronan acid have been immobilized on plasma-modified membranes to obtain antibacterial surfaces [31]. Air plasma treatments have been performed on PS to activate surfaces that have been reacted afterwards with a polyethyleneimine (PEI). A carbodiimide-aided coupling reaction has been then performed between the amino PEI groups and the COOH groups of alginate and hyaluronan acid. Polysaccharide-modified substrates have been used for L-929 mouse fibroblast cell culture and RP62A *Staphylococcus epidermis* bacterial adhesion experiments. Surfaces displayed a cell-resistant effect and the adhesion of bacteria has been reduced by several orders of magnitudes with respect to the untreated substrate.

Carboxymethyldextrans (CMD)s can be conveniently used as model polysaccharide compounds and were attached onto plasma-deposited aminated Teflon (FEP) surfaces obtained both from the deposition of *n*-heptylamine displaying surface amino groups and by using surface aldehyde groups plasma deposited from acetaldehyde [32]. CMD with different molecular weights have been immobilized directly on coatings plasma deposited from *n*-heptylamine or through a polyamine spacer arm via a water-based coupling procedure using 1-ethyl-3-(3-dimethylaminopropyl)carbodiimide (EDC) and *N*-hydroxysuccinimide (NHS) as coupling reagents. XPS analysis indicated that a larger amount of CMD was bound on the surface when a polyamine spacer was used if compared with CMDs directly attached. Colonization of bovine corneal epithelial (BCEp) cells was observed as strictly dependent on the immobilization pathway used for the CMD; in particular, cell attachment and growth were totally inhibited especially when a spacer molecule was introduced between the surface and the CDM while cell attachment was observed on surfaces fabricated directly immobilizing CMDs onto plasma-aminated surfaces. No influence on cell attachment was observed by changing the CMD molecular weight or the carboxymethyl substitution of CMD. Only a complete coverage of the substrate with CMD provided fully cell-resistant surfaces.

Together with their cell-resistant properties, polysaccharide-based coatings provide natural highly hydrated surfaces which are desirable for a large number of biomedical applications. A convenient way to produce hydrophilic films focuses on thin polysaccharide coatings as potential biomaterial surfaces [33] obtained via covalent immobilization (periodate oxidation, aqueous solution) on plasma-modified surfaces displaying surface amine groups. In particular, FEP and organosilicon polymer polytrimethylsilylpropyne (PTMSP) surfaces were modified by means of ammonia plasma discharge and by deposition of *n*-heptylamine [34].

Very recently, plasma micropatterned poly(ethylene terephthalate) (PET) surfaces activated by means of Ar/NH$_3$-fed discharges have been used as primer layers for the photoimmobilization of hyaluronan (hyal) [35]. Hyal has attracted much interest since it represents one of the glycosaminoglycan components of extracellular matrix (ECM) proteins. Hyal matrices have been successfully proved to promote chondrocytes adhesion and proliferation. Micropatterned hyal surfaces induced cell adhesion, cell proliferation, and differentiation of articular knee cartilage chondrocytes. Furthermore, the production of aggrecan and collagen II type has been increased onto modified surfaces [36].

14.2.3
Immobilization of Proteins and Peptides

Several approaches have been developed to drive cell adhesion and plasma-modified surfaces have been proven to show unique cell adhesion properties. Nevertheless, to improve cell adhesion, to obtain a specific growth, and differentiation, different biomolecules, mainly proteins and peptides, have been immobilized on the surface of substrates intended for bioapplication.

The immobilization of whole proteins, if compared with small peptides, shows some advantages related to the possibility of promoting multiple integrin-mediated responses. Integrins mediate signal transduction through interaction with multiple cellular or extracellular matrix ligands. Nevertheless, although the literature contains numerous examples of immobilization of proteins onto various substrates it seems to be much more convenient to immobilize small peptides for several reasons. Proteins, for example, may undergo thermal denaturation, are more expensive, and suffer very often pH-driven modifications [36]. Furthermore, nonspecific adsorption of proteins may cause changes in orientation and conformation, and thus denaturation and inactivation, of the protein, for example due to the hiding of active sites. Peptides, in contrast, can be easily synthesized with a very high degree of pureness, are inexpensive, do not suffer from pH and temperature variation, and exhibit higher stability towards sterilization [37]. It should be highlighted that the immobilization of short peptides, although easier from the experimental point of view, only weakly emulates the multifunctionality of proteins that often carry thousand of amino acid sequences.

Several proteins have been immobilized onto various plasma-modified and native polymeric substrates to improve blood compatibility or to improve cell adhesion and growth. Plasma-modified surfaces have been used to perform chemical immobilization of proteins and among many other examples, serum albumin, gelatin, and collagen have been immobilized onto poly(methyl methacrylate) films plasma treated in O_2 plasma followed by acrylic acid grafting; the immobilization has been performed by using carboxylic surface groups as anchor functionalities [38]. Transferrin and insulin have been immobilized onto NH_3 plasma-treated polyurethane membranes to enhance fibroblast cell growth [39]. An interesting example of co-immobilization of two proteins, in order to improve blood biocompatibility, has been performed with plasma-treated PET substrates [40]. In particular, PET has been exposed to an O_2 plasma treatment and subsequently immersed into an acrylic acid solution to obtain a surface rich in COOH groups; these anchor moieties have been used to co-immobilize insulin [41] which is well known to enhance cell proliferation and heparin which promotes cell attachment.

Because of space limitations, in the next paragraphs we will focus our attention on the immobilization of one specific protein, collagen and collagen-like molecules, since they represent the most abundant protein of the ECM and are widely employed in the biomaterial field.

14.2.3.1 Immobilization of Collagen

Many different types of collagen proteins are present in the ECM that perform different tasks during the development and maintaining of the ECM by themselves or interacting with other ECM proteins, such as fibronectin. Due to its peculiar features and to its natural biodegradable properties collagen is widely used in many biomedical applications [42].

An interesting application dealing with the immobilization of collagen has been reported for plasma-treated poly(lactic acid) (PLA), which is of great importance among biodegradable polymers and has been recently approved by the US Food and Drug Administration for implantation in the human body. PLA does not display any surface functional group suitable for biomolecule immobilization and, therefore, O_2 and NH_3 plasma treatments were employed to graft N- and O-containing groups for collagen anchorage [43]. Poly(D,L-lactide) (PDLLA) has been used as substrate for 3T3 fibroblast cell culture. Cell culture experiments performed on collagen anchored on plasma-modified PDLLA showed that this surface improves the adhesion of 3T3 fibroblasts [44].

Type III collagen has been linked onto a plasma-modified silicon rubber (SR) polymer displaying surface COOH groups for improved cornea ephithelial cell culture. In particular, plasma-induced graft polymerization of acrylic acid was performed after an Ar pre-treatment followed by O_2 exposure to introduce peroxides groups. The covalent binding of poly(acrylic acid) has been performed after the thermal decomposition of peroxides. A water-soluble coupling agent, CMC, has been used to couple collagen amino groups and COOH surface groups. Collagen linked on the surface has given proof, *in vitro*, of its capability to enhance primary rabbit cornea cell adhesion and growth [44].

Furthermore, PET surfaces have been graft polymerized with acrylic acid to introduce COOH groups; afterwards types I and III collagen have been immobilized to improve the adhesion of human smooth muscle cells [45]. PET substrates have been plasma activated using Ar and then exposed to O_2 to allow the covalent binding of poly(acrylic acid) following a procedure similar to that previously described. The effectiveness of the immobilization has been investigated with smooth muscle cells grown in a serum-free culture medium to really evaluate the role of collagen in driving cell adhesion. The adhesion of smooth muscle cells was proved onto plasma-modified films where cells were grown with serum proteins, but was surprisingly suppressed when cells were grown in serum-free medium onto collagen immobilized substrates. Although ECM proteins are very often immobilized to promote adhesion and growth of many different cell types, in this work it was highlighted that types I and III collagen do not seem to be the ideal adhesion proteins to favor smooth muscle cell adhesion in the examined culture conditions.

Plasma-modified surfaces have been used also to immobilize collagen-like molecules (CLMs), synthetic proteins which mimic the properties of natural collagen. These synthetic molecules mimic properties of natural collagen and, in particular, are able to form the triple helical structure (*in vivo*) and therefore may be a valid substitute of collagen in bio-intended applications [46]. FEP substrates have been modified by means of plasma-deposited coatings rich in aldehyde and amine groups and used for CLMs binding to determine whether these molecules are able to stimulate cell binding and tissue colonization. CLMs have been linked onto aldehyde-rich FEP-modified surfaces via cyanoborohydride-aided reaction leading to the formation of a Schiff base between the surface and the molecules. CLMs have been immobilized onto amino group-rich surfaces through a CMD molecule used as a polycarboxylate linker molecule. The CLMs have been linked to the COOH-rich CMD molecule via an EDC/NHS coupling procedure. Bovine corneal epithelial cells were used for cell colonization experiments and these experiments showed that CLMs showed a biological response which may be promising for tissue-interfacing biomedical applications.

14.2.3.2 **Immobilization of Peptides**

Cell adhesion through adsorbed proteins is mediated by specific integrins found on the cell membrane. A number of different short amino acid sequences, which mediate cell-specific adhesion present on ECM adhesive proteins, have been identified and used to modify biomaterial surfaces.

The most common peptide that has been immobilized to enhance cell adhesion and growth of different cell lines is the RGD peptide (Arg-Gly-Asp) that represents the minimum adhesion domain contained in ECM, such as fibronectin and vitronectin [47]. The integrin binding capability of ECM molecules has been mapped to specific oligopeptide sequences within ECM proteins. Furthermore, other peptidic fragments such as YIGSR (Tyr-Ile-Gly-Ser-Arg) which represents the minimum adhesion domain of laminin, a glycoprotein that exercises several biological activities have also been investigated and immobilized onto various substrates [48]. In this review we mainly focus our attention on the immobilization of RGD-containing peptides on plasma-modified surfaces.

RGD and YIGSR have been both immobilized onto plasma-deposited allylamine coatings on polycaprolactone (PCL) and poly(L-lactic acid) (PLLA) to improve AML12 hepatocyte adhesion [49]. PEG chains have been first reacted with the adhesion peptides and then have been coupled with the polyallilamine coating displaying surface amino groups. Modified surfaces, carrying RGD and YIGSR adhesion domains, have been developed to mimic the *in vivo* ECM to enhance hepatocyte adhesion. Proper adhesion assays demonstrated an enhancement of hepatocyte adhesion onto both PCL and PLLA substrates modified with the peptides.

We have reported the immobilization of RGD peptides onto surface carboxylic groups obtained via PE-CVD of acrylic acid (plasma-deposited acrylic acid, pdAA) [50]. The immobilization of RGD has been accomplished trough a bis-amino spacer molecule (*O,O′*-bis(2-aminopropyl)-polyethylene glycol 500) using a water-soluble carbodiimide (EDC)-based reaction. The spacer molecule (SA) bonded to the surface via amide bonds plays the key role of spacing the peptide from the substrate allowing the peptide to easily reach the binding site of the integrins. 3T3 fibroblasts were cultured onto PET substrates modified with RGD and cell adhesion was enhanced on these substrates with respect to native PET and to pdAA substrates. In addition to the improved cell adhesion, fibroblasts cultured on RGD-modified surfaces also displayed a different morphology. In particular, while on native PET well separated, circular or elongated cells were observed, spread cells were assembled in clusters of about 4–6 cells on RGD-modified substrates as shown in Fig. 14.2.

We have also reported the immobilization of RGD peptide onto plasma-modified polyethersulfone (PES) flat membranes for primary human hepatocyte culture to control the adhesion, proliferation, and differentiation of liver cells [51]. PES membranes were modified by PE-CVD processes fed with acrylic acid in order to obtain pdAA surfaces with controlled density of COOH groups.

The modification of the membrane surface with RGD peptide determined specific cellular responses, and therefore hepatocytes cultured on PES-pdAA-SA-RGD membranes displayed an enhanced rate of albumin and urea synthesis, especially when cells were cultivated in the presence of the drug Diclofenac. Furthermore, the ability

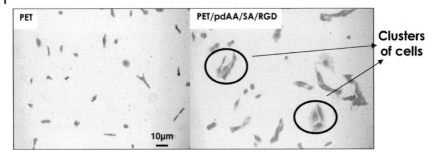

Fig. 14.2 3T3 fibroblasts cultured onto PET and PET/pdAA/SA/RGD; clusters of 4–5 cells are evident only on RGD-modified substrates.

of hepatocytes to eliminate Diclofenac and its metabolites 4′-OHdic and N,5-(OH)$_2$dic resulted in expression at high biotransformation rate when cells were cultured on PES-pdAA-SA-RGD surfaces with respect to unmodified membranes. The immobilization of RGD affected cell attachment to the modified surface inducing functional changes in terms of biotransformation, albumin production, and urea synthesis, as well as in terms of synthesis of total proteins.

Many other peptides containing the RGD sequence have been immobilized onto plasma-deposited substrates. For example, RGDC (RGD-Cys) has been immobilized, via amide bond formation, onto acrylic acid grafted on polytetrafluoroethylene (PTFE) through a low molecular weight spacer molecule [9b]. Human umbilical vein endothelial cells (HUVEC) have been cultured onto RGD-containing peptide surfaces thus revealing an enhanced cell adhesion if compared with the scarcely adhesive PTFE.

14.2.4
Immobilization of Enzymes

Enzymes are highly sensitive specific biological catalysts and are attracting much interest due to their unique properties: they display a high level of catalytic efficiency in mild conditions if compared with chemical catalysis and, furthermore, the specificity of enzyme-aided reactions is extremely high and can be used for separation of optical isomers, for regiospecific reactions, or for substrate separations. Another particular aspect to be considered is that enzymes display their activity in very mild conditions such as low temperature, neutral pH, and ambient pressure conditions, and at the end of the reaction they can be easily biodegraded. All these interesting features make enzymes valuable systems for applications in the food industry as antibacterial surfaces as well as in the pharmaceutical or biomedical field.

Enzymes immobilization has been reported on plasma-modified synthetic and natural substrates, displaying surface anchor groups conveniently used as platforms for chemical enzyme immobilization.

The immobilization of enzymes onto natural and synthetic plasma-treated substrates is very well documented: glucose oxidase [52], papain [53], β-galactosidase [54], and insulin [55] are only a few representative examples.

Glucose oxidase has been immobilized onto various polymeric membranes (polypropylene, PP, PTFE, poly(vinylidene fluoride), PVDF) that have been plasma modified with surface amino groups [56]. Membranes have been treated in NH_3- and N_2-fed glow discharges and the immobilization of the enzyme has been performed via a glutaraldehyde-aided reaction. The activity of the enzyme has been evaluated by using a glucose sensor equipped with enzyme-loaded membranes. The sensor response has been found to be related to the amount of enzyme immobilized and, therefore, to the amount of amino groups grafted by the plasma treatment. Immobilization of glucose oxidase has been also reported on plasma-deposited films obtained from PE-CVD of *N*-vinyl-2-pirrolidone onto poly(ether urethane urea) [57].

Xylose isomerase has been immobilized onto polysulfone (PSU) modified by means of a microwave plasma-deposited allyl alcohol coating rich in hydroxyl groups; in this work the critical role of Ar, as buffer for the discharge, and its contribution to stabilize the plasma has also been evaluated [58]. In particular, the presence of Ar together with allyl alcohol reduced the amount of O-containing functionalities and, therefore, reduced the amount of OH groups suitable for the immobilization of the enzyme. To obtain a support suitable for immobilization a plasma discharge without Ar had to be used. Surface hydroxyl groups have been chemically activated with divinylsulfone to allow the chemical immobilization of the enzyme through its amine groups.

Immobilization of glucose isomerase, a well-known catalyst in the conversion of D-glucose into D-fructose, has been also reported onto plasma-modified PSU surfaces. Amino group-rich surfaces have been obtained using ammonia, *n*-butylamine, and allylamine microwave plasma processes [59]. The activation step for amino groups has been performed with a gluteraldehyde-aided reaction. Glucose isomerase has been successfully immobilized on all plasma-treated samples and the presence of the enzyme, on the polymer surfaces, has been confirmed by FTIR-ATR spectroscopy.

14.2.5
Immobilization of Carbohydrates

Functionalization of synthetic polymers with sugar moieties is emerging as a significant research area. Polymers tailored with pendant carbohydrate derivatives behave as surfaces promoting the adhesion of anchorage-dependent cells such as hepatocytes [60] and have been therefore largely employed in diverse fields of applications ranging from surfaces that promote cell culture [61] to drug delivery systems [62]. Furthermore, since for many biomedical applications biodegradability is a critical issue to be addressed, carbohydrate based polymers appear particularly suitable due to their excellent biodegradability properties [63].

Although carbohydrate-modified substrates have been used to promote the adhesion of several cell lines, in this discussion we will focus our attention on the key role that carbohydrates play with respect to hepatocyte cells. In particular, galactosylated surfaces represent an interesting alternative to surface modification performed with ECM proteins such as collagen or fibronectin, for hepatocyte culture [64], since a specific interaction between the galactose ligand and the asialoglycoprotein receptor present on the hepatocyte cells membrane is well documented [65]. Several studies have shown that surfaces modified with galactose units can improve both hepatocyte attachment and the maintenance of high levels of liver-specific synthetic functions. Among the various examples reported in the literature, PET substrates have been modified by means of plasma grafting of poly(acrylic acid) under UV irradiation to obtain a surface rich in COOH functionalities. A galactose ligand has been linked to the surface and hepatocytes grown onto modified surfaces have shown a round shape and high levels of differentiated functions (urea and albumin synthesis) even comparable to hepatocytes grown on collagen, the substrate normally used to obtain hepatocyte adhesion and growth *in vitro* [66].

The galactose derivative 1-*O*-(6''-aminohexyl)-D-galactopyranoside has been immobilized onto Ar pre-treated PET films via UV-induced plasma graft copolymerization of acrylic acid [67] followed by NHS/EDC coupling. Male Wister rat hepatocytes have been cultured onto modified substrates and both albumin and urea have been expressed at higher levels with respect to collagen-coated PET surfaces.

Recently, we have also reported the immobilization of galactose derivatives on the surface of pdAA coatings, accomplished through a linear bis-amine spacer arm (bNH$_2$PEG) [68]. Galactose molecules have been first oxidized to galactonic acid according to a previously reported procedure, trough a simple oxidation procedure in the presence of a base, iodine, and methanol [69].

Furthermore, we have also immobilized galactonic acid onto plasma-modified PES membranes for primary pig and human hepatocyte culture to control the adhesion, proliferation, and differentiation of liver cells [70]. PES membranes were modified by PE-CVD processes fed with acrylic acid in order to obtain pdAA surfaces with controlled density of COOH groups. The performance of modified membranes was estimated by evaluating the expression of liver-specific biotransformation functions of pig and human hepatocytes. Human liver cells cultured on PES-pdAA-SA-GAL membranes displayed an enhanced albumin production, urea synthesis, and protein secretion for up to 24 days of culture. These modified surfaces supporting long-lasting hepatocyte culture may be employed for the construction of extracorporeal bioreactors mimicking liver-specific functions.

14.3
Conclusions

In this review we focused on the role of plasma processes followed by biomolecule immobilization reactions aimed to achieve specific responses in biological environments. We therefore started from the thesis that the design of biological-inspired materials represents one of the most challenging goals to emulate what biology does *in vivo*; plasma techniques coupled with the chemical immobilization of biomolecules can be successfully used to move towards this target.

The design of biomimetic materials, able to mimic the complex biological processes, surely is one of the most promising strategies for modern biomaterials development. Many efforts have been then devoted to the emulation of the ECM, for example by immobilizing whole proteins (e.g. fibronectin) or adhesion peptides (e.g.

RGD, YIGSR) on plasma-modified surfaces, as we have described. A recent, intriguing approach focuses on the possibility of generating surfaces which display multiple bioactive elements to activate multiple integrin receptor responses on cell membranes. Hybrid RGD/galactose-immobilized PET substrates have been recently developed with the aim of enhancing hepatocyte adhesion and function synergistically [71]; hepatocyte cells stably anchored to the substratum exhibited high levels of liver-specific functions (urea secretion and albumin synthesis) and maintained a limited spreading cell morphology typical of three-dimensional spheroids.

14.4
List of Abbreviations

AA	Acrylic acid
BCEp	Bovine corneal ephithelial cells
CLMs	Collagen-like molecules
CMC	1-Cyclohexyl-3-(2-morpholinoethyl)carbodiimide
CMD	Carboxymethyldextran
ECM	Extracellular matrix
EDC	1-Ethyl-3-(3-dimethylaminopropyl)carbodiimide
FEP	Fluorinated ethylene propylene
FTIR	Fourier transform infrared spectroscopy
FTIR-ATR	Fourier transform infrared spectroscopy in attenuated total reflection mode
HUVEC	Human umbilical vein endothelial cell
Hyal	Hyaluronan
NHS	N-Hydroxysuccinimide
PCL	Polycaprolactone
pdAA	Plasma-deposited acrylic acid
PDLLA	poly(D,L-lactide)
PE-CVD	Plasma-enhanced chemical vapor deposition
PEG	Poly(ethylene glycol)
PEI	Polyethyleneimine
PEO	Poly(ethylene oxide)
PES	Polyethersufone
PET	Poly(ethylene terephthalate)
PLA	Poly(lactic acid)
PLLA	Poly(L-lactic acid)
PS	Polystyrene
PSU	Polysulfone
PP	Polypropylene
PTFE	Polytetrafluoroethylene
PVDF	Poly(vinylidene fluoride)
RGD	Arg-Gly-Asp
RGDC	Arg-Gly-Asp-Cys

SA   Spacer arm
SIMS   Secondary ions mass spectrometry
SR   Silicon rubber
WCA   Water contact angle
XPS   X-ray photoelectron spectroscopy
YIGSR   Tyr-Ile-Gly-Ser-Arg

Acknowledgments

We acknowledge the Italian Ministry of University and Research that has supported our research trough the MIUR-FIRB RBNE012B2K_002 project, and the European Commission for funding our research trough the NMP3-CT-2005–013653 project Livebiomat "Development of new polymeric biomaterials for *in vitro* and *in vivo* liver reconstruction".

References

1 Williams, D.F. (1987) *Prog. Biomed. Eng.*, **4**, 72.
2 Williams, D.F. (1998) in *Advances in Biomaterials* (ed. C.de Putter), Elsevier, Amsterdam, p. 11.
3 Ratner, B.D., Chilkoti, A. and Lopez, G.P. (1990) in *Plasma Deposition, Treatment and Etching of Polymers* (ed. R. d'Agostino), Academic Press, San Diego, CA, p. 463.
4 Ratner, B.D. (1993) *J. Biomed. Mater. Res.*, **27**, 837.
5 Chung, T.S., Loch, K.C. and Goh, S.K. (1998) *J. Appl. Polym. Sci.*, **68**, 1677.
6 d'Agostino, R., Favia, P. and Fracassi, F. (1997) *NATO ASI Series, E: Appl. Sci.*, **346**.
7 d'Agostino, R., Favia, P., Oehr, C. and Wertheimer, M.R. (2005) *Plasma Process. Polym.*, **2**, 7.
8 (a) Ben Rejeb, S., Tatoulian, M., Khonsari, F.A., Durand, F.A., Martel, A., Lawrence, J.F., Amoroux, J. and Le Goffic, F. (1998) *Anal. Chim. Acta*, **376**, 133; (b) Yang, J., Bei, J., and Wang, S. (2002) *Biomaterials*, **23**, 2607.
9 (a) Puleo, D.A., Kissling, R.A. and Sheu, M.S. (2002) *Biomaterials*, **23**, 2079. (b) Baquey, C., Palumbo, F., Portedurrieu, M.C., Legeay, G.,

Tressaud, A. and d'Agostino, R. (1999) *Nucl. Instrum. Meth. Phys. Res. Sec. B: Beam Interact. Mater. Atoms*, **151**, 255.
10 Hsiue, G.H., Lee, S.D., Wang, C.C., Shiue, M.H.I. and Chang, P.C.T. (1993) *Biomaterials*, **14**, 591.
11 Sipheia, R., Martucci, G., Barbarosie, M. and Wu, C. (1993) *Biomater., Artif. Cell. Im.*, **21**, 455.
12 (a) Daw, R., Candan, S., Beck, A.J., Deulin, A.J., Brook, I.M., Macneil, S., Dowson, R.A. and Short, R.D. (1998) *Biomaterials*, **19**, 1717. (b) Detomaso, L., Gristina, R., Senesi, G.S., d'Agostino, R. and Favia, P. (2005) *Biomaterials* **26**, 3831.
13 Lopez, G.P., Ratner, B.D., Tidwell, C.D., Haycox, C.L., Rapoza, R.J. and Horbett, T.A. (1992) *J. Biomed. Mater. Res.*, **26**, 415.
14 Favia, P., Palumbo, F., d'Agostino, R., Lamponi, S., Magnani, A. and Barbucci, R. (1998) *Plasmas Polym.*, **3**, 77.
15 Anderson, J.M. (2001) *Annu. Rev. Mater. Res.*, **31**, 81.
16 Powell, G.M. (1980) in *Handbook of Water-soluble Gums and Resins* (ed. R.L. Davidson), McGraw-Hill, New York, ch. 18.

17 Andrade, J.D., Nagaoka, S., Cooper, S., Okano, T. and Kim, S.W. (1987) *Trans. ASAIO*, **33**, 75.

18 Gombotz, W.R., Guanghui, W., Horbett, T.A. and Hoffman, A.S. (1991) *J. Biomed. Mater. Res.*, **25**, 1547.

19 Bridgett, M.J., Davies, M.C. and Denyer, S.P. (1989) *Biomaterials*, **10**, 411.

20 Grainger, D.W., Okano, T. and Kim, S.W. (1989) *J. Colloids Interf. Sci.*, **132**, 161.

21 Sardella, E., Senesi, G.S., Favia, P. and d'Agostino, R. (2004) *Plasma Process. Polym.*, **1**, 63.

22 Freij-Larsson, C. and Wesslen, B. (1993) *J. Appl. Polym. Sci.*, **50**, 345.

23 Gombotz, W.R., Guanghui, W. and Hoffman, A.S. (1989) *J. Appl. Polym. Sci.*, **37**, 91.

24 Sheu, M.-S., Hoffman, A.S., Ratnwer, B.D., Fweijen, J. and Harris, J.M. (1993) *J. Adhesion Sci. Technol.*, **7**, 1065.

25 Wang, C.-C. and Hsiue, G.-H. (1993) *J. Polym. Sci.*, **31**, 2601.

26 Gong, X., Dai, L., Griesser, H.J. and Mau, A.W.H. (2000) *J. Polym. Sci: B: Polym. Phys.*, **38**, 2323.

27 (a) Marchant, R.E., Yuan, S. and Szakalas-Gratzl, G. (1994) *J. Biomater. Sci. Polym. Ed.*, **6**, 549, (b) Osterberg, E., Bergstrom, K., Holmberg, K., Schuman, T.P., Riggs, J.A. and BurnsN.L. *et al.* (1995) *J. Biomed. Mater. Res.*, **29**, 741,

28 Catterall, J.B., Gardner, M.J., Jones, L.M. and Turner, G.A., (1997) *Glycoconj. J.*, **14** (5), 647.

29 Solchaga, L.A., Dennis, J.E., Goldberg, V.M. and Caplan, A.I. (1999) *J. Orthop. Res.*, **17**, 205.

30 Kingshott, P., StJohn, H.A.W., Chatelier, R.C. and Griesser, H.J. (1997) *Polym. Mater. Sci. Eng.*, **76**, 81.

31 Morra, M. and Cassinelli, C. (1999) *J. Biomater. Sci. Polym. Ed.*, **10**, 1107.

32 McLean, K.M., Johnson, G., Chatelier, R.C., Beumer, G.J., Steele, J.G. and Griesser, H.J. (2000) *Colloids Surf. B: Biointerfaces*, **18**, 221.

33 Griesser, H.J., Chatelier, R.C., Dai, L., StJohn, H.A.W., Davis, T. and Austen, R. (1997) *Polym. Mater. Sci. Eng.*, **76**, 79.

34 Dai, L., St John, H.A.W., Bi, J., Zientek, P., Chatelier, R.C. and Griesser, H.J. (2000) *Surf. Interf. Anal.*, **29**, 46.

35 Barbucci, R., Torricelli.P., Fini, M., Pasqui, D., Favia, P., Sardella, E., d'Agostino, R. and Giardino, R. (2005) *Biomaterials*, **26**, 7596.

36 Elbert, D.L. and Hubbell, J.A. (1996) *Annu. Rev. Mater. Sci.*, **26**, 365.

37 Ito, Y., Kajihara, M. and Imanishi, Y. (1991) *J. Biomed. Mater. Res.*, **25**, 1325.

38 Kang, I.K., Kwon, B.K., Lee, J.H. and Lee, H.B. (1993) *Biomaterials*, **14**, 787.

39 Liu, S.Q., Ito, Y. and Imanishi, Y. (1993) *J. Biomed. Mater. Res.*, **27**, 909.

40 Kim, Y.J., Kang, I.-K., Huh, M.W. and Yoon, S.-C. (2000) *Biomaterials*, **21**, 121.

41 Chytry, V., Letourneur, D., Baudys, M. and Josefonvvicz, J. (1980) *J. Biomed. Mater. Res.*, **14**, 65.

42 Yannas, I.V. and Burke, J.F. (1980) *J. Biomed. Mater. Res.*, **14**, 65.

43 Yang, J., Bei, J. and Wang, S. (2002) *Biomaterials*, **23**, 2607.

44 Lee, S.-D., Hsiue, G.-H., Chang, P.C.-T. and Kao, C.Y. (1996) *Biomaterials*, **17**, 1599.

45 Bisson, I., Kosinski, M., Ruault, S., Gupta, B., Hilborn, J., Wurm, F. and Frey, P. (2002) *Biomaterials*, **23**, 3149.

46 Griesser, H.J., Mc Lean K., Beumer, G.J., Gong, X., Kingshott, P., Johnson, G. and Steele, J.G. (1999) in *Plasma Deposition and Treatment of Polymers, Vol. 544* (eds W.W. Lee,R. d'Agostino and M.R. WertheimerEd), Materials Research Society.

47 Pierschbacher, M.D. and Ruoslahti, E. (1984) *Nature*, **309**, 30.

48 Hirano, Y., Okuno, M., Hayashi, T., Goto, K. and Nakajima, A. (1993) *J. Biomater. Sci. Polym. Ed.*, **4**, 235.

49 Carlisle, E.S., Mariappan, M.R., Nelson, K.D., Thomes, B.E., Timmons, R.B., Constantinescu, A.,

Eberhart, R.C. and Bankey, P.E. (2000) *Tissue Eng.*, **6**, 45.

50 Lopez, L.C., Gristina, R., Ceccone, G., Rossi, F., Favia, P. and d'Agostino, R. (2005) *Surf. Coat. Technol.*, **200**, 1000.

51 De Bartolo, L., Morelli, S., Lopez, L.C., Giorno, L., Campana, C., Salerno, S., Rende, M., Favia, P., Detomaso, L., Gristina, R., d'Agostino, R. and Drioli, E. (2005) *Biomaterials*, **26**, 4432.

52 Cosnier, S., Novoa, A., Mousty, C. and Marks, R.S. (2002) *Anal. Chim. Acta*, **453**, 71.

53 Ganapathy, R., Manolache, S., Sarmadi, M. and Denes, F. (2001) *J. Biomater. Sci. Polym. Ed.*, **12**, 1027.

54 Mohy Eldin, M.S., Bencivenga, U., Portaccio, M., Stellato, S., Rossi, S., Santucci, M., Canciglia, P., Gaeta, F.S. and Mita, D.G. (1998) *J. Appl. Polym. Sci.*, **68**, 625.

55 Kang, I.K., Choi, S.H., Shin, D.S. and Yoon, S.C. (2001) *Int. J. Biol. Macromol.*, **28**, 205.

56 Kawakami, M., Koya, H. and Gondo, S. (1988) *Biotechnol. Bioeng.*, **32**, 369.

57 Danilich, M.J., Kottke-Marchant, K., Anderson, J.M. and Marchant, R.E. (1992) *J. Biomater. Sci. Polymer Ed.*, **3**, 95.

58 Gancarz, I., Bryjak, J., Bryjak, M., Pozniak, G. and Tylus, W. (2003) *Eur. Polym. J.*, **39**, 1615.

59 Gancarz, I., Bryjak, J., Pozniak, G. and Tylus, W. (2003) *Eur. Polym. J.*, **39**, 2217.

60 (a) Hubbell, J.A. (1990) *Trends Polym. Sci.*, **2**, 20, (b) Bahulekar, R., Tokiwa, T., Kano, J., Matsumura, T., Kojima,

I. and Kodama, M. (1998) *Biotechnol. Techniques*, **12**, 721,

61 Kobayashi, K. Sumitomo, H. Kobayashi, A. and Akaike, T. (2001) *J. Macromol. Sci. Chem.*, **A25**, 655.

62 Caneiro, M.J., Fernandes, A., Figneiredo, C.M., Fortes, A.G. and Freitas, A.M. (2001) *Carbohydr. Polym.*, **45**, 135.

63 Metzke, M., Bai, J.Z. and Guan, Z. (2003) *J. Am. Chem. Soc.*, **125**, 7760.

64 (a) Ben ZegevA., Robinson, G.S., Bucher, N.L. and Farmer, S.R. (1988) *Proc. Natl Acad. Sci. USA*, **85**, 2161, (b) Sanchez, A., Alvarez, A.M., Pagan, R., Roncero, C., Vilaro, S., Benito, M. and Fabregat, I. (2000) *J. Hepatol.*, **32**, 242.

65 Park, J.K. and Lee, D.H. (2005) *J. Biosci. Bioeng.*, **99**, 311.

66 Yin, C., Ying, L., Zhang, P.-C., Zhuo, R.-X., Kang, E.-T., Leong, K.W. and Mao, H.Q. (2003) *J. Biomed. Mater. Res.*, **67**, 1093.

67 Ying, L., Yin, C., Zhuo, R.X., Leong, K.W., Mao, H.Q., Kang, E.T. and Neoh, K.G. (2003) *Biomacromolecules*, **4**, 157.

68 Lopez, L.C., Gristina, R., De Bartolo, L., Morelli, S., Favia, P. and d'Agostino, R. (2004) *JABB*, **2**, 211.

69 Moore, S. and Link, K.P. (1940) *J. Biol. Chem.*, **133**, 293.

70 De Bartolo, L., Morelli, S., Rende, M., Salerno, S., Giorno, L., Lopez, L.C., Favia, P. and Drioli, E. (2006) *J. Nanosci. Nanotechnol.*, **6**, 2344.

71 Du, Y., Chia, S., Han, R., Shang, S., Tang, H. and Yu, H. (2007) *Biomaterials*, **27**, 5669.

15
In Vitro Methods to Assess the Biocompatibility of Plasma-Modified Surfaces

M. Nardulli, R. Gristina, Riccardo d'Agostino, and Pietro Favia

In vitro biocompatibily tests of materials intended for biomedical applications (prostheses, implants, biosensors, disposable labwares, catheters, devices, etc.) allow one to obtain preliminary indications on how a material interacts with a well-defined biological environment, with the aim of using it safely in a final application. In this short contribution we review some of the preliminary cytocompatibility tests used, including some results obtained when stable plasma-processed surfaces synthesized in our laboratory are exposed to cell cultures *in vitro*.

Cell culture methods, especially when immortalized cell lines are used, are the most used tests to achieve useful and reproducible results on cell–biomaterial interactions. The observation of the interactions of single cell types with the surface of a material, a study practically impossible to perform in a living organism, offers a formidable way for an in-depth analysis of specific cell–surface interactions. We show also how the correct interplay between the use of molecular and cell biology methods is a unique way to understand more profoundly the subtle and complicated relationships between cell morphology and cell physiology, a fundamental piece of information to gain before starting *in vivo* tests of the material with animals and humans.

15.1
Introduction

Biomaterials have partially answered the need to replace or integrate tissue and organs functionally or metabolically damaged or inactive for some pathological or traumatic damage, and have solved or speeded up many problems in the everyday biomedical practice with wares and devices; new scientific frontiers have also been opened in closely related fields such as tissue engineering and regenerative medicine [1]. The number of applications has grown remarkably in the last few years and the capacity of therapeutic innovations has been involved in the functional restoration of compromised parts of organisms and in the survival and life style improvement of patients.

Advanced Plasma Technology. Edited by Riccardo d'Agostino, Pietro Favia, Yoshinobu Kawai, Hideo Ikegami, Noriyoshi Sato, and Farzaneh Arefi-Khonsari
Copyright © 2008 WILEY-VCH Verlag GmbH & Co. KGaA, Weinheim
ISBN: 978-3-527-40591-6

Prostheses, devices, and scaffolds to be inserted *in vitro, ex vitro,* or *in vivo* in contact with biological fluids have to interact in an optimal way with the biological host, where the nature and the duration of the contact play a paramount role. Biocompatibility is defined as the ability of a material to perform with an appropriate host response in a specific biomedical application [2]. The biocompatibility of a material depends on the series of events occurring at the tissue/material interface, whose outcome must be optimized for each particular use.

Often, an adverse physiological reaction can occur when a foreign material is implanted in the human body; inflammation is the first common reaction of living tissues to an injury, and the living tissue needs this reaction to control, neutralize, or isolate the damaging agent. This process includes a sequence of events that, in the final step, can heal the site where the implant was positioned through the generation of new tissue via native parenchymal cells. An adverse response to the material consists, instead, in the development of connective tissue around it, which leads at last to a fibrous capsule. This final fibrous encapsulation step is known as *foreign body reaction*[3]. The ideal biomaterial should minimize this reaction and promote normal wound healing [4]; with this aim many investigations are pursued to fully understand all phenomena occurring when a material enters in contact with a biological environment, and to develop material surface modification strategies aimed to minimize the reaction and optimize the interaction with the tissue. In the last few decades considerable progresses have been made in understanding and driving tissue/material interactions, which allowed the identification of materials, surface features, and modification strategies for increased integration and biocompatibility. In *in vitro* experiments particular attention is given to observe how biological entities, protein and/or cells, interact with the material under scrutiny; this is usually done by studying adhesion and growth of cells during cell culture time.

When a biomaterial is implanted, or it is placed *in vitro* in contact with a cell culture medium, proteins adsorb on the foreign surface within a few seconds [5]. Such an adsorbed layer, whose distribution and conformation depends also on the duration of the contact, mediates cell–biomaterial recognition and interactions through cell receptors, thus driving the behavior of cells in events like adhesion, growth, physiology, and death, at its surface [6], including, possibly, wound healing and the growth of new tissue.

It is worth noting that *in vivo* the adhesiveness of cells among themselves and/or to a solid substrate is mediated by the extracellular matrix (ECM), a complex network of proteins and glicosaminoglicans synthesized, secreted, and remodeled by cells themselves. The principal function of ECM is, in effect, to provide a mechanical and chemical specific substrate for cell attachment [7]. *In vitro* ECM molecules are synthesized *ex novo* by cell themselves on artificial substrates, when possible, or are adsorbed at the surface of the substrate from the cell culture medium. For this reason, all substrates intended to have cells adhering on them should support protein adhesion [8,9], unless special experimental conditions require media with no proteins added. Since *in vivo* the specific composition of this conditioning layer depends on the biological tissue, one of the specific targets of engineered biomaterial surfaces is the precise, programmed control of the adsorbed protein layer, with the target of driving in a predetermined way cell–materials interactions. Several approaches are used for this purpose; among them, surface modification plasma processes have gained great popularity and efficacy in life sciences since, among many other advantages, they allow one to modify materials surfaces without affecting their bulk [10].

15.2
Surface Modification Methods: Plasma Processes and Biomolecule Immobilization

Non-equilibrium plasmas, also very well known as glow discharges, are employed in different areas of material sciences and nanotechnologies, including microelectronics, semiconductors, the car industry, food packaging, sterilization, and sensors [11–13]. Active species (atoms, radicals, ions, electrons) generated in the gas phase can interact with the surfaces exposed to the plasma in three different kinds of surface modification processes: thin film deposition (PE-CVD, plasma enhanced chemical vapour deposition); ablation (etching); and treatments (grafting of functional groups, crosslinking).

Non-equilibrium plasmas are generated in properly arranged plasma reactors; low-pressure "parallel plate" reactors under controlled conditions (pressure, applied power, gas fluxes, etc.) are mainly utilized, with 13.56 MHz the most commonly used excitation frequency.

Plasma surface modifications occur in non-thermodynamic equilibrium conditions [11,14], and work really close to ambient temperature, allowing one to "tailor" in a controlled way the surface of materials (composition and energy, hardness, adhesion to other materials, biological compatibility, etc.) without altering bulk properties.

In the last few years an enormous range of plasma processes has been observed, in particular in the biomedical field, where such processes are well established and contribute to the achievement of prosthesis, catheters, heart valves, and contact lenses [15]. In this scenario, plasma processes allow one to modify the surface of materials to obtain an optimized response of the biological environment in which they are located.

In several biomedical applications (diagnostic, primary cells culture, prosthesis bio-integration, etc.) it becomes necessary to stimulate adhesion and growth of cells at the surface of conventional materials which would not be able to support it conveniently. In the case of hydrophobic polymers, for example, plasma treatments fed with O_2, NH_3, H_2O, or non-depositing other gases/vapors, allow the grafting of oxygen- and/or nitrogen-containing polar functional groups at the surface of the material, increasing their surface energy and wettability. Consequently, the surface becomes more suitable for the adhesion of protein, thus for adhesion and growth of cells in proper culture conditions [16,17].

The described treatment strategy is similar to the use of PE-CVD processes of functional coatings, whose structure preserves the functional groups present in the starting monomer, which become able to stimulate protein adhesion and cell growth.

The literature offers several examples of this approach, with plasmas fed, for example, with acrylic acid [18–20] or allylamine [21] vapors, to deposit coatings characterized by –COOH and –NH$_2$ groups, respectively.

Plasma-functionalized surfaces can be further modified by covalent immobilization of biomolecules conveniently chosen to provide specific interactions with the cells that come in contact with the materials. This further modification strategy involves the immobilization of peptides, antibodies, enzymes, carbohydrates, anti-thrombotic agents, and other biomolecules through conventional organic reaction synthesis onto plasma-functionalized surfaces [22–25]. In order to preserve the biological activity of the immobilized biomolecules, the conformational mobility should be kept intact; for this purpose, usually the molecule is not directly immo-bilized at the surface of the modified material, but rather it is tied at the surface of the material through a long-chain hydrophilic "spacer arm" molecule.

Certain proteins and saccharides immobilized at the surface of biomedical materials can be recognized by specific integrins, a family of cell membrane surface receptors involved in cell–ECM and cell–cell interactions, and allow specific biolo-gical response. Many papers have been published dealing with short amino acid sequences immobilized with various methods at the surface of polymers; the most commonly investigated is the arginine-glycine-aspartic acid (RGD) oligopeptide, found in the adhesive domains of fibronectin, laminin, collagen, vitronectin, and other ECM proteins [26,27]. In other applications, instead, where it is requested to avoid cell adhesion to substrates and interactions with other biological systems, it is possible to deposit cell-repulsive "non-fouling" coatings [28]. It is also possible to deposit nanocomposite coatings (silver clusters included in an organic matrix) able to release silver ions in water, with recognized antibacterial effect [29]. Another modern application field concerns the use of oxygen and other plasmas [30,31] to sterilize and decontaminate biomedical metals, plastics, and ceramics.

15.3
In Vitro Cell Culture Tests of Artificial Surfaces

Materials to be used in the biomedical field have to be previously tested *in vitro* to establish whether a certain surface can go further toward *in vivo* evaluation with animals and humans.

Cell culture experiments are the more obvious and reproducible methods for *in vitro* testing of materials intended for biomedical purposes. The validity of the cultured cell as a model of physiological functions has been often criticized, simply because *in vitro* experiments cannot reproduce the entire range of cellular responses that occur *in vivo*, due to the lower complexity of cell–cell and cell–matrix interactions *in vitro* compared to living systems. Only *in vitro* cell culture studies, however, allow one to fully control the culture environment, to limit the interactive nature of the biological environment, to isolate and quantify molecules playing a key role in biological responses, to achieve rapidly reproducible results, and, last but not least, to reduce the use of animals in research. For *in vitro* cell culture studies specialized pieces of equipment are required, including a laminar-flow hood, a cell incubator with atmosphere control, refrigerators and freezers, inverted microscopes, and cell culture vessels (e.g. Petri dishes) of proper materials. The choice of culture vessels is determined by different needs (whether cells are grown in monolayer or suspension, the number of the cells to be grown, etc.). Glass Petri dishes, which could be re-used after proper cleaning and sterilization steps, were used in the past. Now disposable polystyrene (PS) dishes and flasks are used now since those intended for cell cultures are usually chemically surface-treated in order to gain hydrophilic surface character to increase cell adhesion and spreading on PS [32–34]. Most cells seeded on native PS, in fact, usually aggregate into clumps, attach poorly to the dish and cease growth, due to the hydrophobic nature of the substrate. Commercial tissue culture polystyrene (TCPS) wares are surface-treated in air corona discharges to create hydrophilic surfaces, or are coated with a thin film of ECM proteins like collagen or fibronectin.

Cell culture experiments on biomaterial surfaces can be performed by using either primary or immortalized cell cultures. Primary cells are originated directly from a living tissue, after isolation of the tissue itself, its mechanical or enzymatic disaggregation, and seeding of the resulting cells in culture vessels. All different cell types that are part of the particular living tissue enter, thus, in a primary cell culture. Primary cells are highly functional, but their reproductive capacity is low; they grow *in vitro* until all the surface available of the substrate is occupied by cells, i.e. until the culture becomes *confluent*. At this stage it becomes necessary to subculture cells into new culture vessels.

Primary cells usually retain their specific functions, and are a very useful tool for studying a specific cell type in a controlled environment. After a certain number of growth passages, depending on the particular cell type, primary cells do not proliferate anymore [35]; further, since they derive directly from a living biological host, their behavior can change strongly from donor to donor, thus leading to low reproducibility in the results. In order to avoid these drawbacks, "immortalized" cell lines could be used. Such highly proliferative cells can be derived from tumor or from primary cells immortalized in some chemical or biological way, as it is amply reviewed in [36]. They always offer a very homogeneous population of cells due to their clonal origin and, consequently, allow biologists to perform highly reproducible experiments.

The phenotype (i.e. the entire set of cell functions) of immortalized cells is often different from that presented from the same cell type *in vivo*, probably due to the lack of several factors that regulate cellular growth, morphology, and specific metabolic functions. In spite of this, the uniformity of cell population, the high proliferation rate, and the possibility of long-term storage allow one to use immortal cell lines in a wider range of experiments.

Differentiation and dedifferentiation characterize the phenotype of cells during their life span both *in vivo* and *in vitro*. The evolution process that lead a mature cell to express its phenotypic properties is known as differentiation; this status can be monitored by analyzing specific marker molecules (proteins, enzymes, etc.; for example, the production of serum albumin is marker for hepatocytes) produced by the cells [37]. The loss of such properties, which can happen in culture conditions, is known instead as dedifferentiation. Decades of work on formation and propagation of primary cells and cell lines has led to a certain understanding of the dedifferentiation process in culture conditions. Sometimes primary cell cultures lose the peculiar features that they show in a living tissue [38], more often this happens also to cell lines, even though this cannot be considered a general rule. For example, HepG2 and Hep3B human hepatoma cell lines retain the biosynthetic capabilities of normal liver cells, and secrete the same proteins usually found in human plasma [39].

In the light of these considerations, it is clear that the best approach for planning *in vitro* tests depends strongly on the situation; although primary cultures offer, in principle, the best tool for a simulation of *in vivo* systems, they often are too expensive and time consuming when compared with cell lines.

At present, a variety of different tests are available for cell cultures; they can assess morphology, membrane integrity, proliferation, and specific functions of cells that enter in contact with artificial materials in materials intended for different biomedical applications (prostheses, implants, biosensors, devices, etc.), as well as the cytotoxicity of the material itself.

In the following sections different methods usually used to test the *in vitro* biocompatibility of biomedical materials will be briefly reviewed, and some results obtained in our laboratory on plasma-processed surfaces will be discussed. The tests are obviously the same for surfaces produced with different procedures; we will rather highlight how some particular characteristics of plasma-modified surfaces influence the behavior of cells.

15.4
Cytotoxicity Analysis

Cytotoxicity is defined, in general, as the ability of a drug or of a material in inhibiting the viability of the cells, i.e. the health of the cells in terms of survival and normal metabolism. The evaluation of cytotoxicity should include counting dead/alive cells over the entire cell population under scrutiny, and assessing their physiological status. Morphological changes of the cells and alterations of their membrane permeability, for example, can be inferred from the behavior (intake and retention, or exclusion) of dyes added to the cell suspension.

Many different cheap, reproducible, and easy quantitative assays are available to test cytotoxicity; the choice depends mainly on the cells themselves and on the particular drug/material under study. We briefly focus now on viability, metabolic, and irritancy assays, the three most common tests used.

15.4.1
Viability Assays

These tests allow one to easily and quickly measure the proportion of viable cells among all those adhering on a substrate [40]. Most viability tests are performed under the microscope, and are sensitive either to the uptake of a dye/stain molecule to which cells are normally impermeable (e.g. trypan blue or naphthalene black), through the broken membrane of dead cells, or to other dye/stain molecules (e.g. diacetyl fluorescein or neutral red) taken up only by living cells [41]. The common protocol for these tests requires that a suspension of the cells is mixed with a solution of the dye molecule, then it is loaded in a cell counting chamber (i.e. a hemocytometer) [42] and the percentage of dead/alive colored cells is measured with the microscope.

15.4.2
Metabolic Assays

Such tests are used to probe the survival of the cell population, defined as the retention of their metabolic or/and proliferative ability. The MTT test is probably the most common metabolic assay, where the yellow water-soluble 3-(4,5-dimethylthiazol-2-yl)-2,5-diphenyl tetrazolium bromide (MTT) molecule is reduced by dehydrogenase enzymes present in the active mitochondria of living cells. MTT is converted into a dark blue compound, formazan, insoluble in water, which precipitates inside the cells. The cell suspension is then lysed in a proper isopropanol-based solution, where formazan is soluble and can be easily spectrophotometrically quantified at 570 nm [43]. The resulting optical density can be correlated with the cell viability.

Figure 15.1 shows MTT test results, as a function of culture time, for HepG2 hepatoma cell lines grown on poly(ethylene terephthalate) (PET) substrates native and modified in a parallel plate reactor with radiofrequency (RF) (13.56 MHz) glow

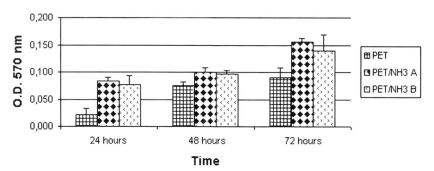

Fig. 15.1 MTT results (570 nm optical density) as a function of culture time for HepG2 hepatoma cell lines seeded at 2×10^4 cell/substrate and grown on PET substrates, native and modified in RF (13.56 MHz) glow discharges fed with NH_3, under the experimental conditions reported below. Pretreatments are needed to limit the hydrophobic recovery [57] of the groups grafted in the NH_3 discharge. The PET/NH_3 B surface is characterized by a slightly higher N/C surface ratio (0.20 vs. 0.18), attesting for a more efficient grafting of N-containing groups. PET/NH_3 A pretreatment: Ar 20 sccm, pressure 350 mtorr, RF power 40 W, 5 min; treatment: NH_3 10 sccm, pressure 200 mtorr, RF power 20 W, 1 min. PET/NH_3 B pretreatment: Ar 10 sccm, pressure 200 mtorr, RF power 30 W, 1 min; treatment: NH_3 10 sccm, pressure 200 mtorr, RF power 20 W, 1 min.

discharges fed with NH_3, in different experimental conditions (see caption). The plasma-treated surfaces result in modified and grafted with different amount of N-containing (e.g. $-NH_2$, $-CN$, etc.) hydrophilic polar functionalities. Such approach is usually pursued to improve cell adhesion on polymer surfaces, or to graft functional groups suitable for biomolecule immobilization. The results evidence that both plasma-treated surfaces support better cell viability compared to native PET, and this is in agreement with quantitative cell-counting data (data not shown), reporting that a higher density of cells is found adhered on plasma-treated with respect to native substrates. No relevant difference in cell viability is found, in effect, between the two different plasma-treated substrates.

The MTT test is rapid, versatile, highly reproducible, and adaptable to large-scale screening; some drawbacks of the MTT assay are known [44,45], mostly related to the fact that MTT can be reduced also by non-mitochondrial enzymes, so the correlation with cell viability becomes harder to be defined. Consequently, MTT results should usually be compared with data from other viability assays.

15.4.3
Irritancy Assays

These tests are used to monitor *in vitro* the inflammation response, the first common reaction to an injury, that a material could trigger in a living tissue. This analysis is an *in vitro* oversimplification of what could happen to cells in a physiological environment, and consists in seeding leukocytes on the substrate under investigation, and in the enzyme-linked immunosorbent assay (ELISA) detection of cytokines, which are usually released from leukocytes under stress conditions [46].

15.5
Analysis of Cell Adhesion

To measure the density of all/selected cells in contact with a given substrate, to control the reproducibility of the experiments, and to make comparative studies, the cell population must be counted in some way. The growth of mammalian cells in culture can be monitored by evaluating the adhesion of the cells at the surface of the substrates, by using a number of parameters related to the increase of cellular biomass over time.

Cell–cell and cell–substrate adhesion are crucial conditions for cell survival, leading to cell spreading at the surface of substrates, cell migration, and differentiated cell functions. There is a correlation between cell adhesion, cell spreading, and cell behavior (e.g. viability, migration, growth, and differentiation); most cells lose rapidly viability, assume a round shape, and undergo apoptosis (i.e. programmed cell death in case of unfavorable environmental conditions [47]) when detached from the ECM, either in living tissues or at the substrate of materials [48]. We have highlighted before that ECM is, *in vivo*, the principal factor of cell adhesion in tissue formation and remodeling, since it provides the mechanical and chemical specific substrate for cell attachment. Therefore, it is very important to investigate how cells adhere *in vitro* to a given substrate intended to be used as biomaterial. To test cell adhesion *in vitro* several protocols can be followed, where most simple ones involve counting cells at regular time intervals.

Cells are trypsinized (i.e. enzymatically detached) from the substrate, harvested in proper vials, then transferred and counted with a hemocytometer chamber. If the suspension of the cells is previously mixed with a proper stain, generally in equal volume, a viability test could be also performed, as described previously.

Flow cytometry allows one to quantify cell density by simultaneously analyzing multiple physical characteristics of single particles, usually cells, as they flow in a fluid stream through a laser beam. The properties measured include relative size, relative granularity or internal complexity, and relative fluorescence intensity. These characteristics are determined using dedicated equipment that records how the cell, suspended in a saline solution and properly labeled with a fluorochrome, scatters incident laser light and emits fluorescence [49]. Flow cytometry is extremely accurate, and is by far more expensive that a hemocytometer chamber.

Indirect measurements of the density of adhered cells can be performed, after cell lysis and proper purification, by estimating the density of the cellular material (total DNA, total proteins) extracted. DNA assays can be performed with fluorescent dyes such as 4',6-diamidino-2-phenylindole (DAPI) [50] and Hoechst 33258 [51], or by measuring the specific absorption of DNA at 260 nm with a spectrophotometer. The total protein content can be determined spectrophotometrically on proteins collected from lysed cells, then stained with Coomassie Blue (Bradford reaction [52]) or with another stains for proteins.

Although the most evident, easy way to check significant differences in cell adhesion on surfaces consists in counting the number of adhered cells, also comparing the different morphologies that a cell can acquire when in contact with substrates characterized by different chemical and morphological surface features allows gaining information on the cytocompatibility of materials. With this aim it is important to observe both the external (degree of spreading, dimensions, elongation, presence of lamellae or filopodia, etc.) and the internal (i.e. the architecture of the cytoskeleton proteins) morphology of cells.

Plasma processes can tune surface chemistry and morphology of materials in a controlled reproducible way; the observation of subtle but significant differences in cell morphology and responses may contribute to rank the cytocompatibility of different plasma-processed substrates.

In order to visualize external details of adhering cells, they must be fixed onto the substrates at different growth times. Paraformaldehyde or gluteraldehyde allow preserving the external cell morphology presented at the moment of the fixation; methanol allows permeabilizing the membrane of the cell to fluorescent dies and/or to antibodies that bind to the internal cell structures and allow their recognition. Fixed cells can be stained (e.g. with Coomassie Blue), then observed with an optical microscope; digital images can be acquired at different magnifications to evaluate the distribution of the many possible single cell morphologies with software for

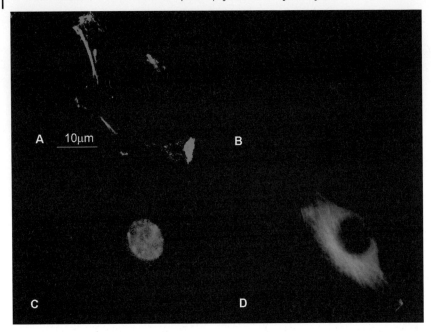

Fig. 15.2 Fluorescence microscopy of a hTERT BJ1 cell seeded on a pdAA coating deposited in a RF glow discharge (Ar 20 sccm, AA vapors 3 sccm, pressure 150 mtorr, RF power 100 W [53]). The cytoskeleton and the nucleus are evidenced. (A) The distribution of F-actin labeled with TRITC (tetramethylrhodamine isothiocyanate)-conjugated phalloidin is detected; (B) tubulin microfilaments labeled with an anti-tubulin monoclonal antibody and a FITC (fluorescein isothiocyanate)-conjugated secondary antibody are revealed; (C) the nucleic acids inside the nucleus are revealed with the use of DAPI; (D) the three pictures are overlapped for a global vision of the cell spread on pdAA.

image analysis. One of the most used software, Image J, can be downloaded from the website of the National Institute of Health (rsb.info.nih.gov/ij/).

The observation of the cytoskeleton by means of immunofluorescence microscopy gives further clues on the adaptation and on the response of cells to different surfaces. For this purpose, the distribution of actin and tubulin in the cell, two main cytoskeletal proteins, is usually investigated, as well as shape, dimension, and location of the nucleus. In Fig. 15.2 different features of the cytoskeleton of a single hTERT BJ1 cell adhering on a plasma-deposited acrylic acid (pdAA; see caption for deposition conditions) [53] coating are shown, where it is evident how the cell is spread and stretched on the cell adhesive surface.

Besides actin and tubulin, other cytoskeleton proteins can be labeled and observed to achieve more specific hints on the formation of specific cell-substrate complexes. For example, vinculin can be visualized to highlight in detail number and size of the focal contacts, i.e. the structures of the cellular membrane specialized in the cell–substrate attachment. This latter analysis is particularly interesting when cells are in contact with topographically structured surfaces.

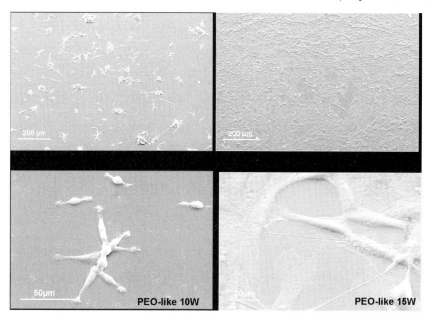

Fig. 15.3 SEM images of 3T3 murine fibroblasts adhered on two different PEO-like coatings plasma deposited under the conditions described below. Both surfaces are cell-adhesive [29], albeit at different extent. It is possible to observe how both cell density and morphology are evidently affected by the chemical nature of the different surfaces. PEO-like 10 W: Ar 5 sccm, DEGDME 0.4 sccm, pressure 400 mtorr, RF power 10 W; PEO-like 15 W: Ar 5 sccm, DEGDME 0.4 sccm, pressure 400 mtorr, RF power 15 W.

Scanning electron microscopy (SEM) is another way to observe cell morphology, with higher resolution with respect to optical microscopy. Obviously, cells need to be fixed, dehydrated, and gold/carbon decorated before being observed, in a much more drastic procedure.

High magnification SEM analysis evidences cellular details such as the protrusions that cells use to probe the surface around them, known as lamellipodia and filopodia, which are difficult to observe at lower magnification. This method of observation can provide three-dimensional topographical information at nanometric resolution, and may reveal important details on the interactions of cells with micro- and nanostructured surfaces. Moreover, the possibility to tilt the substrate in the SEM instrument improves the detection of details. Figure 15.3 show 3T3 murine fibroblasts on two different coatings plasma-deposited in RF glow discharges fed with diethylglycodimethyl ether (DEGDME) vapors run at two different RF power levels. The structure retention of the monomer is not high enough in both poly(ethylene oxide) (PEO)-like coatings to ensure nonfouling properties [29] in this case, so both surfaces result in cell adhesion, albeit at different extents, as is evident in Fig. 15.3. The coating deposited at lower power, in fact, induces adhesion to a lower density of cells, and a lower spreading.

15.6
Analysis of Cell Functions

Cells probe the external surface to which they are attached through their receptor on the cellular membrane. Chemical composition and topography of the surface where cells are grown *in vitro* become their new environment, so they react to it by producing proteins to mimic a sort of ECM habitat similar to that they are used to *in vivo*. Further, they adapt morphology and specific differentiation activities to the new environment: new integrin- and cadherin-mediated adhesion points activate specific signaling pathways inside the cell, that regulate cell migration and gene expression besides cell morphology. To study such a complex behavior and correlate it with surface chemical/morphological features, and to optimize surface features to preserve the optimal cell functions for a specific biomedical application, proteomic and genomic patterns (namely, the set of proteins produced and genes expressed by the cells) should be investigated with molecular biology methods, such as complementary-DNA (cDNA) microarray tests [54]. For example, it has been shown [55] how a chemically well-defined surface obtained by grafting polyacrylic acid on plasma-pretreated PET film under UV irradiation, followed by conjugation of a galactose derivative (1-*O*-(6-aminohexyl)-D-galactopyranoside) to the grafted polyacrylic acid chains is able to maintain both an aggregate morphology and the best physiological performances for hepatocyte human cells. The analysis of specific proteins (collagen, fibronectin, albumun, etc.) synthesized by cells can lead to a better understanding of cell physiology related to the contact with a given material.

Recent results from our laboratory show, in Fig. 15.4, how different substrates, TCPS, PET, polytetrafluoroethylene (PTFE), and pdAA surfaces, modulate the expression of plasmatic fibronectin in HepG2 cells seeded in equal density. The experiments, performed with a monoclonal antibody able to recognize the human plasma fibronectin, show different degree of protein production as a function of the substrates. The expression of fibronectin is an important marker of how much a certain substrate can safely accommodate hepatocytes *in vitro* [56,57].

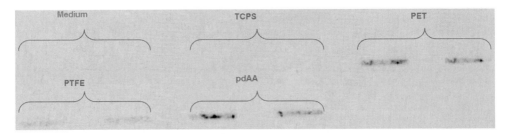

Fig. 15.4 Dot-blot analysis of medium collected from wells where HepG2 on different substrates (TCPS, PET, PTFE, pdAA) were grown. 1 μg of total protein from the supernatant medium, containing both medium and secreted protein, were blotted onto a nylon membrane. Immunodetection with a monoclonal antibody against human plasma fibronectin was performed. pdAA was plasma-deposited on PET under the following experimental conditions: Ar 20 sccm, AA vapors 3 sccm, pressure 150 mtorr, RF power 100 W [53].

15.7
Conclusions

In this review we have seen a panoramic view of the methods usually performed to analyze *in vitro* the biocompatibility of materials, including plasma-processed surfaces, through the evaluations of viability parameters such as cell density, growth rate, cell viability, cytotoxicity, and changes in morphology and physiological behavior. All these tests are necessary, but often not sufficient, to establish whether a surface can go toward *in vivo* evaluations in animals and humans; it is important to stress that "positive" *in vitro* results of the studied material do not automatically means that the tested material can behave in a satisfactory way in the human body.

References

1 Vacanti, C.A. (2006) *J. Cell. Mol. Med.*, **10**, 569.
2 Williams, D.F. (1987) *Progress in Biomedical Engineering*, Elsevier, Amsterdam, p. 4.
3 Anderson, J.M. (2001) *Annu. Rev. Mater. Res.*, **31**, 110.
4 Ratner, B.D. and Bryant, S.J. (2004) *Annu. Rev. Biomed. Eng.*, **6**, 41.
5 Tang, L. and Eaton, J.W. (1995) *Am. J. Clin. Pathol.*, **103**, 466.
6 Tang, L. and Eaton, J.W. (1999) *Mol. Med.*, **5**, 351.
7 Nelson, C.M. and Tien, J. (2006) *Curr. Opin. Biotechnol.*, **17**, 518.
8 Wilson, C.J., Clegg, R.E., Leavesley, D.I. and Pearcy, M.J. (2005) *Tissue Eng.*, **11**, 1.
9 Harrison, C.A., Gossiel, F., Bullock, A.J., Sun, T., Blumsohn, A. and MacNeil, S. (2006) *Br. J. Dermatol.*, **154**, 401.
10 Sardella, E., Favia, P., Gristina, R., Nardulli, M. and d'Agostino, R. (2006) *Plasma Proces. Polym.*, **3**, 456.
11 d'Agostino, R. (1990) *Plasma–Materials Interactions*, Academic Press.
12 d'Agostino, R., Favia, P. and Fracassi, F. (1997) Plasma Processing of Polymers, NATO ASI series E: Applied Science, Kluwer Academic.
13 d'Agostino R., Favia P., Fracassi F. and Palumbo F. (eds) (2003) Book of Abstracts and CD of Full Papers of the 16th Int. Symp. on Plasma Chemistry, ISPC Taormina.
14 Chapman, B. (1980) *Glow Discharge Processes*, Wiley.
15 Favia, P. and d'Agostino, R. (2002) Plasma processed surfaces for biomaterials and biomedical devices: PEO-like, Ag/PEO-like and –COOH functional coatings, micro-patterned surfaces, Le Vide, 203.
16 Pu, F.R., Williams, R.L., Markkula, T.K. and Hunt, J.A. (2002) *Biomaterials*, **23**, 2411.
17 France, R.M., Short, R.D., Duval, E. and Jones, F.R. (1998) *Chem. Mater.*, **10**, 1176.
18 Lombello, C.B., Malmonge, S.M. and Wada, M.L.F. (2000) *J. Mater. Sci. Mater. Med.*, **11**, 541.
19 Dawe, R., Candan, S., Beck, A.J., Devlin, A.J., Brook, I.M., Macneil, S., Dawson, R.A. and Short, R.D. (1998) *Biomaterials*, **19**, 1717.
20 Favia, P., Sardella, E., Gristina, R., Milella, A. and d'Agostino, R. (2002) *J. Photopol. Sci. Technol.*, **15**, 341.
21 Harsch, A., Calderon, J., Timmons, R.B. and Gross, G.W. (2000) *J. Neurosci. Meth.*, **98**, 135.
22 Hersel, U., Dahmen, C. and Kessler, H. (2003) *Biomaterials*, **24**, 4385.
23 Drumheller, P.D. and Hubbell, J.A. (2003) in *Tissue Engineering* (eds B. Palsson, J.A. Hubbell, R. Plonsey and J.D. Bronzino), CRC Press.
24 Favia, P., Palumbo, F., d'Agostino, R., Lamponi, S., Magnani, A. and Barbucci, R. (1998) *Plasmas Polym.*, **3**, 77.

25 Detomaso, L., Gristina, R., Lopez, L.C., Senesi, G.S., d'Agostino, R. and Favia, P. (2004) in *Plasma Processes and Polymers* (eds R. d'Agostino, P. Favia, M.R. Wertheimer and C. Oehr), VCH-Wiley.

26 Temming, K., Schiffelers, R.M., Molema, G. and Kok, R.J. (2005) *Drug Resist Updat.*, **8**, 381.

27 Aumailley, M., Gerl, M., Sonnenberg, A., Deutzmann, R. and Timpl, R. (1990) *FEBS Lett.*, **262**, 82.

28 Balazs, D.J., Triandafillu, K., Sardella, E., Iacoviello, G., Favia, P., d'Agostino, R., Harms, H. and Mathieu, H.J. (2005) in *Plasma Processes and Polymers* (eds R. d'Agostino, P. Favia, M.R. Wertheimer and C. Oehr), Wiley-VCH, 351.

29 Sardella, E., Gristina, R., Senesi, G.S., d'Agostino, R. and Favia, P. (2004) *Plasma Process. Polym.*, **1**, 63.

30 Lerouge, S., Fozza, A.C., Wertheimer, M.R. and Yahia, L.H. (2000) *Plasmas Polym.*, **5**, 31.

31 Rossi, F., DeMitri, R., DosSantos Marques, F., Bobin, S. and Eloy, R. (2003) Proceedings 16th Int. Symp. on Plasma Chemistry (ISPC-16), University of Bari, 22–27 June 2003, Taormina (I), ORA/PRO 63917.

32 Ertel, S. Chilkoti, A., Horbett, T.A. and Ratner, B.D. (1991) *Biomater. Sci. Polym. Ed.*, **3**, 163.

33 Curtis, A.S.G.Forrester, J.V. and Clark, P. (1986) *J. Cell Sci.*, **86**, 9.

34 McFarland, C.D., Mayer, S., Scotchford, C., Dalton, B.A., Steele, J.G. and Downes, S. (1999) *J. Biomed. Mater. Res.*, **44**, 1.

35 Alberts, B., Johnson, A., Lewis, J., Raff, M., Roberts, K. and Walter, P. (2002) *Molecular Biology of the Cell*, 4th edn, Garland Publishing.

36 Obinata, M. (2007) *Cancer Sci.*, **98**, 275.

37 Zola, H. (2000) *J. Biol. Regul. Homeost. Agents*, **14**, 218.

38 Goldman, B.I. and Wurzel, J. (1992) *In Vitro Cell Dev. Biol.*, **28A**, 109.

39 Knowles, B.B., Howe, C.C. and Aden, D.P. (1980) *Science*, **20**, 497.

40 Freshney, R.I.(2000) *Culture of Animal Cells*, 3rd edn, Wiley-Liss, New York.

41 Castañoa, A. and Gómez-Lechónb, M.J. (2005) *Toxicol. In Vitro*, **19**, 695.

42 Haskard, D.O. and Revell, P.A. (1984) *Clin. Rheumatol.*, **3**, 319.

43 Mossman, T.J. (1983) *Immunol. Methods*, **65**, 55.

44 Liu, Y., Peterson, D., Kimura, H. and Schubert, D. (1997) *J. Neurochem.*, **69**, 581.

45 Liu, Y. (1999) *Prog. Neuro-Psychopharmacol. Biol. Psychiat.*, **23**, 377.

46 Grandjean-Laquerriere, A., Laquerriere, P., Guenounou, M., Laurent-Maquin, D. and Phillips, T.M. (2005) *Biomaterials*, **26**, 2361.

47 Hale, A.J., Smith, C.A., Sutherland, L.C., Stoneman, V.E., Longthorne, V.L., Culhane, A.C. and Williams, G.T. (1996) *Eur J Biochem.*, **236**, 1.

48 Nelson, C.M. and Bissell, M. (2006) *Annu. Rev. Cell Dev. Biol.*, **22**, 287.

49 Robinson, J.P. (2004) *Flow Cytometry Encyclopedia of Biomaterials and Biomedical Engineering*, (eds G.E. Wnek and G.L. Bowlin), Marcel Dekker Co., pp. 630–640.

50 Stuart, K.R. and Cole, E.S. (2000) *Methods Cell Biol.*, **62**, 291.

51 Poot, M., Hoehn, H., Kubbies, M., Grossmann, A., Chen, Y. and Rabinovitch, P.S. (1994) *Methods Cell Biol.*, **41**, 327.

52 Bradford, M. (1976) *Anal. Biochem.*, **72**, 248.

53 Detomaso, L., R Gristina, G., Senesi, S., d'Agostino, R. and Favia, P. (2005) *Biomaterials*, **26**, 3831.

54 Shoemaker, D.D., Schadt, E.E., Armour, C.D., He, Y.D., Garrett-Engele, P., McDonagh, P.D. and Loer, P.M. (2001) *Nature*, **409**, 922.

55 Chao, Yin, Lei, Ying, Peng-Chi, Zhang, Ren-Xi, Zhuo, En-Thang, Kang, Kam, W. Leong and Hai-Quan, Mao. (2003) *J. Biomed. Mater. Res. A*, **67**, 1093.

56 Midwood, K.S., Mao, Y., Hsia, H.C., Valenick, L.V. and Schwarzbauer, J.E. (2006) *J. Invest. Dermatol. Symp. Proc.*, **11**, 73.

57 Briggs, S.L. (2005) *J. Wound Care*, **14**, 284.

16
Cold Gas Plasma in Biology and Medicine
E. Stoffels, I.E. Kieft, R.E.J. Sladek, M.A.M.J. Van Zandvoort, and D.W. Slaaf

16.1
Introduction

Non-equilibrium plasma is a particular medium. It contains energetic charged species with temperatures of 30 000 K and higher, which coexist with neutral gas that is much colder – at room temperature or at most at a few hundred degrees. Since plasmas are usually ignited by applying voltage to a gas, electrons and ions can directly gain energy from the electric field. This energy is dissipated in elastic/inelastic collisions with neutrals, irradiated and/or transferred to heat and transported outside. Sustaining an enormous temperature difference between charged and neutral species requires a specific way of plasma operation, so that energy transfer between energetic particles and ambient gas is hindered. This situation is readily achieved at low pressure, because collisions in a rarefied medium are scarce. Low-pressure (0.0001–0.01 atm) discharges are of major importance in lighting technology and in processing of solid-state surfaces (coating, etching, cleaning, and activation). Their exceptional chemical activity allows virtually any surface modification, while their low background temperature makes them suitable for treatment of heat-sensitive surfaces. Surface treatment for the sake of biomedical technology has become a well-established technique. Applications include plasma coating of artificial (bone) implants with biocompatible layers (hydroxyapatite, diamond, etc.) [1,2] and surface micropatterning to control/promote cell adhesion [3]. Plasma treatment has even facilitated an advanced technology for preparing wound dressings that aid healing of chronic wounds [4]. Since these techniques involve low-pressure discharges, they are limited to *ex vivo* treatment of prostheses and scaffolds. Another large potential of non-equilibrium plasmas is bacterial decontamination of air and medical/surgical equipment. For the latter application, both low-pressure and atmospheric plasmas can be used, but only *ex vivo* treatment has been considered [5–7].

In the newest trends in medical and non-medical plasma processing, increasingly more use is made of atmospheric non-equilibrium discharges. They provide more flexibility in sample handling, (usually) faster procedures, and allow substantial cost saving, because expensive vacuum equipment is not needed. Much effort has been

Advanced Plasma Technology. Edited by Riccardo d'Agostino, Pietro Favia, Yoshinobu Kawai, Hideo Ikegami, Noriyoshi Sato, and Farzaneh Arefi-Khonsari
Copyright © 2008 WILEY-VCH Verlag GmbH & Co. KGaA, Weinheim
ISBN: 978-3-527-40591-6

invested in developing suitable plasma sources, which have similar properties to low-pressure discharges in terms of high chemical activity and low gas temperature. At present, it is possible to operate stable atmospheric glows at low voltage/power and at close to ambient temperature. This technology has given rise to an innovative concept in biomedical plasma applications: *in vivo* treatment with atmospheric plasmas (plasma surgery) and plasma-aided disinfection. This new direction has led to further refinement of plasma devices, because for medical purposes the requirements on the source quality are stricter than ever before.

In vivo treatment with gas plasmas is not completely unknown in medical sciences. Hot ionized jet (argon plasma coagulation, or APC device), developed by the company ERBE-MED [8], is successfully applied for non-contact coagulation of wounds, ulcers and for nonspecific tissue removal. The main role of plasma in this technique is to supply thermal energy (heat) that will denaturate and desiccate the bleeding tissue. Usage of this plasma is especially popular in endoscopic procedures (e.g. coagulation of ulcers in gastrointestinal system [9]), treatment of wounds and infections in the oral and nasal cavity, and many others. Another approach involves a discharge created by short electric pulses applied to a physiological liquid: water is locally heated by the electric current, and breakdown occurs in the vapor [10]. This version is somewhat less aggressive than the APC plasma. The pulsed discharge can be used in a variety of medical procedures, including spine surgery [11] and even treatment of cardiovascular lesions (only in research [12]).

Recently, the biomedical team at Eindhoven University of Technology has introduced another concept of *in vivo* plasma treatment [13], which intentionally makes use of the non-equilibrium properties of the plasma. The aim is not to denaturate the tissue, but to operate under the threshold of thermal damage and to induce a specific response or modification. A suitable plasma source for such applications should operate (preferably) at body temperature, and display only a modest chemical activity.

The most obvious recipe to obtain a cold plasma is to take care that the dissipated power (or rather power density) is low. For example, a power density of 10^4–10^5 $W\,m^{-3}$ (typical for low-pressure plasmas) will result in a low gas temperature. Atmospheric plasmas usually do not operate under such conditions, but some special designs may make it possible. When designing a cold atmospheric source one has to minimize the power input, while maximizing the energy losses (power outflux) to the surrounding. The following discharge types can be considered:

- High-frequency (HF) discharges: when the excitation frequency is above ion-plasma frequency (typically >1 MHz), only electrons will be heated by the HF field, while ions will remain relatively cold. Energy transfer from electrons to heavy particles is exceptionally inefficient, so that gas heating is limited. In addition, HF discharges (>100 kHz) are recommended for medical applications, because they neither induce electric disturbances in the nervous system nor affect cardiac function.
- Transient discharges: one can operate a plasma using trains of short electric pulses, so that the duty cycle remains low. Furthermore, if the electric pulse duration is shorter than

10^{-6} s, there is not enough time for energy transfer from electrons to gas atoms/molecules.

- Plasmas with large surface-to-volume ratio: in this situation energy losses by thermal conduction will keep the gas temperature in the plasma volume low. Such plasmas can be realized in many geometries: microplasmas with dimensions of about 1 mm, planar parallel-plate discharges with a small (mm) interplate distance, etc. Microplasmas consume very little electric power, so they are convenient and inexpensive. In the latter case (parallel-plate), the electrodes are often covered with a dielectric layer (so-called dielectric barrier discharges [14]). The insulating layer charges during plasma operation, and reduces electric field and current in the discharge.
- Flowing plasmas: when gas flow is applied through the discharge zone, the temperature remains low because of convective cooling. If necessary, afterglow (downstream) treatment can be performed.

Of course, any combination of the above principles can be used. The Eindhoven team has chosen the following: a radiofrequency (RF) microplasma with gas cooling. For the sake of *in vivo* treatment, a small plasma device (*plasma needle*) has been invented [15]. Plasma needle is a point-like discharge, generated at the tip of a metal wire, enclosed in a tube with a low gas flow. This geometry offers exceptional flexibility and precision in treatment of biological samples.

In this chapter basic features of the *plasma needle* will be described (Section 16.3). The electrical characteristics, gas temperature, thermal fluxes towards the treated surfaces, and production of active radical species will be discussed. Special attention will be given to the interactions of the plasma needle with liquid samples. These are very important from the point of view of biomedical applications, because treatment of cells and tissues is always performed in an aqueous environment. Section 16.4 contains some facts on bacterial deactivation using the plasma needle. Several methods of sample preparation will be shown, and plasma efficiency in killing various kinds of bacteria will be discussed. Special attention will be paid to the intended usage of cold plasma in treatment of dental caries. Finally, Section 16.5 will deal with interactions of the plasma needle with living cells in culture and with arterial tissue (*ex vivo*) maintained in a bioreactor. Cell reactions will be described and potential applications of cold plasma treatment in (micro)surgery will be indicated.

16.2
Experiments

To create a small-sized cold atmospheric plasma, 13.56 MHz RF voltage is applied and a plasma is generated at the end of a sharp metal pin electrode (needle). The (grounded) surrounding serves as a remote counter-electrode. Formerly, plasma

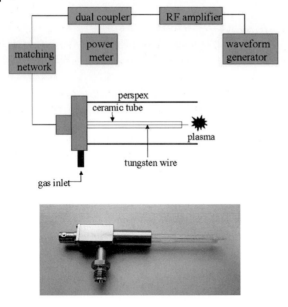

Fig. 16.1 The experimental setup (top) and an image of the portable plasma needle device (bottom).

needle was operated in many geometrical configurations, e.g. closed "torch" or plasma-box. At present, a convenient and flexible design has been introduced. A schematic and an image of the flexible plasma device are shown in Fig. 16.1. The needle has a diameter of 0.3 and it is inserted in a 5 cm long Plexiglas tube with inner diameter 0.8 cm. A helium flow of up to 2 L min^{-1}, regulated by a mass flow controller (Brooks series 5850E), is directed through the tube.

For the plasma generation a waveform generator (Hewlett Packard 33120A) and an RF amplifier (Amplifier Research 75AP250RF) are used. The power is monitored using a dual directional coupler and an Amplifier Research PM 2002 power meter. The power meter is placed in between the amplifier and the matching network to determine forward and reflected power. The voltage is measured using a voltage probe P6150A from Tektronix, inserted in the electric throughput of the needle. The probe is a 1000 times attenuator with 100 MΩ resistance and 3 pF capacitance.

Gas-phase plasma properties have been studied using optical emission spectroscopy (OES) [16], mass spectrometry (MS), and Raman scattering. All these techniques have been used to determine plasma composition; OES and MS have provided data on gas temperature. For OES, a Jobin-Yvon H25 spectrometer has been used, supplied with 1500 lines nm^{-1} grating. The image produced by the spectrometer has been obtained with an intensified charged coupled device (iCCD) camera (Andor DH534) at a cooling temperature of $-35\,^{\circ}$C. The exposure time is set at 0.7 s, and the gate width at 2000 ns. Plasma radiation has been imaged using a quartz lens with a focal length of 0.25 m. Mass spectrometric measurements have been performed using a molecular beam mass spectrometer (MBMS) developed by HIDEN Analytical Ltd. In this system species from the plasma are sampled using a triple stage differentially pumped beam inlet system and subsequently detected with the HIDEN EQP mass/energy analyzer. The setup for Raman scattering is more complex; an exact description is given elsewhere.

Plasma interactions with various (non-living) objects have been visualized either optically or by monitoring electric properties of the discharge (electric power [16]). Temperature of objects brought into contact with the plasma has been determined by thermocouples of type PT-100 (platinum) resistance temperature detectors [17]. Energy flux from the plasma to the surface has been monitored using a home-built thermal probe [18], consisting of a sensor (metal platelet) attached to a thermo-couple and provided with data acquisition system that allows recording time-dependent signals. The principle of the thermal probe is very simple: when the plasma is ignited, temperature of the sensor rises, and after plasma extinction the sensor cools down. From the dynamical response of the sensor and calibration against another energy source (e.g. laser) the energy from the plasma collected by the sensor is calculated.

Diffusion of plasma-produced species into liquid samples has been studied by means of laser-induced fluorescence in combination with fluorescent probes [19]. Major attention has been given to reactive oxygen species (the ROS family, consisting of O, OH, HO_2, H_2O_2, NO and their derivatives), because of their role in various biological processes. For detection of ROS, 5- (and 6)-chloromethyl-2',7'-dichloro-dihydrofluorescein diacetate acetyl ester (CM-H_2DCFDA) has been employed. This is a commonly known indicator of ROS formation and oxidative stress in living cells. Upon reaction with ROS, a stable multiple-conjugated double bond system is created, which displays fluorescent properties. The exact protocol of probe preparation is given at www.probes.com. For plasma treatment, liquid samples of the probe with a volume of 0.4 mL are prepared in 96-well plates with a flat bottom of nucleon surface. After plasma exposure the samples are examined with a microplate fluorescence reader FL600 of Bio-Tek. The fluorescence of the oxidized CM-H_2DCFDA probe is excited at 485 nm by a quartz halogen lamp (P/N 6000556S), supplied with a 485/20X filter. Fluorescent emission is collected at about 530 nm using a 530/25M filter. To obtain quantitative data, the probe has been calibrated using NO radicals, produced by NOR-1, an NO releaser. It is assumed that the reaction efficiency of NO and ROS with the probe is comparable.

To demonstrate the antibacterial properties of the plasma needle, deactivation tests on *Escherichia coli* (Gram negative) and *Streptococcus mutans* (Gram positive) have been performed. The *E. coli* samples have been handled according to the protocol described elsewhere [17]. A 50 μL droplet of the bacterial suspension has been pipetted on nutrient agar media in a sterile Petri dish. The initial number of bacteria is 10^5–10^6. After plasma treatment, dishes have been incubated under culture conditions, at 37 °C for 24–48 h. Colony-forming units (CFUs) on agar plates have been counted to determine the number of survivors. Alternatively, thin films of *E. coli* have been prepared by spreading 0.1 mL suspension with 10^8 bacteria per mL onto a nutrient agar medium in a sterile Petri dish [20]. Plasma has been applied locally to the dish; the sample has been incubated for 24 h and plasma-induced damage to the bacterial population has been assessed visually.

Fluorescent staining has been extensively used both in studies of bacteria and of mammalian cells. Various assays are available for assessment of cell viability and activity. The principle of fluorescent staining is the same as in a popular plasma diagnostics, laser-induced fluorescence: fluorescent products can be illuminated by a laser or a lamp and their broad-band emission can be collected using appropriate filters. Stained cells have been studied using a normal fluorescence microscope, a confocal laser microscope, a two-photon laser fluorescence microscope, or a micro-plate reader. The latter is an automated device to record space-averaged fluorescence from a standardized cell sample (one well in a multi-well array); it is equipped with a lamp and a set of filters. The information is by far not so detailed as the one provided by microscopy, but one can obtain quantitative data very fast. A micro-plate reader is ideal for parameter scans, and in cases when large numbers of experimental points are required for statistics.

The simplest and most powerful diagnostic method is the viability assay: cell tracker green (CTG), and SYTO 9 or SYTO 13 together with propidium iodide (PI). CTG reacts with cytoplasm thiols in a glutathione S-transferase mediated reaction and is converted into a cell-impermeant fluorescent product. During normal cell functioning, thiol levels are high and glutathione S-transferase is abundant; the cells stain bright green. Reduction of CTG emission in comparison to control samples can be due to reduction of living cell population (cell death) and/or to decrease of enzymatic activity in living cells. In contrast to CTG, SYTO 9 or 13 does not provide information about cell activity. It passes through the membrane and stains DNA and RNA in both dead and alive cells with a green fluorescent marker. Thus, SYTO can be used to determine the total number of cells regardless of their condition. PI is a specific stain to detect dead cells. Similar to SYTO, PI binds to the DNA and RNA and turns into red fluorescent product, but since it is a membrane-impermeant molecule, it can penetrate only into cells with damaged membranes. As explained further in the text, membrane-permeable cells are dead (necrotic). Furthermore, dual staining SYTO + PI allows one to count dead and alive cells separately, because PI binds to nucleic acids stronger than SYTO, so necrotic cells would display only red PI fluorescence, while alive ones would remain green. In this particular study, dual staining SYTO 9 + PI has been used to detect dead *S. mutans* bacteria in an "artificial cavity model", which has been constructed to study the penetration depth in plasma treatment. This model consists of two parallel microscope glasses covered with a cultured *S. mutans* biofilm. The glasses are separated by a spacer, so that a slit of 1 mm between them is created. Plasma is applied to the edge of the model and it creates a semi-circular "dead zone" between the glasses, which is assessed with a fluorescence microscope.

Interactions of the cold plasma with living mammalian cells have been studied to unravel basic cell responses and to search for ways of cell removal without inducing accidental cell death (necrosis). The following cell types have been used:

- fibroblasts: Chinese hamster ovarian cells (CHO-K1) [21], 3T3 mouse fibroblasts.

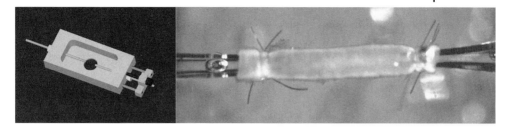

Fig. 16.2 Left: perfusion reactor in which isolated blood vessels can be maintained; right: detailed picture showing how the vessel is mounted.

- Muscle cells: rat aortic smooth muscle cells (A7r5) [22].
- Endothelial cells: bovine aortic endothelial cells (BAEC) [22].

For the above cells, viability staining with CTG and PI has been applied.

Furthermore, reactions of cells in a tissue have been studied on a well-defined and reproducible tissue sample. An artery section obtained from a Swiss mouse (carotid and uterine arteries) has been chosen, because, for arterial tissue, the structure and the responses to various damage factors are well known. The structure of a mouse artery is relatively simple: it is thin (lumen diameter is about 300 μm) and it has only few cell layers. However, it is a good example to study, because it contains all essential elements which constitute other tissues. The outer part (tunica adventitia) is a layer of connective tissue, containing collagen and fibroblast cells, the middle part (media) is a layer of smooth muscle cells reinforced with elastin fibers, and the inner part (intima) is a layer of endothelial cells and sub-endothelial connective tissue. Arteries have been excised from euthanized mice and mounted in a perfusion reactor (see Fig. 16.2) under a flow of physiological buffer at 40 mmHg pressure. SYTO 13 + PI has been applied to check cell viability and tissue has been observed using the two-photon laser fluorescence microscope.

16.3
Plasma Characteristics

The experiments show that plasma needle is indeed a nondestructive source. The power dissipated in the plasma is generally low – it ranges from 10 mW to several watts; the peak-to-peak RF voltage varies from 160 V (ignition threshold) to about 700 V. A voltage–power curve is shown in Fig. 16.3. The typical size of the plasma glow is less than 1 mm, but it increases in size with increasing forwarded power. Plasma can operate in two modes. In the unipolar one the grounded surroundings are remote (few millimeters away from the powered needle); the glow is confined at the needle tip. When the plasma needle is brought closer to a surface, a mode transition occurs: the glow expands and increases in brightness. Typical images of the glow in two modes are shown in Fig. 16.4. The change in plasma appearance and properties is

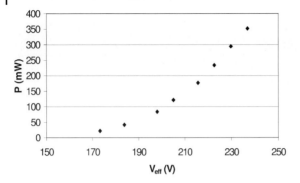

Fig. 16.3 Dissipated plasma power as a function of applied voltage (unipolar mode).

easy to understand. The electrode gap (needle-to-ground distance) determines the local electric field (E). At an imposed (RF) voltage V, the field scales approximately as V/d. The electric field determines the ionization rate, electron/ion densities, conductivity σ, and eventually the power consumption in the plasma ($P = \sigma E^2$). It is observed that the consumed power increases rapidly at short distances (Fig. 16.5). This mode of plasma operation is most effective in surface treatment, because most of the plasma energy and active species produced in the glow can be deposited on the treated object. Measurements performed using the calibrated thermal probe have shown that the power transfer efficiency (the fraction of plasma power that has been absorbed by the surface) is close to one at a distance of less then 2 mm.

In order to demonstrate that the plasma needle does not cause thermal damage to treated objects, temperatures of the gas as well as of various treated surfaces have been determined. A summary of results is given in Fig. 16.6. Optical emission spectroscopy (OES) provides the rotational temperature from N_2 bands. The data originate from a zone that has the highest emission intensity; this is also the hottest

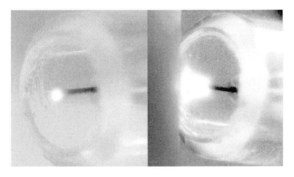

Fig. 16.4 Images of the plasma glow. Left: unipolar mode (plasma concentrated at the needle tip); right: bipolar mode (plasma in contact with skin surface).

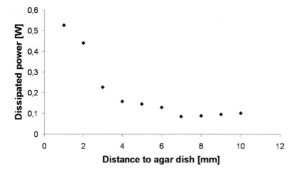

Fig. 16.5 Dissipated plasma power as a function of plasma–surface distance (bipolar mode with Petri dish as a counter-electrode).

part of the plasma. Therefore considerably high temperatures are found. Mass spectrometry samples the neutral gas density (n) from a downstream zone (2–4 mm away from the needle). The temperature is calculated according to $p = nk_B T$, applied assuming constant (atmospheric) pressure. Because this is an afterglow zone, the corresponding temperatures are lower than obtained by OES. Nevertheless, the trends given by both methods are quite consistent. It is evident that gas heating occurs only at high power input. For the thermocouple measurements, a thermocouple has been immersed about 0.5 mm under the liquid surface and the distance between the needle and the liquid has been varied. This is not a gas-phase method; in fact, it is more relevant from the point of view of biomedical applications, because it provides information about the heat that the sample will encounter. Surface temperature depends on the thermal capacity and conductivity of the surface, plasma power, and exposure time, but generally heating is tolerable.

Gas in the plasma consists mainly of helium. From Raman scattering measurements the admixture of air into the helium flow (by diffusion from the sideways) has

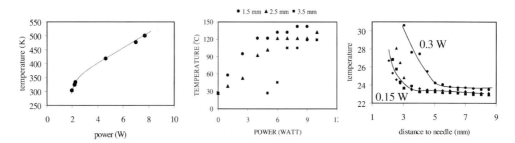

Fig. 16.6 Left: OES temperature versus applied power; middle: mass spectrometric (MS) data versus power at various needle-to-MS orifice distances; right: immersed thermocouple data versus needle-to-water distance at two powers.

been estimated to be about 0.5%. However, this small admixture is responsible for the chemical activity of the plasma. OES shows the presence of many molecular species in the plasma. MS data indicate that positive ions like N^+, N_2^+, and O^+ are more abundant than He^+. Two regimes of plasma operation can be distinguished: at the voltages just above the ignition threshold (120 V), plasma emission is very weak and the spectrum features mainly emission lines of molecular species. Above 150 V a sudden increase in plasma intensity is observed; especially helium lines become much more pronounced (more than 100 times more intense than below 150 V). Typical spectra of these two operation regimes are displayed in Fig. 16.7. Note that in the low-power regime N_2, N_2^+, and OH lines are relatively strong as compared to the helium line at 587.6 nm. The chemical activity of the plasma is already sufficient for tissue treatment, while the voltage/power is too low to cause any tissue damage.

During treatment of living tissues the active plasma zone is never in direct contact with individual cells. The latter ones are covered with up to 1 mm physiological liquid (saline) to prevent them from desiccation. Thus, plasma–cell interactions always take place through the gas–liquid interface. Active radicals from the plasma diffuse and dissolve in the liquid, reach the cells and undergo reactions with organic cell constituents. Detection of oxygen-containing radicals (ROS) produced by the plasma needle has been performed using a fluorescent probe. In Fig. 16.8 the ROS concentration induced by plasma treatment is shown as a function of exposure time. As expected, a linear dependence is found. From these data the radical flux towards the surface can be deduced ($10^{18}\,m^{-2}\,s^{-1}$). The concentration in the solution lies in the μM regime, which is close to the natural level of active radicals produced by

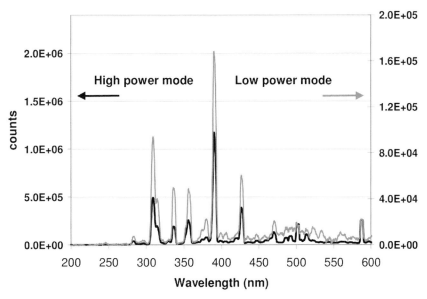

Fig. 16.7 Spectra of the plasma in low- and high-voltage operation regime (normalized at the 587.6 nm helium line).

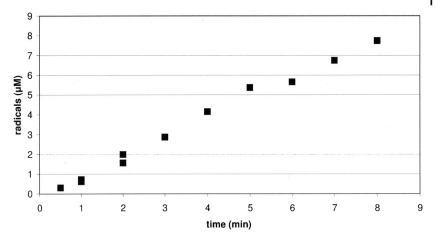

Fig. 16.8 ROS radical concentration in plasma-treated liquid samples, determined using the fluorescent probe. Plasma power is about 100 mW and needle-to-sample distance 1 mm.

a healthy organism. Of course, at higher levels (mM), radicals cause so-called oxidative stress and cell damage [23]. In μM concentrations, oxygen radicals are indispensable in fighting bacterial infections. Therefore one can expect that plasma needle will have mild disinfecting properties, i.e. it will deactivate bacteria without damaging the body cells.

16.4
Bacterial Inactivation

Bactericidal properties of many plasma sources are well-documented in the literature [19,20]. Plasma needle combines these properties with a mild character, which will allow an *in vivo* application. The medical application of plasma treatment is straightforward: plasma needle can be used for various local disinfections. Noncontact disinfection of wounds and dental cavities would be of great value in dermatology and dentistry. In this section the efficiency of killing micro-organisms will be demonstrated.

Escherichia coli is the first test object, because this species is easy to culture and fairly resistant to chemical damage factors. Droplets of bacterial suspension containing 10^7–10^8 colony forming units in 50 μL fluid have been treated at low plasma powers (100 mW), plated out, and incubated overnight. The survival curve (number of colonies as a function of treatment time) for this experiment is shown in Fig. 16.9. Because of a substantial thickness of the droplet (1–2 mm), several tens of seconds are needed for deactivation. The so-called *D*-value, the time required to deactivate one decade (90%) of the bacteria, is about 40 s. When thinner samples are treated,

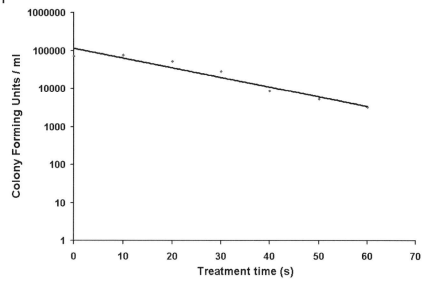

Fig. 16.9 Number of colony forming units as a function of plasma treatment time for droplet samples of *E. coli* at 100 mW, 1 mm needle-to-sample distance.

deactivation times are substantially shorter. Treatment of thin (0.1 mm) *E. coli* films grown in Petri dishes results in fast formation of characteristic voids at the incidence of the plasma needle. The voids, up to 1 cm in diameter, are completely free of bacteria, while the adjacent regions are not influenced – the void borders were sharp. It was argued that bacterial destruction is caused by short-living plasma species and not, for example, by heat. A parameter study has been conducted to optimize the conditions for real *in vivo* treatment. Figure 16.10 shows the void sizes and corresponding numbers of destroyed bacteria as a function of treatment time. One can see that the curves saturate; similar effect was observed when plasma power was increased – the void reaches a certain size and does not increase further. Thus, usage of high powers and prolongation of treatment time is not necessary, because it does not improve the efficiency of disinfection. Furthermore, plasma treatment is effective even at quite long needle-to-sample distance: voids are induced even at 8 mm away from the needle. These findings are very encouraging for *in vivo* applications, where one strives to use low powers and large distances, to suppress all possible side effects.

Plasma is a gaseous medium, of which one expects a certain ability to penetrate irregular cavities and fissures. This inspired the research on plasma disinfection of dental cavities. Usage of plasma would have many essential advantages. First of all, it would allow saving plenty of healthy dental tissue, which is usually removed in conventional preparation of cavities (drilling). In many cases (e.g. beginning caries) tissue removal is not necessary at all – after disinfection the lesion may cure by itself. Usage of plasma on dental plaque and early caries would greatly improve oral

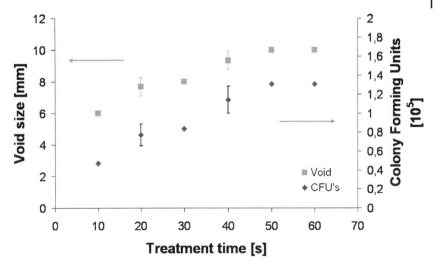

Fig. 16.10 Size of the void in *E. coli* sample and the corresponding amount of destroyed bacteria as a function of treatment time at 180 mW and 1 mm.

hygiene. The bacterial species that is predominantly responsible for caries is the biofilm-forming Gram positive facultative anaerobe *Streptococcus mutans*. A bio-film has a dense structure: living bacteria are embedded in a matrix of their own secretions, dead bacteria, nutrients, etc., so from the point of view of sterilization it is a difficult object. Preliminary tests on deactivation of *S. mutans* bio-films have been already performed. It appears that deactivation of thin films (0.1 mm) can be easily achieved, but for 0.3–0.6 mm films only partial deactivation is possible. On the other hand, such thickness is quite extreme – plaque as thick as a sub-millimeter slab should be rather removed mechanically. Penetration depth has been studied on a model consisting of two parallel glass plates with a slit of 1 mm; the inner surfaces of the plates are covered with a *S. mutans* film. The advantage of this model is the simplicity and experimental convenience: since the glass plates are transparent, bacterial survival can be directly diagnosed using fluorescent staining and micro-scopy. Plasma is applied at the edge of the glass, with the needle coplanar with the plate and pointing into the slit. At the incidence of the needle, a semicircular dead zone is produced on both plates. This can be seen in Fig. 16.11, where the dead bacteria are stained with PI. The diameter of the dead zone is 1 to 1.5 cm. This penetration depth is more than enough in treatment of dental cavities; however, this simplified cavity model suffers from some inadequacies. In real cavities typical slit dimensions may be smaller and they may vary dependent on the position within the tooth. *In vivo* experiment is the only means to verify the effectiveness of plasma therapy and to assess the post-treatment tissue repair. One can state for sure: no damage to healthy tissue will occur – in the worst case (insufficient disinfection), the cavity will have to be treated in the conventional way.

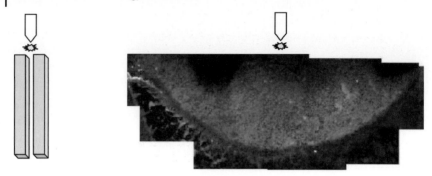

Fig. 16.11 Plasma treatment of *S. mutans*. Left: stack of two plates with bacterial films on inner surfaces (side view); right: dead part of the film (PI-stained light grey zone) as a result of treatment (top view) at 110 mW, 1 mm distance.

16.5
Cell and Tissue Treatment

Besides bacterial disinfection, plasma needle offers a possibility to perform sophisticated modification of body cells. Plasma interactions with eukaryotic cells are much more complicated than with bacteria. The cells are quite resistant, and can have many interesting responses of self-defense. Some of these responses are of interest for tissue treatment. For example, we have established that plasma can modify cells without causing membrane rupture. Membrane rupture results in so-called accidental cell death (necrosis), where the cytoplasm is released. The latter is poisonous to the tissue, so the body launches a defense mechanism known as inflammatory reaction. This occurs in any type of surgery: conventional surgery (incision), electrosurgery, cryosurgery, laser treatment, and burning/coagulation. The organism can cope with some necrosis, but it always results in bulk tissue damage, delayed healing, and no neat tissue repair (e.g. scar formation). Large necrotic sections pose a serious danger and must be removed. So far there has been no alternative to necrosis, because no appropriate technique has been available. Plasma needle introduces a new concept in surgery, which is based on operating without inflammation. In cold plasma treatment, necrosis can be avoided, and modification can be limited only to a small group of target cells. This high precision will allow local surgery, e.g. treatment of extremely vulnerable membranes and embryos. Desired effects are controlled cell removal and manipulation.

The major effect of plasma on living cells is cell detachment. Usually, cells in culture form a two-dimensional quasi-tissue: they are attached to each other and to the substrate by cell adhesion molecules (CAMs). An untreated sample, showing an array of elongated fibroblast cells is shown in Fig. 16.12. Plasma treatment at moderate power levels (100–200 mW) results in a temporary loss of contact between the cells and their surroundings (neighboring cells and substrate matrix). The cells assume a rounded shape, like droplets (Fig. 16.12) and eventually they loosen from

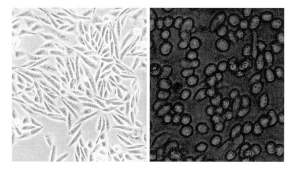

Fig. 16.12 Left: untreated sample of CHO-K1 fibroblasts; right: rounded (detached) cells after plasma treatment.

the substrate and float in the medium. The area of reach (region in which cells are detached) is a circle from 0.1 mm to about 1 cm. The border of this region is very sharp. Treatment time of about 2 s per spot is sufficient to induce this effect; longer treatment results in a larger area of reach. Detachment is of general nature – it has been observed on all studied types of cells, fibroblasts, smooth muscle, endothelial and epithelial cells – the extent of the phenomenon is in all cases about the same. The cells restore the contact within several hours after exposure to the plasma. This is a typical time scale for synthesis of proteins and repair of minor damage in viable cells. Apparently, plasma does not have a drastic negative effect on cell viability and activity. This is also indicated by a low occurrence of necrosis: under proper treatment conditions (low power) no necrotic cells are found. Typically, necrosis can occur at the power higher than 300 mW and exposure time of 30–60 s; in such cases it can be caused by heat and/or desiccation (long exposure to gas flow).

Cell detachment may be a result of interactions of plasma radicals with the cell membrane: the reactive oxygen species can oxidize and thus break the CAMs. This process need not result in damage to the cell interior. Alternatively, plasma species can act like messengers: they convey certain stimuli, which will alert the cell and trigger a mechanism to move away from a dangerous area. Again, in this process the stimulus remains under the threshold of damage: if the cell can "escape" in time, it will remain intact. Summarizing, cell detachment is a reversible effect that allows removing or mobilizing cells. This will also enable one to extract cells from the tissue to prepare a native graft. In absence of necrosis, no inflammation in the body will be induced.

Tissue treatment is the next logical step in preparation of the plasma surgical technique. In tissues, a complication arises due to the presence of extracellular matrix. Here some preliminary facts on plasma interaction with arterial tissue are given. So far, *ex vivo* treatment of carotid and uterine mouse arteries has been performed. Arteries are mounted in a perfusion reactor and filled with liquid at a transmural pressure of 40 mmHg. The first outcome is that the transmural pressure remained constant, thus no artery perforation took place. At low plasma powers

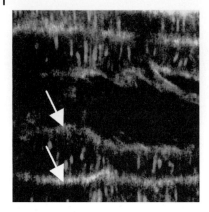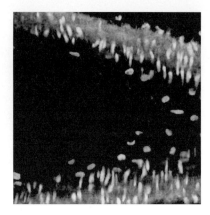

Fig. 16.13 Arterial sample from a Swiss mouse, a two-photon fluorescence micrograph showing a part of the medium. Left: untreated, arrows indicate elastin fibers, smooth muscle cells (SMCs) are stained with SYTO 13. Right: after 1 min treatment with 450 mW (1 mm away from the sample); some elastin bands are broken, SMCs have lost their alignment.

(100 mW) the effects are limited only to adventitia. Occasional necrosis in fibroblasts has been observed, but the collagen matrix remains intact. A treated sample is shown in Fig. 16.13. In this sample, some smooth muscle cells and elastin fibers are damaged. However, one can also see that the cells change their shape and orientation: from elongated and aligned before to rounded and chaotic after treatment. This might indicate that cell detachment, similar to that observed for cells in culture, can also occur in the tissue. In these experiments treatment conditions have been too harsh, so that necrosis has been abundant. However, in treatment of cardiovascular diseases (atherosclerotic plaque) one aims at removing large amounts of tissue, so occasional necrosis should not be a great hinder. In fact, plaque removal with the needle might be much less destructive and patient-friendly than mechanical methods or laser/spark tissue ablation.

The fact that plasma influences only living cells and not the extracellular matrix is not necessarily a hindrance. The most important feature is that plasma can detach/mobilize cells without killing them. When necessary, non-living extracellular matrix can be separated mechanically (by cutting) after the cells have been driven away from the area. The idea of operating without inflammation is valid as long as no cell necrosis occurs. At present one cannot foresee whether this technique will be applicable in large-scale operations, but microsurgery seems feasible. First experiments on small test animals are in preparation.

The remaining question in plasma–cell interactions is the role of UV radiation. Spectra of the plasma needle (Fig. 16.7) show little emission at 305 nm, due to the OH radical. Some vacuum UV of helium transitions can be also present, but it is expected to be optically thick under atmospheric conditions. In general, atmospheric plasmas are poor UV sources, and short-wavelength (<200 nm) radiation is usually cut off by aqueous media in which cells and tissues are maintained. Furthermore, the influence of pure UV irradiation (without plasma) on cells has been studied, and necrosis has been the only observed effect [24]. Therefore, one can tentatively assume that cell reactions induced by the needle are of chemical, and at higher powers, of thermal, nature. However, it is not unthinkable that in other types of plasma sources (e.g. air plasmas) UV radiation will play a role.

16.6
Concluding Remarks and Perspectives

Non-thermal plasma can be seen as a well-controllable source of chemical, thermal, and radiative energy that can be transferred to tissue. It can have various effects on the tissue: it can operate above the threshold of cellular damage and necrosis, or under very mild conditions where no necrosis occurs. Both modes of operation may have their specific medical applications. This chapter describes a particular source: a plasma needle that is cold, compact, moderately active, and nondestructive. It can be applied to living tissues without deep damage and pain sensation. Most likely, the plasma technique will not be used for bulk surgery. The major advantage is the finesse: superficial action and minimum damage to the tissue. Several medical applications can be already identified: disinfection of wounds and dental cavities, and microsurgery. Although it will take much effort to implement cold plasma treatment in clinical practice, current studies provide very interesting fundamental facts and give confidence in the success of plasma surgery.

References

1 Cheang, P. and Khor, K.A. (1996) *Biomaterials*, **17** (5), 537.
2 Freitas, R. A. Jr. (2003) *Nanomedicine, Vol. IIA: Biocompatibility*, Landes Bioscience, Georgetown, TX.
3 d'Agostino R., Favia P., Oehr Ch. and Wertheimer M. (eds) (2005) *Plasma Processes and Polymers*, Wiley-VCH.
4 Haddow, D.B., Steele, D.A., Short, R.D., Dawson, R.A. and MacNeil, S. (2003) *J. Biomed. Mater. Res.*, **64A**, 80.
5 Laroussi, M. and Leipold, F. (2004) *Int. J. Mass Spectrom.*, **233**, 81.
6 Laroussi, M. (2002) *IEEE Trans. Plasma Sci.*, **30**, 1409.
7 Philip, N., Saoudi, B., Crevier, M.-C., Moisan, M., Barbeau, J. and Pelletier, J. (2002) *IEEE Trans. Plasma Sci.*, **30**, 1429.
8 See: www.erbe-med.de.
9 Stoppino, V., Cuomo, R., Tonti, P., Gentile, M., DeFrancesco, V., Muscatiello, N., Panella, C. and Ierardi, E. (2003) *J. Clin. Gastroenterol.*, **37**, 392.
10 Stalder, K.R., McMillen, D.F. and Woloszko, J. (2005) *J. Phys. D: Appl. Phys.*, **38**, 1728.
11 See: www.arthrocare.com.
12 Slager, C.J., Essed, C.E., Schuurbiers, J.C., Bom, N., Serruys, P.W. and Meester, G.T. (1985) *J. Am. Coll. Cardiol.*, **5**, 1382.
13 Stoffels, E., Kieft, I.E., Sladek, R.E.J., Van derLaan, E.P. and Slaaf, D.W.

(2004) *Crit. Rev. Biomed. Eng.*, **32**, 427.

14 Nersisyan, G. and Graham, W.G. (2004) *Plasma Sources Sci. Technol.*, **13**, 582.

15 Stoffels, E., Flikweert, A.J., Stoffels, W.W. and Kroesen, G.M.W. (2002) *Plasma Sources Sci. Technol.*, **11**, 383.

16 Kieft, I.E., Van derLaan, E.P. and Stoffels, E. (2004) *New J. Phys.*, **6**, 149.

17 Sladek, R.E.J., Stoffels, E., Walraven, R., Tielbeek, P.J.A. and Koolhoven, R.A. (2004) *IEEE Trans. Plasma Sci.*, **32**, 1540.

18 Stoffels, E., Sladek, R.E.J., Kieft, I.E., Kesten, H. and Wiese, H. (2004) *Plasma Phys. Contr. Fusion*, **46**, B167.

19 Kieft, I.E., Van Berkel, J.J.B.N., Kieft, E.R. and Stoffels, E. (2005) Radicals of plasma needle detected with fluorescent probe, in *Plasma Processes and Polymers* (eds d'Agostino, Favia, Oehr and Wertheimer), Wiley-VCH, p. 295.

20 Sladek, R.E.J. and Stoffels, E. (2005) *J. Phys. D: Appl. Phys.*, **38**, 1716.

21 Kieft, I.E., Broers, J.L.V., Caubet-Hilloutou, V., Ramaekers, F.C.S., Slaaf, D.W. and Stoffels, E. (2004) *Bioelectromagnetics*, **25** (5), 362.

22 Kieft, I.E., Darios, D., Roks, A.J.M. and Stoffels, E. (2005) *IEEE Trans. Plasma Sci.*, **33**, 771.

23 Halliwell, B. and Gutteridge, J.M.C. (1999) *Free Radicals in Biology and Medicine*, University Press, Oxford.

24 Sosnin, E.A., Stoffels, E., Erofeev, M.V., Kieft, I.E. and Kunts, S.E. (2004) *IEEE Trans. Plasma Sci.*, **32**, 1544.

17
Mechanisms of Sterilization and Decontamination of Surfaces by Low-Pressure Plasma

F. Rossi, O. Kylián, and M. Hasiwa

17.1
Introduction

Decontamination of medical device surfaces is an increasing concern worldwide. Recent statistics show that nosocomial infections are responsible for several thousands of death each year in Europe, and have a significant impact on health costs. Their major sources are micro-organisms (bacterial spores, viruses, molds, or yeasts) but other complications come also from biomolecules (e.g. pyrogens). The risk linked to the transmission of Creutzfeldt–Jacob disease (CJD) by contaminated surgical instruments has also been reported recently [1]. Instruments entering the human body must be therefore sterilized and decontaminated by removal or inactivation of any harmful biological residuals, especially in the invasive surgery and dental praxis. Therefore much effort has been devoted in the last years to develop and validate techniques of sterilization and decontamination of wide range of materials. In principle ideal technique should fulfill several requirements:

- Efficiency of treatment (low treatment time at low cost).
- Universality of method. Ideal decontamination method should be applicable for wide range of objects from the point of view of their composition and properties (e.g. heat sensitive polymeric materials or corrosive materials) as well as of their shapes (e.g. hollow substrates).
- Safety of treatment both for the operator and the environment – i.e. optimal method should be non-toxic without any side effects or dangerous byproducts.

One promising way how to reach theses objectives is the treatment of surfaces by plasma as demonstrated in this chapter. To reach this objective we will first present the different commonly used techniques together with their advantages and drawbacks. Subsequently, the principles of low-pressure plasma treatment and a review of the results presented in the literature will be made. Finally, selected recent results related to the bacterial spore sterilization, depyrogenation, and plasma–protein interaction will be presented and discussed.

Advanced Plasma Technology. Edited by Riccardo d'Agostino, Pietro Favia, Yoshinobu Kawai, Hideo Ikegami, Noriyoshi Sato, and Farzaneh Arefi-Khonsari
Copyright © 2008 WILEY-VCH Verlag GmbH & Co. KGaA, Weinheim
ISBN: 978-3-527-40591-6

17.1.1
Overview of Sterilization and Decontamination Methods

In the following, we will distinguish between the sterilization processes, which must kill micro-organisms (bacteria, spores, viruses) and decontamination processes for which the objective is to remove or inactivate biological contamination (pyrogens, proteins, biomolecules).

Typical examples of materials which are reused in the hospital environment and thus require repeated sterilization and decontamination are major surgical instrument trays, minor surgical kits, respiratory sets, fiber optics (endoscopes, angioscopes, bronchioscopes, etc.). Typical instruments reused in a dental environment require repeated sterilization as well; representative examples of them are, for instance hand-pieces, dental mirrors, plastic tips, model impressions, and fabrics.

17.1.1.1 Current Cleaning and Sterilization Processes

All the operations of sterilization must be preceded by a cleaning process, which has the function to remove major contamination, by simple mechanical and chemical action. It is less studied and until recently less controlled than the sterilization operation itself, because the risk of complication and infection is generally higher for micro-organisms. The principal products for cleaning and disinfection and their relative mechanisms are presented in Table 17.1.

The principal methods of sterilization and their relative mechanisms are presented in Table 17.2. One major sterilization process in present use is that which employs ethylene oxide (EtO) gas in combination with Freon-12 (CCl_2F_2) at pressures up to 3 bar in a special shatter-proof sterilization chamber. This process, in order to achieve effective sterile levels, requires exposure of the materials to the gas for at least one to

Tab. 17.1 Different types of disinfectants and antiseptics.

Classes	Examples	Modes of action
Phenol and phenolic compounds	Carbolic acid, hexachlorophene, cresol, phenol, orthophenylphenol	Disruption of plasma membrane through protein denaturation
Alcohols	Ethanol, isopropanol (50–70% aqueous solution)	Lipid solvents and protein denaturants
Surfactants	QACs; soaps, detergents	Disruption of plasma membrane through charge interactions with phospholipids Disruption of plasma membrane through charge interactions with lipoproteins
Halogens	Iodine	Reaction with tyrosine to inactivate proteins
Alkylating agents	Formaldehyde, glutaraldehyde, ethylene oxide	Denaturation of proteins and nucleic acids through attachment of methyl or ethyl groups to these molecules
Heavy metals	Mercury, silver, copper	Protein denaturant

three hours. Perhaps the chief drawback to this system, however, is its dangerous toxicity. Ethylene oxide is a highly toxic material dangerous to humans. It was recently declared a carcinogen as well as a mutagen. It requires a very thorough aeration process following the exposure of the medical materials to the gas in order to flush away toxic EtO residues and other toxic liquid by-products like ethylene glycol and ethylene chlorohydrin. Unfortunately, it is a characteristic of the gas and the process that EtO and its toxic byproducts tend to remain on the surface of the materials being treated. Accordingly, longer and longer flush (aeration) times are required in order to lower the levels of these residues absorbed on the surface of the materials to a safe operational value.

A number of other approaches for performing sterilization have also been employed. One such process is high-pressure steam autoclaving. It requires high temperature (110–140 °C) and is not suitable for materials which are affected by either moisture or high temperature, e.g. sensitive to corrosion and sharp-edged metals, plastic-made devices, polymers, etc., employed by the hospital and the dental care communities.

Another approach utilizes radioactive sources. The gamma sterilization approach is expensive and requires large installations. The use of radioactive sources moreover requires expensive waste disposal procedures, as well as stringent radiation safety precautions. The sterilization technique also presents problems because of radiation-induced molecular changes of some materials, which, for example, may render flexible materials brittle, e.g. catheters. However, this method is particularly adapted for sterilization after manufacturing and packaging (implants).

A newer and already well-established sterilization technique is based on the use of low-temperature plasma (Sterrad system, Johnson & Johnson) combined with hydrogen peroxide. The J&J system offers a short cycle, low temperature, and wide compatibility with materials [2], but the use of high concentration hydrogen peroxide requires high safety precautions. Moreover, the Sterrad system is more an improved H_2O_2 sterilization device than a true plasma sterilization method: it was demonstrated that the use of plasma *has no effect* on the sterilization kinetics, but was

Tab. 17.2 Mechanisms of classic methods of sterilization.

Sterilization methods	Modes of action
Ethylene oxide	Alkylating agent
	Affects essential components of the cell: denaturation of proteins and nucleic acids
Heat sterilization (moist heat)	Denaturation of proteins, inactivation of enzymes, oxidizing agent, alterations of lipids
	Spores: denaturation of DNA, inhibition of the germinal system
Irradiation by gamma rays and electron beam radiation	Interferences with DNA, RNA (strand breakage, base damage)
	Inhibition of DNA replication and repair, and protein synthesis

necessary to desorb and decompose the hydrogen peroxide adsorbed on the surfaces, thereby reducing the treatment duration.

In the pharmaceutical industry depyrogenation and decontamination are a major concern, but the operation is generally applied to glass vials, which are not as temperature sensitive as medical devices. In general, depyrogenation in industry is made in a heat tunnel between 260 and 350 °C, which is expensive in energy and floor space (cooling). The case of PrP raises new demands which are not solved since the only processes validated for PrP inactivation are based on steam autoclaving at 134 °C for 18 min, or treatment with concentrated NaOH solution.

This brief summary indicates that a method of decontamination and sterilization efficient at low temperature, without effect on the substrates treated, and not using toxic products has still to be developed: this consideration led many groups to study the plasma sterilization based on the use of non-toxic gases.

17.1.1.2 Low-Pressure Plasma-Based Method

Low-pressure plasma sterilization has been the object of various general studies which indicated different potential mechanisms of its action [3–7]. Generally speaking, non-equilibrium discharges combine several advantages. Firstly, the non-equilibrium discharges are effective sources of number of chemically active particles (e.g. O, N, H, OH) in their ground state, exited to higher energetic levels as well as ionized that can effectively etch, modify or, in the case of ions accelerated in the plasma sheath, sputter the contaminant deposited on the surface. Since the gases used for the sterilization and decontamination purposes are usually common gases without biocidal effects (e.g. O_2, N_2, Ar, H_2) the operation is non-toxic, safe, and environmentally friendly. Secondly, the reactive species produced in the plasma usually have short life times and disappear quickly after switching off the discharge. Thirdly, non-equilibrium plasmas are sources of high-energy UV/vacuum ultraviolet (VUV) photons that can inactivate or destroy micro-organisms or biomolecules. Compared to the classic UV/VUV treatment the UV photons can be brought to the appropriate site by its emitting atom or molecule so that shadowing effect is lowered in this case [8]. Finally, one of the main advantages of plasma treatment is that it can be performed at low temperatures desired for the treatment of plastic objects.

From the above, it follows that different mechanisms can contribute to the sterilization and decontamination during plasma processing. Naturally, the role and the contributions of possible mechanisms can vary with discharge parameters, with treated biomaterial and it depends also on the particular experimental configuration as is shown schematically in Fig. 17.1.

17.2
Bacterial Spore Sterilization

The idea of employing non-equilibrium plasmas for bacterial spore sterilization is relatively old [9] and there exists extensive literature devoted to this topic (e.g. comprehensive reviews of Lerouge *et al.* [6] and Moisan *et al.* [7]). However, the

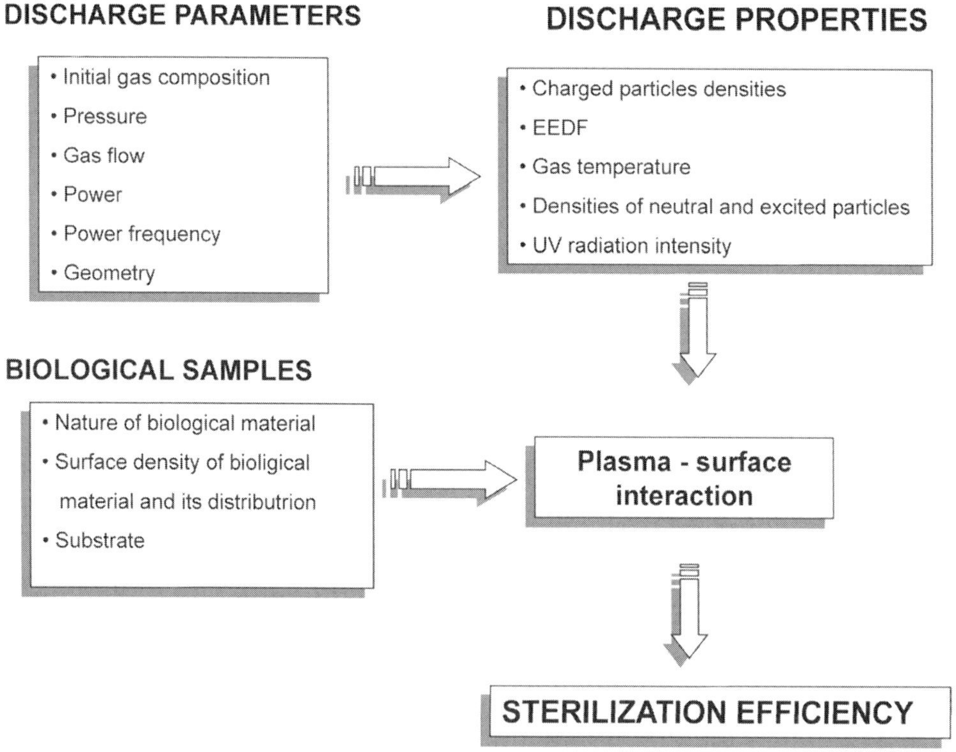

DISCHARGE PARAMETERS

- Initial gas composition
- Pressure
- Gas flow
- Power
- Power frequency
- Geometry

DISCHARGE PROPERTIES

- Charged particles densities
- EEDF
- Gas temperature
- Densities of neutral and excited particles
- UV radiation intensity

BIOLOGICAL SAMPLES

- Nature of biological material
- Surface density of bioligical material and its distributrion
- Substrate

Plasma - surface interaction

STERILIZATION EFFICIENCY

Fig. 17.1 Parameters influencing sterilization efficiency.

mechanism of spore killing is still unclear and the subject of controversy. Two main mechanisms have been proposed:

- Destruction of spores' genetic material induced by UV/VUV radiation penetrating through spores protecting coats. This mechanism has been emphasized by several groups under several discharge configurations. However, there is still discussion about the most effective spectral range for spores killing. For example Moisan *et al.* [7] and Moreau *et al.* [8] suggested that the most important for the spores destruction is UV radiation in the spectral range 200–300 nm, contrary to the results of Sholoshenko *et al.* [10] and Feichtinger *et al.* [11] who stated that radiation with wavelengths shorter than 200 nm plays a crucial role. Another study with synchrotron radiation by Munakata *et al.* [12] on different types of *Bacillus subtilis* showed that DNA was the major target for the inactivation of the spores. VUV radiations of 125–175 nm were effective in killing the spores, and distinct peaks of the sensitivity were seen with all types of the spores. Insensitivities at 190 and 100 nm were common to all types of spores tested, indicating that these wavelengths were particularly absorbed by the outer layer materials. The VUV peaks center at 150 nm were more efficient for the spores defective in recombination repair mechanisms, while the far-UV peaks at around 235 and 270 nm were more efficient for spores deficient in removal mechanisms of spore photoproducts. Thus, the UV action spectra were explained by three factors: the penetration depth of each radiation in a spore, the efficiency of producing DNA damage that could cause inactivation, and the repair capacity of each type of spore.

- Erosion of micro-organisms through their etching by active particles has been identified as the dominant mechanism of spores' destruction for example by Lerouge *et al.* [3] and Nagatsu *et al.* [13]. These authors according to their results suppose that the main process leading to spore death is their etching and that the sporicidial action of the plasma is not greatly affected by UV/VUV radiation. Moreover, the mechanism of membrane destruction has been evidenced in the case of atmospheric plasma discharge by Laroussi *et al.* [14] and Mendis *et al.* [15].

Moreover, the kinetics of spore destruction varies between authors, some finding a 2- or 3-phase destruction [7,8,16], others finding only 2 [11,17,23].

17.3
Depyrogenation

Bacterial endotoxins, a main component in the bacteria cell wall, are fever-causing molecules, most often found on medical devices. Called pyrogens, these compounds are composed of lipopolysaccharides (LPS) that have to be detected and quantified according to ISO 10993 "Biological Evaluation of Medical Devices". LPS surface contamination is hard to remove. This is due to the fact that it is insensitive to pH changes and thermostable [18]. For instance, a 3 log reduction is obtained in 40 min at 200 °C, or 1 min at 250 °C. Although the possibility of employing non-equilibrium plasmas for pyrogen destruction has been proposed already in 80′ by Peeples *et al.* [19].

17.4
Protein Removal

The removal of proteins and residues from surgical instruments has been studied by Whittaker *et al.* [20]. Those results showed that surgical instruments were still contaminated by protein residues after typical decontamination procedure. Those residues identified by energy dispersive X-ray analysis and scanning electron microscopy (SEM) were strongly adherent on metallic surfaces but could be removed by a radiofrequency (RF) Ar–O$_2$ plasma discharge. More specifically, Baxter *et al.* [21]

have shown that elimination of PrP was possible with an Ar–O$_2$ RF plasma. However, in both cases, a detailed mechanism of the interaction of plasma with protein is still missing. In order to find out which gas mixture is more efficient, extensive studies of protein removal rate were performed for bovine serum albumin (BSA) and collagen films. In this chapter, recently obtained results are going to be presented.

17.5
Experimental

17.5.1
Experimental Setup

The experimental setup for sterilization tests described in this chapter consists of a home developed double inductively coupled plasma (ICP) source (Fluxtran) [22]. Fluxtran is a novel inductively coupled plasma source enabling the treatment of hollow substrate with high aspect ratio. This ICP plasma source consists of two oppositely faced RF coils placed close to the field emission dielectric windows of the processing chamber. The coils are connected in parallel to the power supply, thus current flows in the same direction in both of them. Such configuration enables one to generate magnetic flux perpendicular to the substrate in the entire space between the coils. This results in the creation of an electric field in the whole substrate volume that enables uniform plasma treatment, even in small cavities.

The discharge chamber used for the sterilization tests is pumped by the primary pump and is connected to the inlet system composed of four mass flow controllers (maximum flow 10 sccm) connected to the gas lines (O$_2$, H$_2$, N$_2$, and Ar). The background pressure that can be reached in this configuration is 4 mtorr. Bacterial samples to be sterilized are inserted to the discharge chamber by means of the load lock mounted on the diagnostic window flange. The samples are placed in the system in the direction of the X-axis under the RF coils. In all experiments described below, the treatment was done in pulsed mode to ensure that the temperature did not increase beyond 60 °C.

The experimental setup used for the depyrogenation as well as for the protein removal studies is based on a microwave discharge and consists of a stainless steel chamber (a cylinder 200 mm in diameter and 380 mm in length) with several windows for *in situ* diagnostics. The plasma is excited in the 0.1 to 1 torr pressure range by a microwave power supply working at 2.45 GHz. Microwaves are introduced in the chamber through a silica window placed at the extremity of a circular 100 mm waveguide. The microwave circuit includes the microwave supply (2 kW), a circulator, a three-stub impedance matching system, and a rectangular–circular waveguide transition. The chamber is pumped by primary pump and a roots blower allowing a base vacuum of 2 mtorr. The gas flows are controlled by mass flow controllers connected to gas lines (O$_2$, H$_2$, N$_2$, and Ar). More details are given in [23]. It should be noted that in all reported experiments, the temperature inside the processing chamber measured by means of IR pyrometry never went beyond

$60\,°C$, thus much lower than the temperature necessary for thermal removal of LPS activity [18].

17.5.2
Biological Tests

The sterilization tests have been performed in different gas mixtures using *G. stearothermophilus* spores coated on steel discs (declared spore population 2.5×10^6) as the biological indicators. In this study, a statistical approach has been used for the estimation of the sterilization efficiency. The samples have been treated at different conditions and for different times and afterwards have been incubated for 7 days at $60\,°C$ and then checked for sterility (i.e. no positive growth of spore colonies is observed). For each experiment, 5 to 10 samples were treated, giving statistics about the sterilization probability in these conditions. On the basis of the results obtained it is then possible to derive the probability that a sample treated in certain mixture for a given time is sterile or not.

It must be stressed that this procedure does not give a kinetic of the decrease of colony forming units (CFU) as a function of time but only a probability that the initial spore population is completely inactivated.

The plasma–spore interaction has been studied from two points of view:
- Efficiency of spore killing has been investigated in order to find the best sterilization conditions.
- Morphological changes of spores exposed to the plasma have been studied by SEM.

17.5.3
Pyrogen Samples Detection

The samples to be sterilized were plastic 24-well plates covered with $100\,\mu L$ of LPS (Sigma Aldrich) diluted in de-pyrogenized water having different LPS concentrations ranging from 10 ng of LPS per mL to 0.01 ng per mL. The well plates were introduced to the plasma chamber and exposed to the O_2-H_2 (50:50) plasma. Afterwards the bioactivity of LPS after treatment has been measured by the enzyme-linked immunosorbent assay (ELISA) test described below.

In order to have a better insight into the pyrogen inactivation process additional tests have been performed with Lipid A (Sigma), the pyrogenic part of LPS. In these experiments has been compared effectiveness of Lipid A deactivation estimated by ELISA test with mass loss during the plasma processing measured by a quartz crystal microbalance (QCM). In both sets of experiments $10\,\mu L$ of Lipid A water solution $(0.1\,\mathrm{mg\,mL^{-1}})$ have been deposited on the 24-well plate (in the case of the biological test) or on a gold-coated quartz crystal for the QCM study.

The estimation of pyrogen amount on a solid surface is difficult to perform. The operation normally done consists in washing the surface with pyrogen-free water and measuring the pyrogenic content of the solution by the rabbit test (measure of the body temperature increase of the animal after injection) or with the Limulus Amebocyte Lysate Assay (LAL test). We have used instead the whole blood test developed at JRC and University of Konstanz [24]. The test is based on immune defense mechanisms based on cytokine release (e.g. IL-1β, TNF-α), from erythrocytes and leukocytes in the presence of a foreign substance in the human organism. By adding LPS (from gram-negative bacteria) or lipoteichoic acid or LTA (from gram-positive bacteria) or their pyrogenic compound (Lipid A) to human whole blood and measuring the cytokines released by a sandwich-ELISA, we are able to simulate the human reaction to immune-stimulating principles *in vitro*. Interestingly, it is possible to measure the biological response to LPS directly on different surfaces, thus avoiding the washing step necessary with the other methods. The measurement is done in a physiological environment, in the relevant species and donor independent. Positive and negative controls are included, the detection limit is much lower than the rabbit or LAL tests, and this method can be used for the detection of different pyrogenic substances (e.g. LTA, PGN or zymosan) [25].

In the experiments reported below, the results are presented as the relative LPS activity as measured by the ELISA test, referred to the value before treatment.

17.5.4
Protein Removal Tests

Protein films from collagen and BSA (Sigma Aldrich, Germany) were deposited from a distilled water solution on a quartz crystal covered with a gold film. A 100 μL BSA solution (1 mg mL^{-1}) or 5 μL of collagen solution (1 mg mL^{-1}) was pipetted onto each of the coupons and allowed to air dry. The quartz crystal is placed in a QCM (Leybold, Germany) in the microwave chamber described above and treated in different gas mixtures in the near post discharge.

17.6
Results

17.6.1
Sterilization

Discharges in O_2-containing mixtures are commonly used for sterilization, since they provide large amounts of UV radiation important for bacteria DNA destruction as well as relatively large amounts of oxygen atoms for etching of biological materials. In the present study, time evolution of sterilization efficiency as well as its dependence on initial mixture composition has been investigated. First, experiments in different O_2–N_2 mixtures have been performed. The results show that the best sterilization results are achieved in O_2 rich mixtures, as is demonstrated in Figs. 17.2 and 17.4 (complete sterilization has been observed in the mixture 95% O_2 : 5% N_2 within 5 min). Comparing this result with optical emission spectroscopy (OES) (see Fig. 17.3) it can be concluded that under our experimental conditions, overall

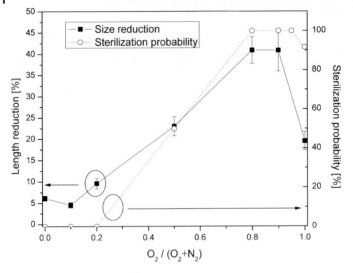

Fig. 17.2 Sterilization probability and spores' size reduction as a function of O_2-N_2 dischange mixture composition (power 500W, 20% DC, 5 ms time-on, pressure 100 mtorr, flow 10 sccm).

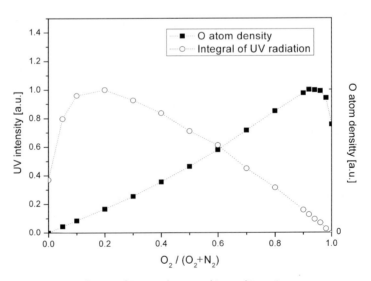

Fig. 17.3 Dependencies of O atom density and integral intensity of UV radiation in the spectral range of 200–300 nm on the O_2–N_2 discharge mixture composition (power 500W, pressure 100 mtorr, flow 10 sccm).

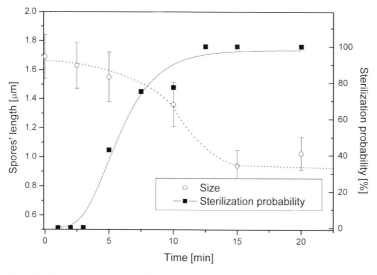

Fig. 17.4 Time evolution of sterilization probability and mean spores' length (power 500 W, 20% DC, 5 ms time-on, pressure 100 mtorr, flow 10 sccm, pure oxygen).

sterilization efficiency is more related to O atom density than UV radiation intensity, i.e. sterilization is not related to DNA damage as in UV sterilization but more likely to the etching of spores.

In order to verify the crucial role of etching on sterilization of spores, SEM analysis of treated samples has been performed. As can be seen in Fig. 17.2, significant changes of spore dimensions (about 40% of their size) occur in discharges with a high amount of atomic oxygen. On the other side, almost no variations of spore sizes have been observed in mixtures with low O atom concentration. The importance of O atoms for the spores' etching was cofirmed by experiments caried out in Ar-O_2 mixtures (Fig. 17.5). As can be seen also in this case, the significant reduction of spore dimensions is observed only in mixtures providing a high density of O atoms.

Concerning the time evolution of spore size reduction it has been found that in O_2 discharge significant reduction starts after 5 min of plasma treatment. As can be seen in Fig 17.4, this time corresponds to the time when complete sterilization of spores has been achieved according to the biological tests. Agreement of SEM results with biological ones confirms that in our case the dominant process of death of spores is their etching.

17.6.2
Depyrogenation

As mentioned above, plasma-based depyrogenation is a relatively new topic, and therefore it is naturally important to demonstrate its applicability. For this reason

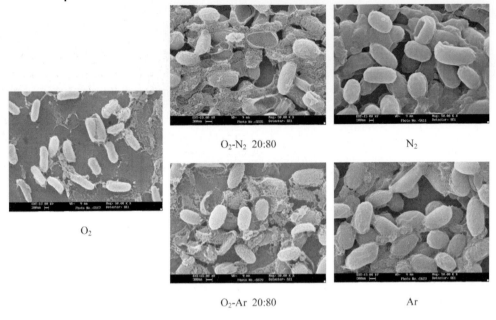

O$_2$-N$_2$ 20:80   N$_2$

O$_2$

O$_2$-Ar 20:80   Ar

Fig. 17.5 SEM images of treated spores in different gas mixtures (power 500 W, 20% DC, 5 ms time-on, pressure 100 mtorr, flow 10 sccm, treatment time 10 min).

preliminary tests have been performed focused on the estimation of the time scale of plasma treatment and its limitations. Results of experiments made in a microwave plasma with an O$_2$–H$_2$ (50%:50%) gas mixture are shown in Fig. 17.6. It can be seen that 1 log reduction of the biological activity of LPS occurs in 20 to 200 s. An interesting fact from the point of view of possible applications is that time evolution of LPS deactivation is dependent on the kind of treated pyrogens, which is probably due to their different chemical composition. The fastest depyrogenation has been observed for *E. coli* O111 LPS when almost 2 log reduction of its bioactivity has been observed after 5 min of plasma treatment. By dry heat a similar decontamination would be obtained at 170 °C after 100 min, which clearly demonstrates the interest of the plasma discharge.

In order to ensure that variations of depyrogenation efficiency observed for different LPS are not caused by variations of plasma composition, optical emission spectra of the discharge have been recorded simultaneously with the pyrogen treatment. Nevertheless, no significant variations of intensities of main detected spectral lines and bands have been observed during the treatment and therefore changes of depyrogenation rate can be attributed to the LPS type only. We also checked that neither the microwave radiation (without plasma) nor the temperature reached during the experiments were responsible for the LPS loss of activity. However, we found that the efficiency of depyrogenation is dependent on the initial LPS contamination level (see Fig. 17.7).

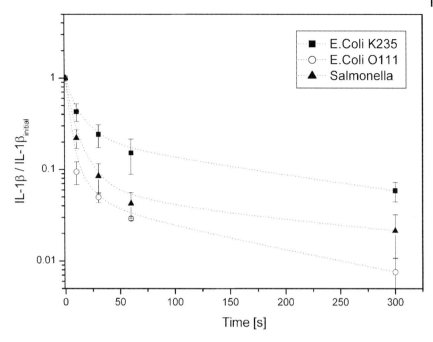

Fig. 17.6 Time evolution of bioactivity of different LPS after plasma treatment (applied power 1000 W, 100 mtorr, 100 sccm, O_2–H_2 50:50, deposited amount 0.1 ng of LPS).

In order to have a better insight in the mechanisms responsible for de-pyrogenation, a comparison between mass loss during plasma treatment and biological activity were made with QCM experiments on Lipid A thin films. Figure 17.8 shows the relative mass loss due to etching of a Lipid A film deposited on a quartz crystal in O_2, O_2–H_2 and Ar–H_2 (50:50) mixtures. It can be seen that H_2–O_2 mixture is much more efficient that O_2 or Ar–H_2 in removing the Lipid A deposit, 80% being etched after 10 min of treatment instead of about 20% in the case of O_2 and Ar–H_2 mixture. However, Fig. 17.8 shows that the biological activity of the Lipid A evolves identically in both hydrogen-containing mixtures.

17.6.3
Protein Removal

Figure 17.9 shows the time evolution of BSA and collagen film samples (normalized to the initial thickness) treated in different plasma gas mixtures (microwave, 1000 W, 120 mtorr, 100 sccm). It can be seen that the etching rate observed varies strongly with the gas used, the highest values being obtained for H_2–O_2 50:50 mixture. For these conditions, the removal rate is more than 5 times higher than that obtained with a pure Ar discharge. It can also be seen that the removal rate of the collagen and BSA for

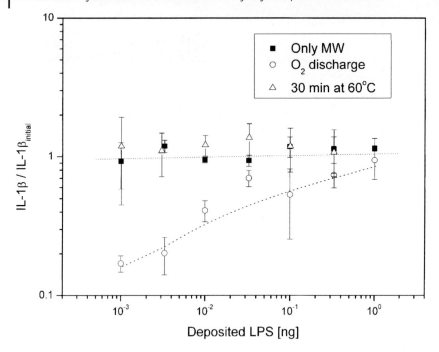

Fig. 17.7 Bioactivity of LPS exposed only to microwave radiation (300 W, 100 sccm O_2, 100 mtorr, 5 min), exposed to the O_2 plasma radiation (300 W, 100 sccm O_2, 100 mtorr, 5 min) and bioactivity of LPS heated for 30 min at 60 °C.

H_2–O_2 follows initially the same variation, followed by a slowing of the removal rate for BSA after 15 min.

17.7
Discussion

17.7.1
Plasma Sterilization

Comparison of the obtained sterilization results with the results of OES (Figs. 17.2 and 17.3) and with the results of the SEM study of treated and untreated samples leads to the conclusion that under the experimental conditions employed, the sterilization efficiency is controlled by the etching and erosion of the spores and not by UV radiation.

In order to explain the contrasting findings in terms of UV sterilization effects, we have suggested following simplified kinetics mechanism [26]. For simplicity we assume that there are two kinds of spores with given initial quantities. The first group

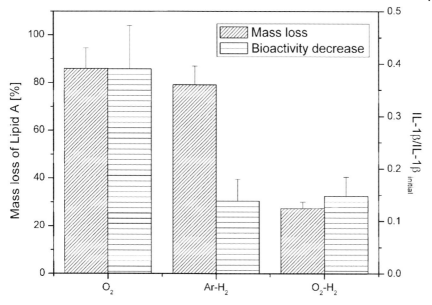

Fig. 17.8 Comparison of the mass loss of Lipid A and decrease of its bioactivity after plasma treatment in O_2, Ar–H_2 (50:50), and O_2–H_2 (50:50) discharges (applied power 1000 W, 100 mtorr, 100 sccm, treatment time 10 min).

of spores has genetic material that can be directly reached and destroyed by the UV/VUV radiation (top layer). The second type of spores represents spores shielded from UV radiation either by other spores or by some residuals (e.g. cell debris). In order to enable UV radiation to destroy DNA of these spores it is necessary primarily to remove the shielding material (top spores) either by etching or by photo-desorption. If we denote the first group of spores as S_1 and the second group as S_2 we can write down equations governing sterilization kinetics:

$$\frac{d[S_1]}{dt} = -k_1 \cdot UV \cdot [S_1] + k_2[X][S_2]$$
$$\frac{d[S_2]}{dt} = -k_2[X][S_2]$$

where [X] is the amount of etching agent and k_1 and k_2 are rate constants of UV radiation killing of spores and removing of shielding material respectively. Although the proposed schema is very simplified it can describe qualitatively the trend of the experimental results reported in the literature. Figure 17.10 shows the comparison of the model with the experiment from [23], with $k_1[UV]/k_2[X]$ ratio equal to 13.3 and $[S_2]/[S_1]$ equal to 5×10^{-4}. The calculation models a 2-step kinetics, which takes into account the production of etching radicals and UV emission. Moreover, the suggested schema allows better insight to the nature of spore sterilization and consequently to explain the basic features of the experimental results. In order to demonstrate this fact

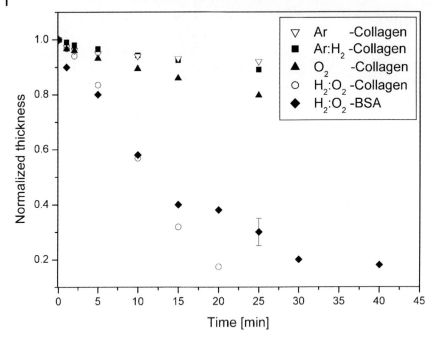

Fig. 17.9 Time evolution of collagen and BSA films thickness, normalized to initial value, in different microwave plasma gas mixtures (1000 W, 120 mtorr, 100 sccm).

several calculations have been performed using as input parameters different values of densities of spores S_1 and S_2 as well as rate constants k_1, k_2 and densities of erosion agent and UV radiation intensity. The results of simulations can be summarized as follows:

- The kinetics of bacterial killing can be divided into two phases. In the first phase, direct destruction of spores by VU/VUV radiation is dominating. The second phase is driven mainly by the efficiency of removal (etching) of shielding material.
- The total number of bacterial spores does not change the shape of survival curves if other parameters are kept constant (Fig. 17.11(a)).
- Ratio of uncovered and shielded spores influences significantly the time needed to reach complete sterilization. As can be seen in Fig. 17.11(b) it is mainly due to the different amount of surviving spores after the first sterilization phase. An increasing number of shielding spores increases the number of spores surviving the first phase but has no influence on the slope of the second phase that is driven mainly by the efficiency of removal of shielding material.

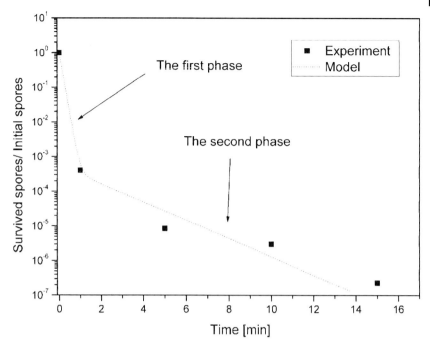

Fig. 17.10 Comparison of the proposed model with the experimental results. Experimental data are taken from [23] ($k_1[UV]/k_2[X] = 13.3$, $[S_2]/[S_1] = 5 \times 10^{-4}$).

- Increasing efficiency of erosion and removal of shielding material leads to significant increase of total sterilization efficiency (Fig. 17.11(c)). In this case the first phase has almost the same course for wide range of k_2 but the second phase is rapidly accelerating by increasing destruction rate k_2 of shielding material.

- An increase of killing rate of S_1 spores induced by higher UV radiation decreases the time at which the second sterilization phase starts. Nevertheless, the time needed for complete sterilization remains almost independent on the UV radiation effect since the second phase of sterilization is driven by the removal of the shielding material (Fig. 17.11(d)).

As can be seen from the analysis of numerical simulation there is one important conclusion – although the UV/VUV radiation is in the proposed schema the *only* mechanism leading directly to spore death, the sterilization efficiency is governed mainly by the amount of shielding material and efficiency of its removal. In other words, in the case of samples with no shielding material and without overlapping of

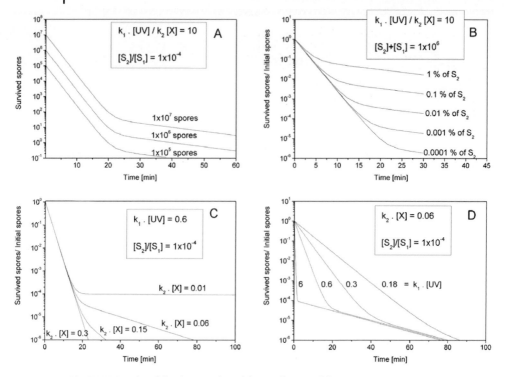

Fig. 17.11 Results of the theoretical model. (a) Influence of the spore number, (b) influence of the overlapping degree; (c) influence of the shield material removal efficiency; (d) effect of UV/VUV radiation intensity.

spores, the time needed for complete sterilization is dependent only on the UV/VUV radiation intensity. The presence of shielding material or spores overlapping causes the duration for complete sterilization to be more related to etching than to the UV/VUV effect. This effect is a possible explanation of the different experimental results mentioned above.

Using the presented numerical schema it is also possible to describe the presented experimental results. For the modeling, we have used the measured quantities of UV radiation intensity and O atom concentration as input parameters. Moreover, it has been assumed that number of covered bacterial spores S_2 represents 1–10% of the total amount of treated spores.

The comparison of the model with the experimental results is represented in Fig. 17.12. As can be seen, the sterilization probability estimated on the basis of experiments as well as log reduction of spore population predicted by the model after 10 min of plasma treatment follows the same trend – i.e. faster sterilization is achieved for higher amount of oxygen in the initial discharge mixture. Moreover, depending on the mixture composition three regions can be distinguished:

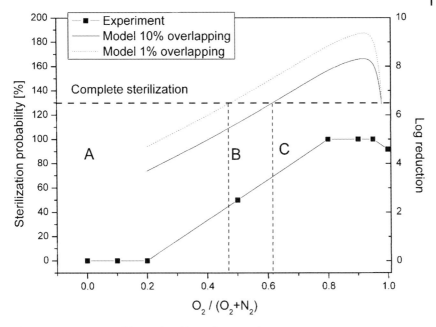

Fig. 17.12 Comparison of the predicted log reduction and probability to obtain complete sterilization after 10 min of plasma treatment in O_2–N_2 mixtures (ICP plasma source, 500 W, 20% DC, time-on 5 ms, 10 sccm, 100 mtorr, 10 min).

- The region between pure nitrogen discharge and discharge with 40% of oxygen, in which the complete sterilization is never reached after 10 min of the treatment.
- The region between 40 and 55% of oxygen, where complete sterilization is only reached for the lowest level of spores overlapping (i.e. some samples are after 10 min of treatment completely sterile and some are not, depending on the initial spore distribution). This corresponds to the experimental results, since for 50% oxygen, a 50% probability of obtaining full sterility has been estimated.
- More than 55% of oxygen mixture finally assured complete sterility after 10 min for all of the samples as has been observed experimentally.

Although, as it has been demonstrated, the proposed model gives realistic results, it has to be noted that more systematic study should be performed on large sets of experimental results for different conditions. Moreover, this simplified reaction schema could not explain the third phase of the survival curves reported by several authors (e.g. Moisan *et al.* [7]).

17.7.2
Depyrogenation

Our results clearly show that depyrogenation of surfaces can be reached at temperatures lower than 60 °C for durations which depend on the type of LPS and the level of contamination. These results must be compared to the thermal treatment used to obtain similar results, such as 30 min at 180 °C for 1 log reduction reported in [18] which clearly show the interest of plasma treatment. The mechanisms of action of the plasma is not clear yet, our results showing that the loss of biological activity can be due to a chemical etching mechanism (O_2–H_2 mixture), or a modification of the different chemical groups forming the Lipid A structure in particular the long pentaacyl, tetraacyl, and hexaacyl chains responsible for the pyrogenic activity [27].

17.7.3
Protein Removal

As in the case of LPS removal, the H_2:O_2 gas mixture leads to a major removal efficiency, 10 times and 5 times larger than a pure Ar and O_2 plasma respectively. The collagen films seem to lead to a linear decrease of thickness as a function of time for the different plasma compositions tested, while BSA leads to a 2-phase etching profile, as in the case of Lipid A (Fig. 17.9). The values of removal rate for H_2–O_2 are of the same order of magnitude at 10 min treatment for collagen, BSA, and Lipid A, while they differ by a factor of 2 for Ar–H_2.

The explanation for the 1- or 2-phase behavior observed for collagen and BSA/Lipid A respectively is related to the enrichment of the surface with non-volatile inorganic compounds as the treatment proceeds [28].

17.8
Conclusions

The applicability of low-pressure plasma treatment for sterilization of bacterial spores, for reduction of the biological activity of pyrogens, and removal of proteins has been demonstrated.

For sterilization, a theoretical model of spore killing kinetics has been introduced in order to explain discrepancies regarding estimation of the most important process leading to the spore death. According to this model, it has been shown that the main feature that governs sterilization efficiency is the presence of materials shielding spores from the direct UV/VUV radiation and the efficiency of its removal. Although the suggested model describes relatively well the experimental results, it is still noticeably simplified in comparison with the real situation and it is therefore to be elaborated according to the new experimental results.

In the case of depyrogenation, it has been demonstrated that low-temperature discharge can significantly inactivate different kinds of pyrogens within several minutes of plasma treatment. It has been also shown that the decrease of LPS biological activity is caused primarily by the plasma and not by increased temperature, low pressure, or microwave radiation. Besides this, two possible pathways of depyrogenation have been proposed based either on the removal of the LPS contamination, the H_2–O_2 plasma being the most efficient in this case, or on a chemical reaction of Lipid A with hydrogen radicals, leading to a breaking of the C–C chains of the Lipid A structure. This kind of mechanism might be particularly interesting for a deactivation process without damaging the underlying substrate.

Finally, we demonstrated the removal of two proteins films, collagen and BSA, at low temperature in typical durations of 20–30 min. The most efficient gas mixture is again H_2–O_2, as in the case of Lipid A. A 1- or 2-phase mechanism is observed depending on the cases, and this phenomenon must be related to the progressive enrichment of the surface on no-volatile compounds. However, our first results with H_2–O_2 show that, initially, there is not a major difference for the removal rate between BSA, collagen, and Lipid A. Further experiments are on-going for the analysis of the chemical mechanisms operating during etching and/ or de-activation.

The results obtained confirm that a plasma discharge can be operated efficiently to sterilize and decontaminate surfaces from biological products. It can be seen though that a multi-step process will be necessary since the optimum conditions are different for sterilization and decontamination. These results also give us good reasons to think that the removal and/or deactivation of PrP on surfaces is possible without damaging the substrates treated.

Acknowledgments

This work was supported by the EU growth project Steriplas (GRD1-19999-10584), and the FP6 2005 NEST project "Biodecon". Support from CSMA Ltd (UK) is gratefully acknowledged.

References

1 Brown, P., Preece, M., Brandel, J.-P., Sato, T., McShane, L., Zerr, I., Fletcher, A., Will, R.G., Pocchiari, M., Cashman, N.R., d'Aignaux, J.H., Cervenáková, L., Fradkin, J., Schonberger, L.B. and Collins, S.J. (2000) *Neurology*, **55**, 1075–1081.

2 Okpara-Hofmann, J., Knoll, M., Dürr, M., Schmitt, B. and Borneff-Lipp, M. (2005) *J. Hosp. Infect.*, **59**, 280–285.

3 Philip, N., Saoudi, B., Crevier, M.C., Moisan, M., Barbeau, J. and Pelletier, J. (2002) *IEEE Trans. Plasma Sci.*, **30**, 1429–1436.

4 Lerouge, S., Wertheimer, M.R., Marchand, R., Tabrizian, M. and Yahia, L'H. (2000) *J. Biomed. Mater. Res.*, **51**, 128–135.

5 Lerouge, S., Wertheimer, M. and Yahia, L'H. (2001) *Plasma Polym.*, **6**, 175–188.

6 Lerouge, S., Fozza, A.C., Wertheimer, M.R., Marchand, R. and Yahia, L'H. (2000) *Plasma Polym.*, **5**, 31–46.

7 Moisan, M., Barbeau, J., Moreau, S., Pelletier, J., Tabrizian, M. and Yahia, L'H. (2001) *Int. J. Pharm.*, **226**: 1–21.

8 Moreau, S., Moisan, M., Tabrizian, M., Barbeau, J., Pelletier, J., Ricard, A. and Yahia, L'H. (2000) *J. Appl. Phys.*, **88**: 1166–1174.

9 Menashi, W.P. (1968) US Patent 3 383 163.

10 Soloshenko, I.A., Tsiolko, V.V., Khomich, V.A., Schedrin, A.I., Ryabtsev, A.V., Bazhenov, V.Yu. and Mikhno, I.L. (2000) *Plasma Phys. Rep.*, **26**, 792–800.

11 Feichtinger, J., Schulz, A., Walker, M. and Schumacher, U. (2003) *Surf. Coat. Technol.*, **174–175**, 564–569.

12 Munakata, N., Saito, M. and Hieda, K. (1991) *Photochem. Photobiol.*, **54**, 761–768.

13 Nagatsu, M., Terashita, F., Nonaka, H., Xu, L., Nagata, T. and Koide, Y. (2005) *Appl. Phys. Lett.*, **86**, 1–3.

14 Laroussi, M., Mendis, D.A. and Rosenberg, M. (2003) *New J. Phys.*, **5**, 41.1–41.10. Laroussi, M. and Leipold, F. (2004) *Int. J. Mass Spectrom.*, **233**, 81–86.

15 Mendis, D.A., Rosenberg, M. and Azam, F. (2000) *IEEE Trans. Plasma Sci.*, **28**, 1304–1306.

16 Moisan, M., Barbeau, J., Crevier, M.-C., Pelletier, J., Philip, N. and Saoudi, B. (2002) *Pure Appl. Chem.*, **74**, 349–358.

17 Schneider, J., Baumgärtner, K.M., Feichtinger, J., Krüger, J., Muranyi, P., Schulz, A., Walker, M., Wunderlich, J. and Schumacher, U. (2005) *Surf. Coat. Technol.*, **200**, 962–966.

18 Hecker, W., Witthauer, D. and Staerk, A. (1994) *PDA J. Pharm. Sci. Technol.*, **48**, 197–204.

19 Peeples, R.E. and Anderson, N.R. (1985) *J. Parenteral Sci. Technol.*, **39**, 9–15.

20 Whittaker, A.G., Graham, E.M., Baxter, R.L., Jones, A.C., Richardson, P.R., Meek, G., Campbell, G.A., Aitken, A. and Baxter, H.C. (2004) *J. Hosp. Infect.*, **56**, 37–41.

21 Baxter, H.C., Campbell, G.A., Whittaker, A.G., Jones, A.C., Aitken, A., Simpson, A.H., Casey, M., Bountiff, L., Gibbard, L. and Baxter, R.L. (2005) *J. Gen. Virol.*, **86**, 2393–2399.

22 Colpo, P. and Rossi, F. (2001) European Patent EP 1126504.

23 Rossi, F. (2004) Plasma sterilisation: mechanisms overview and influence of discharge parameters, in *Plasma Processes and Polymers* (eds R. d'Agostino, P. Favia, M.R. Wertheimer and C. Oehr), Wiley-VCH.

24 Hasiwa, M.,Kullmann, K.,von Aulock, S. and Hartung, T. (2007), *Biomaterials* **28**, 1367–1375.

25 Hasiwa, M., Kylián, O., Hartung, T. and Rossi, F. *J. Endefox Reas.*, sumitted for publication.

26 Rossi, F., Kylián, O. and Hasiwa, M. (2006) Plasma Process. Polym., **3**, 431–442.

27 Erridge, C., Bennett-Guerrero, E. and Poxton, I.R. (2002) *Microbes Infect.*, **4**, 837–851.

28 Ceconne, G., Gilliland, D., Kylián, O. and Rossi, F. (2006) *Ctech J. Phys.*, **56**, B672–B677.

18
Application of Atmospheric Pressure Glow Plasma: Powder Coating in Atmospheric Pressure Glow Plasma

M. Kogoma and K. Tanaka

18.1
Introduction

In 1987 at ISPC-8, Kogoma and Okazaki presented a method to develop an atmospheric pressure glow plasma (APG or APGD) which has a homogeneous and low-temperature glow region. They used a dielectric barrier discharge to create a short length pulse discharge and they used He mixed with monomer gases to realize the lowest starting voltage in the system. Short length pulse discharges will prevent the temperature from rising and He gas can extend the glow discharge period before transferring to an arc discharge at atmospheric pressure. For the application of APG, many kinds of discharge systems were developed [1–7].

Some previous studies have reported powder treatment with glow plasma at a low pressure [8–10]. However, since handling of powders is difficult, we consider that treatment of powders with low pressure glow plasma is impractical. Therefore, an atmospheric pressure glow discharge system has a strong advantage over a low-pressure plasma system. So we tried to use atmospheric pressure glow (APG) plasma for powder treatment [4–7]. Recently, we have been trying to develop thin-film deposition systems on the surfaces of many kinds of powders, such as organic pigments, TiO_2, SiO_2, ZrO_2 magnetic powders, and phosphorescent powders for plasma display panels. We use two methods to make the film surfaces: PCVD and Absorb-Dry method. In this chapter, we will first introduce typical powder treatments to produce ultrathin silicon oxide films on the powders in APG plasma. Second, we will show an application to the TiO_2 fine powder coating with SiO_2 to inhibit the photocatalytic ability of cosmetics powders.

18.2
Development of Silica Coating Methods for Powdered Organic and Inorganic Pigments with Atmospheric Pressure Glow Plasma

Some kinds of pigments that are used for cosmetics give rise to a skin irritation problem when they contact the skin directly. To solve this problem, it is often more

Advanced Plasma Technology. Edited by Riccardo d'Agostino, Pietro Favia, Yoshinobu Kawai, Hideo Ikegami, Noriyoshi Sato, and Farzaneh Arefi-Khonsari
Copyright © 2008 WILEY-VCH Verlag GmbH & Co. KGaA, Weinheim
ISBN: 978-3-527-40591-6

desirable to deposit a silica thin film, which has chemical stability and does not spoil the tones of the pigment owing to its colorless transparency. Silica coating so far has been carried out by hydrolysis of tetraethoxysilane (TEOS) in hot TEOS solution. This method, however, takes over 2 or 3 weeks to hydrolyze TEOS completely. (Hereafter we call this method the *Wet method*.) Moreover, if hydroxide groups that are byproducts during hydrolysis of TEOS remain in the silica coating, they may cause damage to human skin.

To oxidize TEOS completely for a very short time at a low temperature, we think that plasma oxidation is the best way. In an earlier study, we showed that a silica film was able to be deposited on the surface of Fe_2O_3 powder (red) by APG plasma CVD in a short time [5,6]. In this study, we used Fe_3O_4 (black), FeOOH (yellow), and Lithol Rubine BCA (red) (Fig. 18.1) powders. These are not as stable as Fe_2O_3 and can be deteriorated easily by heating or by attack of active species such as oxygen radicals. Therefore they need a low temperature (less than 100 °C) treatment and some way to protect them from active species in the plasma. Then we investigated and developed silica deposition methods for each pigment.

18.2.1
Experimental

A schematic of the discharge tube is shown in Fig. 18.2. The high voltage electrode was an inner aluminum tube (outer diameter = 6.5 mm) and the ground was a stainless mesh (length = 255 mm) wrapping over the outer tube (inner diameter = 16 mm). The discharge area was cooled by water and its temperature was kept under 100 °C during discharge. The pigment powder was blown up through the inner tube and was treated by plasma while it fell down between the inner and the outer tube. The powder was treated only by TEOS vapor/N_2O/He plasma and silica was deposited on the surface of the powder directly. We called this method the *Dry method*. The discharge conditions were as follows: discharge frequency was 13.56 MHz, discharge power was 250–300 W, and treatment time was 15–60 min. Silica-coated powder was evaluated by X-ray photoelectron spectroscopy (XPS), transmission electron microscopy (TEM), energy dispersive X-ray analysis (EDX), and infrared (IR) analysis.

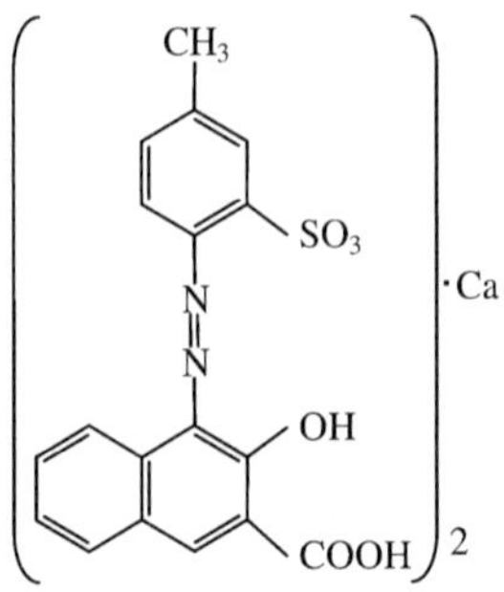

Fig. 18.1 The structural formula of Lithol Rubine BCA.

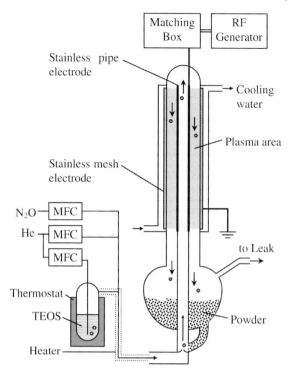

Fig. 18.2 Schematic of discharge tube.

First, we tried to deposit silica on Fe_3O_4 powder by the Dry method. However, it would not circulate in the discharge tube because of its very low fluidity caused from magnetism. In the case of using Wet method, the silica coat generally improved the pigment powder's fluidity. Thus, Fe_3O_4 powder was treated by the Wet method for 7 days with 7 or 28 wt% TEOS in ethanol; hydrolysis silica with half-dissociated TEOS was then deposited on the powder surface. Then the hydrolysis silica with half-dissociated TEOS was oxidized by N_2O/He plasma which produces oxygen atoms. We called this method the *Partial Dry method*.

18.2.2
Results and Discussion

TEM images of the Fe_3O_4 powder treated only by the Wet method are shown in Fig. 18.3(a) and (b). According to these images, a white film of thickness 1–3 nm is wrapped around each Fe_3O_4 particle. In the part of the white film, a peak assigned to Si was measured by EDX. And, as shown in Fig. 18.4, a peak of SiO_2 appeared in the IR spectra of the treated powder. We considered that the powder was coated with silica homogeneously. The ratios of carbon to all elements (Fe, O, Si, and C) in the powder oxidized by N_2O/He plasma were measured by XPS. The dependence of the carbon

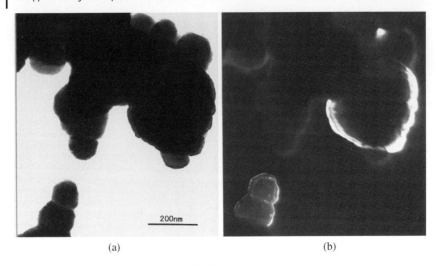

(a)   (b)

Fig. 18.3 TEM images of Fe_3O_4 coated with silica by means of Wet method: (a) transparent; (b) black reflection.

content on treatment time is shown in Fig. 18.5. The ratio decreased as treatment time increased, but carbon was not oxidized completely. The carbon measured by XPS was not only that on the top surface but also that in bulk near the surface, since the measurement depth of XPS is about 100 nm. Moreover, only top surfaces contact the skin directly. Therefore, even if the powder retains some carbon in bulk, we considered it would not give any bad influence to the skin. The plasma did not change the color of the powder. We supposed that the silica coat with half-dissociated TEOS acted as a protective film.

Since FeOOH powder has good fluidity, we tried to treat FeOOH by the Dry method. But its color became reddish by discharge, even though the discharge area was cooled. Oxygen radicals in the plasma probably oxidized the powder. To protect the powder from oxygen radicals, we deposited a protective film on the powder surface by TEOS/He plasma for a start. After a fixed time, with adding N_2O into

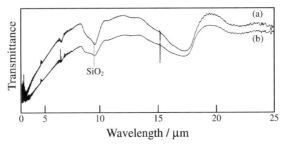

Fig. 18.4 IR spectra of Fe_3O_4 powder treated by Wet method with (a) 8% and (b) 2% TEOS solution.

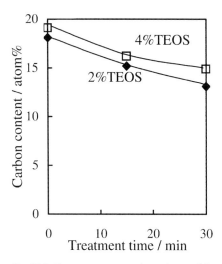

Fig. 18.5 The treatment time dependence of the carbon content for different TEOS concentrations. The flow rates of He and N_2O were 1500 and $10\,cm^3\,min^{-1}$, respectively. The discharge power was 200 W.

plasma, $TEOS/N_2O/He$ plasma caused deposition and oxidation at the same time. We called this method the *2 Step-Dry method*.

By use of the 2 Step-Dry method, the change of color could be prevented. This suggested that plasma-polymerized TEOS film acted as a protective film and that FeOOH was oxidized by oxygen radicals during the treatment with $TEOS/N_2O/He$ plasma. C 1 s spectra of untreated powder, of the powder treated only by TEOS/He plasma for 10 min, and of the powder treated by $TEOS/N_2O/He$ plasma for 20 min after 10 min TEOS/He plasma treatment are shown in Fig. 18.6(a)–(c), respectively. In the C 1 s spectrum of the powder treated only by TEOS/He plasma, a peak assigned to C–O or C=O is found. This peak, however, disappeared after treatment with N_2O. So we considered that almost all ethoxy groups of half-dissociated TEOS were oxidized. From a TEM image and EDX data of treated FeOOH powder, it was confirmed that silica film was formed on the surface homogeneously in the same way as for Fe_3O_4 powder.

Lithol Rubine BCA (red color pigment; BCA) powder has good fluidity and it was estimated that it would be decomposed by oxygen radicals. So, first we tried to treat it by means of the 2 Step-Dry method. But BCA became black although it was treated only by TEOS/He plasma. The C 1 s spectra of untreated powder and of the powder treated by the 2 Step-Dry method are shown in Fig. 18.7(a) and (b), respectively. Since BCA has phenyl groups, the $\pi-\pi^*$ shake-up peak was found in the C 1 s spectrum of untreated BCA. But the treatment with the 2 Step-Dry method eliminated the shake-up peak, as shown in Fig. 18.7(b). Thus, a protective film has to be formed in some other way without a plasma. So we used the following method. The powder was put into 4 or 8 wt% TEOS–ethanol solution and absorbed TEOS. It was left overnight to

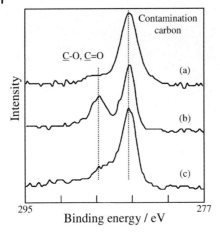

Fig. 18.6 C$_{1s}$ spectra of FeOOH powder: (a) untreated, (b) treated by TEOS/He plasma and (c) treated by TEOS/N$_2$O/He plasma. The plasma conditions were as follows: flow rates of He and N$_2$O were 500 and 10 cm^3 min^{-1}, respectively; supply rate of TEOS was 10 mg min^{-1}; discharge power was 300 W.

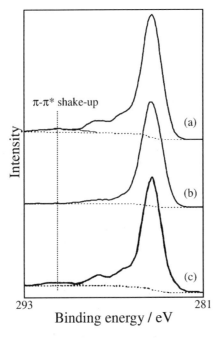

Fig. 18.7 C$_{1s}$ spectra of BCA powder: (a) untreated, (b) treated by 2 Step-Dry method and (c) treated by Absorb and Dry method.

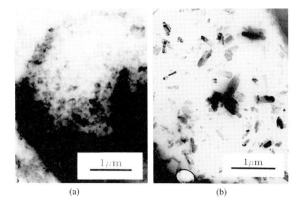

(a)   (b)

Fig. 18.8 TEM images of the compounds with (a) untreated BCA and (b) silica-coated BCA.

evaporate the ethanol. Then this powder that absorbed TEOS was oxidized by $N_2O/$ He plasma. We called this method the *Absorb and Dry method*.

The absorbed TEOS acted as a protective film and so the change of the color was prevented. Moreover, the shake-up peak remained, as shown in Fig. 18.7(c). Next, we made compounds used for lipsticks with BCA to examine the effect of silica coating. Figure 18.8(a) and (b) show TEM images of the compounds with untreated and silica-coated powder, respectively. The unclear contours of the untreated powder in Fig. 18.8(a) suggested that the powder was crushed to pieces during blending into compounds. But the contours of the silica-coated powder were very clear. Therefore, we considered that silica coating also improved the mechanical properties of the powder. And this result affected color brightness especially. Usually, a compound with a pigment powder whose original form remains shows high color brightness because the diffused reflection of incident light on it is very low. Therefore, the compound with silica-coated powder showed high brightness.

In the case of using the Partial Dry method, one could perhaps obtain the same result. But the Partial Dry method needs longer treatment time than the Absorb and Dry method. We tried to use the Absorb and Dry method for Fe_3O_4 powder too. However, the Fe_3O_4 powder condensed in TEOS–ethanol solution and we could not treat it by plasma. So the Absorb and Dry method cannot be used for every powder.

18.2.3
Conclusion

The APG plasma CVD treatment demanded moderate fluidity of powder to circulate in the discharge tube. One of the ways to satisfy this demand was to coat the powder with silica by means of hydrolyzing TEOS in hot TEOS solution (several percent) for

1 week. For the pigments weak in plasma, we found that making a protective film on their surface prevented their deterioration. The 2 Step-Dry method was suited for the pigments that are weak in plasma oxidation. The Partial Dry method and the Absorb and Dry method were suited for such weak pigments.

18.3
Application to TiO$_2$ Fine Powder Coating with Thin Film of SiO$_2$ to Quench the Photosensitive Ability of the Powder

TiO$_2$ is a UV-reflective white pigment powder that is used frequently for cosmetics. But TiO$_2$ is a photosensitive catalyst, and a photosensitized powder can easily attack the human skin surface and oxidize the oil in human sweat as a squalene (2,6,10,15,19,23-hexamethyl-2,6,10,14,18,22-tetracosahexaene). Oxidized squalene molecules produce many kinds of peroxy organic compounds that elicit allergies or cause cancers. We tried to develop the SiO$_2$ coating methods on TiO$_2$ powder pigments by means of the Absorb and Dry method to prevent the oxidation of squalene oil on UV irradiation.

18.3.1
Experimental

The diameter of TiO$_2$ (anatase 80% + rutile 20%, $15 \, \mathrm{m^2 \, g^{-1}}$, Toho Titanium Corporation) particles is about 100 nm. First, the TiO$_2$ powders were pre-treated with TEOS–ethanol mixture to adsorb TEOS on the surface of particles. Next, after the ethanol dried (about 2 days), SiO$_2$ as an inorganic super-fine layer was deposited on the TiO$_2$ surface by plasma oxidation in O$_2$/He. The amount of deposited SiO$_2$ can be controlled by the concentration of TEOS in the solvent ethanol during the adsorbing process. The discharge reactor consists of a quartz glass tube and a stainless steel electrode, which is connected with a matching network and a radiofrequency RF) generator (13.56 MHz), and a copper mesh as an outer electrode connected to earth. In the reaction tube, an ultrasonic homogenizer, which is different from Fig. 18.2, was also installed to flocculate powder into the discharge zone. The RF power was 2500 W. The reaction gases were He and O$_2$. The gas flow rates were $10 \, \mathrm{L \, min^{-1}}$ and $100 \, \mathrm{mL \, min^{-1}}$, respectively. In the reaction tube, fine particles were introduced on an ultrasonic horn and were flicked strongly. The powder was carried into the mixture gas and went through the plasma area. Treated powder was carried into the separation trap, and finally collected in the powder pool. The treating rate is about 100 g per 50 min for the powder. The surface of TiO$_2$ coated by SiO$_2$ was analyzed by XPS. Pure squalene oil mixed with a small amount of TiO$_2$ samples in a Pyrex glass bottle which contained air was irradiated for an hour by Xe-UV light from the outer side of the glass bottom. Then the UV-irradiated squalene oil in the bottle was heated for an hour at 100 °C. Finally, vaporized products in the bottle were sampled and measured by GC/MS (Shimazu, QP5050).

18.3.2
Results and Discussion

18.3.2.1 XPS Analysis

Figures 18.9 and 18.10 show C 1 s and O 1 s spectra of TiO_2 of untreated TiO_2, 10% TEOS adsorbed TiO_2, and 10% TEOS adsorbed and plasma oxidation TiO_2, measured by ESCA. In Fig. 18.9(a) the peak shows untreated TiO_2 powder that contains a small amount of carbon impurity. After TEOS was adsorbed on the powder surfaces, another peak arose on the shoulder of the main peak (Fig. 18.9(b)), which should be assigned to the carboxyl groups in the dissociated and half hydride TEOS molecule. In Fig. 18.9(c), the carbon contamination peak is decreased by plasma oxidation of the layer due to the purification of SiO_2.

In Fig. 18.10, sharp peaks of O 1 s were obtained in the spectra. In the spectrum of Fig. 18.10(a), the single peak (529.9 eV) is assigned to TiO_2. In the spectrum of Fig. 18.10(b), the peak (532.5 eV) that can be assigned to SiO_2 newly appeared, while the peak assigned to TiO_2 decreased. Therefore, it was found that the surface was coated with SiO_2 but the surface was not coated completely. This means that the surface of TiO_2 was still partly covered by half-dissociated TEOS molecules. In the spectrum of Fig. 18.10(c), the peak assigned to TiO_2 almost disappeared and the shape of the peak looked like that of pure SiO_2. So, the surface in this case was perfectly coated with an inorganic SiO_2 layer without any organic contamination.

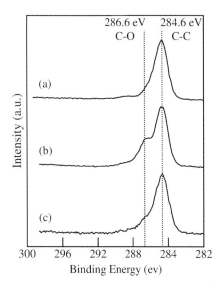

Fig. 18.9 C₁ₛ XPS spectra: (a) untreated TiO₂; (b) 10 wt% TEOS adsorbed TiO₂; (c) sample (b) with plasma oxidation.

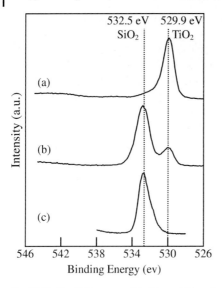

Fig. 18.10 O_{1s} XPS spectra. (a), (b), and (c) as in Fig. 18.9.

18.3.2.2 TEM Analysis of Powder

In the TEM image (Fig. 18.11), we can see the ultrathin layer that covers the TiO_2 bulk particle surfaces without any defects or holes. The thickness is about a few nanometers. It seems that the homogeneous inorganic layer interrupts the penetration of middle-sized chain molecules such as squalene. The layer should be amorphous SiO_2 because inside of the layer we can see no crystal lattice.

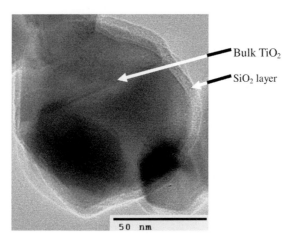

Fig. 18.11 TEM image of TiO_2 particles treated with SiO_2 in the atmospheric pressure glow plasma.

18.3.2.3   GC/MS Spectrum of the Vapor from UV-Irradiated Squalene Oil That Mixed With the Powders

Figure 18.12 shows the GC/MS spectrum of Xe-UV-irradiated squalene with TiO_2 and without TiO_2 in the atmosphere. The following designations of (a), (b), and (c) are the same as those above. The GC/MS spectrum of UV-irradiated squalene with (a) indicates significant amounts of ethyl alcohol, acetone, and many kinds of organic oxides. These compounds were generated by UV-oxidation reaction of squalene by TiO_2 catalyst reaction. The spectrum of UV-irradiated squalene with (b) has only a larger peak of ethyl alcohol and a small peak of acetone, while the peaks from all the other peroxides have disappeared. The peak of ethyl alcohol decreased drastically but the quantity of the detected ethyl alcohol is still harmful to human skin. The spectrum of UV-irradiated squalene of sample (c) has no signal. It seemed that we could attain the SiO_2 protection film, which prevents the oxidation of squalene oil even with UV irradiation. The squalene molecule has eight methyl groups in its structure. Because the naked TiO_2 surface has strong oxidation reactivity, untreated TiO_2 will react with squalene and cut the main chains of the molecules, producing many kinds of organic oxide products such as $C_5H_{10}O$, as shown in Fig. 18.12. On the other hand, sample (b) has only a few lower molecular number oxides such as acetone or ethanol. Probably, the surface of sample (b) has been partly covered with SiO_2 from hydrated TEOS molecules on the TiO_2. So, the surface of the sample (b) should have many small pits of which the sizes are smaller than a nanometer. A molecule as large as squalene itself cannot penetrate inside of the SiO_2 layer through the pits. Only the small-sized parts in the end of the main chain such as methyl groups can creep inside the layer, and then react with TiO_2 surfaces to produce low molecular number oxides. But sample (c) is completely covered by a tight quartz layer. No hydrocarbon molecules can get through the hard layer. This is the reason why we have no signals in the GC/MS spectrum of sample (c).

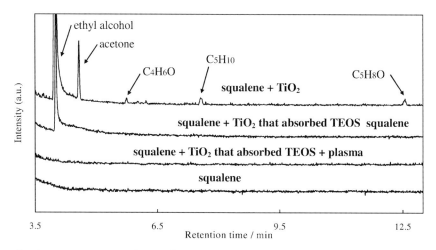

Fig. 18.12 GC/MS spectra of UV-irradiated squalene with treated and untreated TiO₂.

18.3.3
Conclusion

We attained an ultrathin silica layer on TiO_2 single particles by means of atmospheric pressure glow plasma using TEOS adsorption and plasma oxidation method. The attained silica protection layer interrupted perfectly the oxidation of squalene oil mixed with treated TiO_2 powder with UV irradiation. The Absorb and Dry method can be applied not only for TiO_2 but also for many kinds of powders that should provide strong protection from the penetration of gases or molecules.

A powder treatment system such as that shown in Fig. 18.2 has many advantages compared with low-pressure glow plasma powder treatment systems, for example very simple apparatus and ease of cleaning the inside of the reactor without any breaking of complicated vacuum seals. Moreover, because only radical reactions will take place in the reactor without any ion-damaging effect in such high-pressure non-thermal plasma reactor, it can treat very soft materials such as biomedical powders or organic pigments.

Acknowledgments

The authors thank Professor Dr. Chihiro Kaito (Ritumeikan University) for his help with the TEM images and Professor Dr. F. S. Howell (Sophia University) for checking the English.

References

1 Kanazawa, S., Kogoma, M., Moriwaki, T. and Okazaki, S. (1988) *J. Phys. D: Appl. Phys.*, **21**, 838.

2 Yokoyama, T., Kogoma, M., Kanazawa, S., Moriwaki, T. and Okazaki, S. (1990) *J. Phys. D: Appl. Phys.*, **23**, 374.

3 Yokoyama, T., Kogoma, M., Moriwaki, T. and Okazaki, S. (1990) *J. Phys. D: Appl. Phys.*, **23**, 1125.

4 Ogawa, S., Takeda, A., Oguchi, M., Tanaka, K., Inomata, T. and Kogoma, M. (2001) *Thin Solid Films*, **386**, 213.

5 Mori, T., Okazaki, S., Inomata, T. and Kogoma, M. (1995) *Proc. 8th*

Symp. Plasma Science and Materials, 52.

6 Mori, T., Okazaki, S., Inomata, T., Takeda, A. and Kogoma, M. (1996) *Proc. 9th Symp. Plasma Science and Materials*, 7.

7 Kogoma, M., Tanaka, K. and Takeda, A. (2005) *J. Photopolym. Sci. Technol.*, **18**, 277.

8 Kobayashi, T., Terada, T. and Ikeda, S. (1989) *OCCA Chester Conference paper*, 252.

9 Park, S.H. and Kim, S.D. (1994) *Polym. Bull.*, **33**, 249.

10 Tsugeki, K., Yan, S., Maeda, H., Kusakabe, K. and Morooka, S. (1994) *Mater. Sci. Lett.*, **13**, 43.

19
Hydrocarbon and Fluorocarbon Thin Film Deposition in Atmospheric Pressure Glow Dielectric Barrier Discharges

F. Fanelli, R. d'Agostino, and F. Fracassi

Atmospheric pressure dielectric barrier discharges in both filamentary and glow regime attract significant research interest in the field of surface processing of materials. In particular, intense efforts to evaluate the utilization of dielectric barrier discharges (DBDs) in plasma- enhanced chemical vapor deposition (PECVD) of thin films are in progress. In this contribution an overview on the basic aspects of thin film deposition by filamentary and glow DBDs is presented along with our latest results on deposition of hydrocarbon and fluorocarbon films in glow DBDs fed with He–C_2H_4, He–C_3F_6, and He–C_3F_8–H_2 gas mixtures. The effect of several process parameters, such as feed composition and excitation frequency, on both discharge operational mode and coating composition has been investigated.

19.1
Introduction

DBDs are considered nowadays a promising alternative to low-pressure plasmas for surface processing of materials since they couple the advantage of atmospheric pressure operation to non-equilibrium conditions. Even if the main benefit of DBDs appears to be the avoidance of expensive evacuation systems, their utilization could be a decisive advantage also in processing of high outgassing materials such as polymeric membranes, paper, fabrics, synthetic and natural fibers, and rubber.

Many recent studies deal with the PECVD of thin films by DBDs, a research field traditionally dominated by low-pressure plasmas. In particular great research efforts have been devoted to evaluate if DBDs can actually be advantageous compared to low-pressure plasmas in deposition of hydrocarbon-, fluorocarbon-, and SiO_2-like coatings.

It is well known that the DBDs employed in the field of surface processing of materials very often operate in a filamentary regime [1,2] (filamentary dielectric barrier discharges, FDBDs). A more recent development, still at the laboratory stage, is the modification of surfaces, including thin film deposition in spatially homogeneous, diffuse dielectric barrier discharges. Under particular experimental conditions, in fact, a homogeneous discharge regime, the so-called glow dielectric

Advanced Plasma Technology. Edited by Riccardo d'Agostino, Pietro Favia, Yoshinobu Kawai, Hideo Ikegami, Noriyoshi Sato, and Farzaneh Arefi-Khonsari
Copyright © 2008 WILEY-VCH Verlag GmbH & Co. KGaA, Weinheim
ISBN: 978-3-527-40591-6

barrier discharge (GDBD), can be obtained [3–9]. In the field of PECVD processes, GDBDs attract significant research interest because they are expected to be more suitable for homogeneous coating deposition than FDBDs. A limitation to the development of new applications of DBD in the homogeneous regime is the lack of an adequate process control on the industrial scale.

In this contribution, which mainly deals with the PECVD of hydrocarbon and fluorocarbon coatings in GDBDs, a wide and detailed description of the use of DBDs for thin films deposition is reported. The following issues are considered: the two possible discharge operational modes, namely filamentary and glow regimes; the most common electrode configurations and gas injection systems; the state of the art in the deposition of hydrocarbon and fluorocarbon thin films both in filamentary and glow regime (in comparison with low pressure PECVD); and the recent results of thin film deposition by means of GDBDs fed with He–C_2H_4, He–C_3F_6, and He–C_3F_8–H_2.

19.2
DBDs for Thin Film Deposition: State of the Art

DBD surface treatments, often misleadingly called "corona treatments", have been widely utilized to change the surface properties of polymers, e.g. to promote the wettability, printability, and adhesion on polymer surfaces. Among the various plasma applications, filamentary DBDs in air appeared to be very convenient for adhesion and wettability improvement [10–13] even though other feed gases were also employed for the grafting of nitrogen- or fluorine-containing functionalities [3,5,8,14–17]. This was initially successfully achieved in low-pressure plasma reactors but, later on, progress in purging and sealing techniques allowed the utilization of reactors operating at atmospheric pressure.

PECVD of thin films, which for many years was a typical low-pressure technique, is also receiving increasing attention in recent years, both in filamentary and glow DBD regime, since there is an industrial need of technologies which do not require vacuum equipment in order to simplify the processes and to lower apparatus costs compared to the low-pressure counterpart. The aim of the numerous studies involved in this research is currently the achievement a reliable process control also necessary for industrial scale-up.

19.2.1
Filamentary and Glow Dielectric Barrier Discharges

DBDs are usually classified in two main categories: filamentary discharges (FDBD) and glow discharges (GDBD).

The filamentary regime was the first discharge operational mode reported in the literature and it is still the most popular in the plasma community. When the applied voltage is high enough to cause breakdown in a gas at atmospheric pressure, a large number of microdischarges are observed and the discharge assumes a filamentary structure [1,2]. Microdischarges, extensively investigated by many authors [1,2,18], are randomly and evenly distributed over the entire electrode surface. As a consequence of the independent developing of microdischarges, the discharge current is formed by a high number of current peaks of different amplitude and duration in the range of a few nanoseconds, which is the typical microdischarge lifetime.

Recent investigations concerning the treatment of polymers by FDBDs reported, under particular experimental conditions, a permanent damage of the polymer surface, in the form for instance of *"crater formations"*, due to microdischarges occurrence [10,11]. Filaments may in fact be too strong and/or they may ignite repeatedly at the same position in different cycles, so that a stationary filament pattern is formed. In deposition processes, this phenomenon can have detrimental effect on homogeneity and final properties of the growing coating. Generally, such a drawback can be overcome by a careful choice of process parameters (i.e. input power) and by moving the sample at a controlled speed as in continuous treatment apparatus. Moreover, special power supplies have been developed to generate repetitive pulse trains resulting in improved statistical distribution of the microdischarges across the surface, a decisive prerequisite for more uniform treatment [1,2]. The careful choice of the sample speed inside the discharge region and of the power delivered to the discharge allows one to achieve a compromise condition in which the substrate residence time in the discharge is short enough to limit the surface exposure to microdischarges and in the meantime to get the required thickness of the coating.

In recent years, the generation of large radius dielectric barrier discharges, in which no evidence of microdischarges is observed, has been reported. Since 1988, Okazaki, Kogoma and co-workers at Sophia University in Tokyo have published a pioneering work in this field using sinusoidal voltages in different gases with and without additives [3]. They showed that some specific conditions were needed to obtain a stable and homogeneous discharge [3–5]. Electrical measurements allowed one to observe that the atmospheric pressure glow discharge (APGD) is characterized by a periodical discharge current signal, with the same period of the applied voltage, formed by only one peak per half-cycle. Massines and co-workers widely contributed to this discussion; their investigations on the appearance of glow regimes in pure gas plasmas (i.e. helium and nitrogen) are considered of great significance within the scientific community [6–9].

Up to now it is difficult and tricky to reliably control homogeneous glow discharges at atmospheric pressure. For instance, changes of the electrode configuration or small variations in electrical parameters (i.e. frequency and voltage) along with the introduction of a reactive gas, as required in deposition processes, can drastically affect discharge stability and behavior and cause a transition into a more stable filamentary discharge mode.

Generally for each DBD apparatus with its architectural and electrical characteristics, it is possible to determine the frequency–voltage conditions that define the GDBD existence domain for a pure gas, such as helium or nitrogen, that are usually utilized as the main gases in deposition processes. The addition of a reactive monomer generally narrows the GDBD operational window and, in fact, if the concentration of reactive gas in the feed exceeds a certain threshold limit, dependent on the main gas and on the monomer molecular structure, glow generation is prevented and the transition to a filamentary discharge can be observed. Therefore, the operational window of the glow regime should be carefully evaluated in order to achieve a confident control of the deposition process.

19.2.2
Electrode Configurations and Gas Injection Systems

Different electrode configurations can be used for the generation of a DBD. For PECVD, parallel plate reactors are usually used, even if recent studies report the utilization of surface electrode and coplanar DBD geometries [1,2]. According to the basic definition of DBD, they are characterized by the presence of at least one dielectric layer in the current path between the metallic electrodes. The gas injection can be tailored to the electrode geometry and to the process to be performed. Typical configurations are sketched in Fig. 19.1. The basic geometry employed in deposition processes corresponds to a typical parallel plate configuration in which both the electrodes are covered by a dielectric layer (Fig. 19.1(a)). Discharge gap widths of a few millimeters are typically employed, while the areas of the electrodes (usually with rectangular shape) can range from several mm^2 to a few m^2 depending on the substrate and on the process to be performed. A careful choice of dielectric material (e.g. glass, ceramics, and polymers) should be made since it is directly exposed to the plasma and therefore could be responsible of substrates contamination. This system requires a lateral gas injection; the feed gas is introduced in the discharge zone by means of an inlet slit and it is pumped through an outlet slit in the opposite side [19–21].

A possible alternative to this approach is a parallel plate geometry in which the gas is injected from above (Fig. 19.1(b)) [22]; in particular the feed enters into the plasma region through a slit between two identical high voltage (HV) electrodes and it could be sideways pumped by means of two opposite outlet slits.

The DBD apparatus can also consist of a HV electrode formed by a cylindrical metallic rod contained into a coaxial outer dielectric tube and a wide grounded electrode that can be covered with a dielectric plate (Fig. 19.1(c)) [23]. The gas is injected through the slit of a gas diffuser faced with the discharge region.

Usually in these configurations, the ground electrode can scan forward and backward at a controlled speed, in order to deposit homogeneous thin films onto large-area substrates. It should be considered, in fact, that a lateral gas injection can result in variation of coating thickness and chemical composition as a function of the gas residence time into the discharge area. Moreover, as previously discussed, in the case of a FDBD sample movement allows one to limit the surface exposure to microdischarges and hence to reduce surface inhomogeneity and damage.

Roll-to-roll systems can be used for thin film deposition onto flexible substrates, such as polymeric webs (Fig. 19.2). This approach is usually employed if DBD treatments have to be performed as continuous in-line processes. These apparatuses are similar to corona treaters which have been utilized for decades for web treatment.

(a)

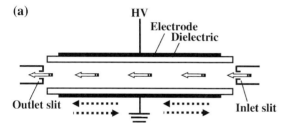

(b)

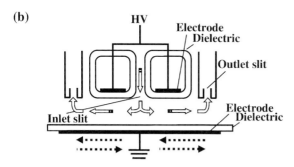

(c)

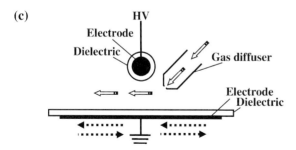

Fig. 19.1 Typical electrode configurations and gas injection systems used for PECVD by means of DBDs.

19.2.3
Hydrocarbon Thin Film Deposition

Low-pressure PECVD of hydrogenated amorphous carbon (a-C:H) thin films has been widely studied utilizing several monomers. Coatings with well-defined characteristics, such as hydrogen concentration and sp^3/sp^2 carbon ratio, were deposited [24–30]. The effect of several process parameters such as reactive feed gas, bias voltage, power, and pressure was accurately investigated. For instance, Kobayashi *et al.* [31,32] obtained from ethylene-fed plasmas a polymer-like coating with a strong tendency to oxidize after preparation; moreover at low pressure, they observed

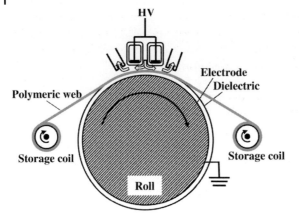

Fig. 19.2 Basic design of a system for continuous surface treatment of polymeric webs.

powder formation. The results were exploited to investigate the hypothesis of free radical polymerization that mainly occurs in the gas phase.

One of the first investigations of atmospheric pressure deposition in hydrocarbon-containing discharges was published in 1979 by Donohoe and Wydeven [33] and concerned the polymerization of ethylene diluted in helium, in a uniform pulsed discharge. In this pioneering work a soft polymer with a low cross-linking degree was obtained. The study of high molecular weight oligomers in the gas phase allowed the investigation of powder formation phenomena.

The polymerization of ethylene in helium-containing APGDs was also reported by Yokoyama *et al.* [5]. The effect of ethylene concentration and discharge current was studied; an increase of the deposition rate up to $1 \, \mu m \, h^{-1}$ was observed as a function of discharge current. A polyethylene-like film, similar to those deposited with low-pressure PECVD, was obtained but unfortunately a complete investigation of coating properties was not performed.

Many other studies concerning hydrocarbon deposition in DBD from several monomers and gas mixtures, frequently in a filamentary regime, have been published. Goossens *et al.* [34,35] studied C_2H_4 deposition in mixtures with helium or argon in a filamentary regime. A deposition rate of about $100 \, nm \, min^{-1}$ was reported with 1 slm of reactive gas mixed with 10 slm of inert gas. A polyethylene-like coating was obtained even though some oxygen was bound to the polymer network, probably originating from post-deposition reaction due to atmospheric exposure. Recently, Girard-Lauriault *et al.* [36] reported a novel material, nitrogen-rich plasma-polymerized ethylene (PPE:N), deposited using atmospheric pressure DBDs fed with nitrogen (about 10 slm) and ethylene (about 10 sccm). The nitrogen content in the film was easily and reproducibly controlled by varying C_2H_4 concentration in the feed gas. The chemical properties, surface morphology, and mechanical properties of a-C:H films deposited by methane-fed DBD were compared by Liu *et al.* [37] at medium and atmospheric pressure. At atmospheric pressure and in filamentary mode, the discharge led to the deposition of a soft polymer-like a-C:H film, with rough surface due to microdischarges. According to these results, it appeared difficult to obtain hard films with smooth surfaces at high pressure. Klages *et al.* [38] investigated this aspect utilizing highly diluted mixtures of hydrocarbons (CH_4, C_2H_4, and C_2H_2) and argon or nitrogen. Their results showed that high hardness values, similar to those of diamond-like carbon (DLC) films, were not achieved, and that films with high monomer structure retention could be obtained from appropriate precursors. To our knowledge, the only work reporting the deposition of diamond-like coatings by DBDs was published by Bugaev *et al.* [39]. Best properties were obtained using methane or acetylene in mixtures with hydrogen; the coatings appeared to be similar to the DLC films obtained by traditional methods (e.g. low-pressure PECVD), but the films grown in parallel plate configuration had a significant quantity of defects which could be correlated to the existence of the filamentary DBD.

19.2.4
Fluorocarbon Thin Film Deposition

Low-pressure PECVD of fluorocarbon films has been widely studied since the 1970s. Numerous studies succeeded in tuning chemical composition (i.e. F/C ratio and cross-linking degree) and morphology of the deposited coatings, hence in tailoring their thermal, mechanical, and electrical performances, as well as their wettability [40–50]. The plasma polymerization of fluorocarbons is even more complicated than that of hydrocarbons because of the significant etching–deposition competition [40–42]. It was demonstrated that the choice of the monomer is of primary importance, since it is the source of reactive fragments and film precursors in the plasma. Usually in saturated fluorocarbon (e.g. CF_4, C_2F_6, and C_3F_8) fed plasmas low deposition rates are observed, hydrogen addition increases the deposition rate allowing F/C ratio and crosslinking degree variation [42,46]: less fluorinated and more cross-linked coatings can be produced by increasing the H_2 content of the feed.

In contrast to the huge number of publications on low-pressure fluoropolymer deposition, only a few articles dealing with fluoropolymer deposition by DBDs have been published up to now and, in particular, the possibility of controlling the chemical properties of the coatings, an important feature of low-pressure plasma deposition processes, has not been clearly demonstrated yet.

One of the first investigations on deposition in fluorocarbon-containing APGDs was reported by Yokoyama *et al.* [5] utilizing tetrafluoroethylene in mixtures with high flow rates of helium. A deposition rate up to approximately 2 μm h^{-1} was reported as a function of the monomer flow rate and of the discharge current, and X-ray photoelectron spectroscopy (XPS) analyses showed a F/C ratio ranging between 1.4 and 1.7. After this fundamental study, Kogoma *et al.* investigated the use of this process for the deposition onto fine silica particles [51] and in the inner surface of commercially available polyvinylchloride tubes for biomedical applications [52].

Thyen *et al.* [53] studied the deposition of Teflon-like coatings in a filamentary DBD fed by C_2F_4–N_2–Ar. Soft and smooth coatings with F/C ratio of 1.6 were deposited with a fairly high deposition rate of about 100–200 nm min^{-1}. Vinogradov *et al.* [54,55]

investigated the deposition of thin films in FDBDs fed by Ar–fluorocarbon mixtures. Several fluorocarbons were employed, i.e. CF_4, C_2F_6, $C_2H_2F_4$, C_3F_8, C_3HF_7, and c-C_4F_8. *In situ* investigation of the plasma phase was performed by means of UV and Fourier transform infrared (FTIR) absorption spectroscopy, and UV–visible emission spectroscopy. The effect of oxygen or hydrogen addition to the gas feed was evaluated on the deposition rate and film composition. The morphological investigation showed that smooth films could be obtained using feeds with low hydrogen content.

19.3
Experimental Results

19.3.1
Apparatus and Diagnostics

The experimental apparatus consists of a parallel plate electrode system (5 mm gap) contained in a Plexiglas box (Fig. 19.3). Each electrode (30×30 mm^2 area) is covered by a 0.635 mm thick Al_2O_3 plate (CoorsTek, 96% purity, 9.5 relative permittivity). The plasma is generated by applying an AC high voltage at 2.8 kV$_{p-p}$ in the frequency range 10–30 kHz with a power supply composed by a variable frequency generator (TTi TG215), an audio-amplifier (Outline PA4006), and a high-voltage transformer (Montoux).

The electrical characterization was performed with a digital oscilloscope; the voltage applied to the electrodes was measured by means of a high-voltage probe, while the current was evaluated by measuring the voltage drop across the 50 Ω resistor in series with the ground electrode.

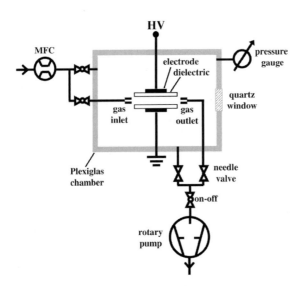

Fig. 19.3 Schematic of the experimental apparatus.

The gas flow rates are controlled by MKS electronic mass flow controllers (MFC) and the system pressure is monitored by means of a MKS 122 baratron. In order to avoid overpressure, the Plexiglas enclosure is slightly pumped with a rotary pump (Pfeiffer). The feed gas is introduced in the interelectrode zone through an inlet slit and pumped by means of an opposite outlet slit, and therefore a longitudinal gas injection is realized.

Glow discharges were fed with $He–C_2H_4$, $He–C_3F_6$, and $He–C_3F_8–H_2$ gas mixtures (Air Liquide, He C; Air Liquide, C_2H_4 N35; Zentek, C_3F_6, 99% purity; Zentek, C_3F_8, >99% purity).

Deposition processes of hydrocarbon films were preformed at a He flow rate (ϕ_{He}) of 2 slm with ethylene concentration ranging from 0.1 to 0.5%, while deposition of fluorocarbon films was performed at a He flow rate (ϕ_{He}) of 4 slm and a fluorocarbon concentration of 0.01%; the H_2-to-C_3F_8 flow rate ratio was varied from 0 to 2.

The plasma phase was studied by means of optical emission spectroscopy (OES) in the UV–visible region, while coating characterization was performed with FTIR spectroscopy, XPS, scanning electron microscopy (SEM), and static water contact angle (WCA) measurements. Film thickness was evaluated on substrates partially masked during the deposition with an Alpha-Step® 500 KLA Tencor Surface profilometer, at different positions inside the interelectrode gap, i.e. as a function of the gas residence time. In order to compare the results obtained under different experimental conditions, the average value of the film thickness, and hence of the deposition rate, was considered in the region 10–20 mm from the gas entrance inside the discharge area; at each experimental point an error bar corresponding to the minimum and the maximum values of the measured deposition rate was associated [21].

19.3.2
Deposition of Hydrocarbon Films by Means of He–C_2H_4GDBDs

In order to clarify the discharge operational mode and evaluate the existence domain of the glow DBD, the first step of the investigation was the electrical characterization of the discharge. It resulted that for all the concentrations of ethylene in helium explored, microdischarges appeared at excitation frequencies lower than 9 kHz. At frequency higher than 10 kHz and voltage lower than 3.0 kV_{p-p}, a periodical discharge current signal can be observed up to 0.5% of ethylene in the feed gas. The current signal showed the same period of the applied voltage and it is composed by only one peak per half cycle, as for a typical GDBD in helium [7]. The higher the ethylene concentration, the higher the breakdown voltage and the discharge current amplitude (Fig. 19.4).

As expected, due to the lateral gas injection, the thickness of the deposit increased with the distance from the gas admission point inside the discharge area (i.e. with the residence time of the gas). At 20 kHz, the deposition rate increased from 22 to 40 nm min^{-1} on the average, for ethylene concentration of 0.1 and 0.5%, respectively.

Figure 19.5 shows the normalized FTIR absorption spectra of deposits as a function of ethylene concentration. The main spectral features are the broad band

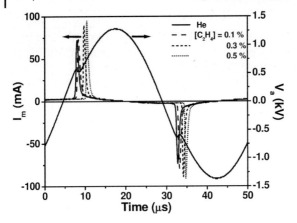

Fig. 19.4 Applied voltage (V_a) and measured current (I_m) at 20 kHz and 2.8 kV$_{p-p}$ for pure helium and three different values of [C_2H_4] in the feed (0.1, 0.3, and 0.5%).

between 3000 and 2800 cm^{-1} due to CH stretching vibration modes and the two signals between 1480 and 1370 cm^{-1} from CH_2 and CH_3 bending. The intensity and structure of the CH stretching band could be consistent with an amorphous and polymer-like structure with mainly sp^3 carbon bonded to hydrogen [29,57–61]. Some oxygen, reasonably due to O_2 and H_2O contaminations in the deposition chamber and/or to coating oxidation after atmospheric exposure, is also contained in the coating as OH (3450 cm^{-1}), C=O (1720 cm^{-1}) and C–O (1090 cm^{-1}) groups [58,60,61]. As the ethylene concentration increases, the C=O band disappears and the OH and C–O signals become less pronounced.

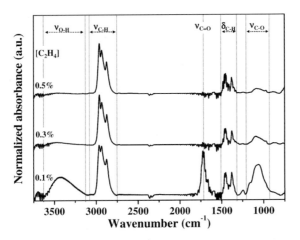

Fig. 19.5 Normalized FTIR spectra of films deposited as a function of ethylene concentration in the feed ($f = 20$ kHz, $V_a = 2.8$ kV$_{p-p}$, [C_2H_4] = 0.1–0.5%).

In agreement with FTIR results, XPS analyses show a decrease of oxygen uptake, from 8 to 0.8%, as the feed content of ethylene passes from 0.1 to 0.5%; at the same time the WCA varies from 75 to 90°.

In order to investigate the stability of the coatings, the deposits obtained at 0.5% of C_2H_4, 20 kHz, and 2.8 kV_{p-p} were also analyzed after one month of aging in air. Only a slight increase of the FTIR OH and C–O bands was detected compared to the as-deposited films, and the oxygen uptake increased from 0.8 to 2.5% after 1 month of aging.

By increasing the frequency, i.e. the number of current pulses per unit of time, a growth of the discharge power and of the average deposition rate from 20 up to 80 nm min^{-1} was detected (Fig. 19.6(a)); at all the excitation frequencies explored the deposition rate shows the same trend as a function of the position inside the discharge area (Fig. 19.6(b)). The FTIR spectra were not appreciably affected by frequency and voltage variation and XPS analyses showed that oxygen concentration was always lower than 3% under all the experimental conditions explored. The coating morphology (SEM observation) is of good quality and no significant evidence of powders or defects was found at the different excitation frequencies examined.

OES investigation allowed the identification of the main emitting species in the plasma. Pure helium GDBD is characterized by intense emissions from helium and air impurities, i.e. nitrogen, oxygen, and water [62,63], ethylene addition reduces all the emission intensities in particular those of oxygen-containing species which completely disappear, while CH (4300 Å system) and C_2 (Swan system) emissions appear.

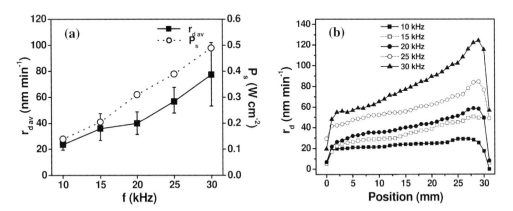

Fig. 19.6 (a) Input power and average deposition rate as a function of excitation frequency at a fixed voltage of 2.8 kV_{p-p} and (b) deposition rate profile as a function of the position in the discharge area at different excitation frequencies ($\phi_{He} = 2$ slm, $\phi_{C2H4} = 10$ sccm, $[C_2H_4] = 0.5\%$, $f = 10$–30 kHz).

19.3.3
Deposition of Fluorocarbon Films by Means of He–C$_3$F$_6$ and He–C$_3$F$_8$–H$_2$GDBDs

The electrical characterization of the discharge allowed the evaluation of the existence domain of a GDBD for C$_3$F$_6$- and C$_3$F$_8$-containing feed. A glow discharge was obtained, at excitation frequencies higher than 15 kHz and applied voltages lower than 3.0 kV$_{p\text{-}p}$, for concentrations up to 0.01 and 0.025% for C$_3$F$_6$ and C$_3$F$_8$, respectively. Under these conditions, in fact, a periodical discharge current signal, composed by only one peak per half-cycle [5], was observed; otherwise a filamentary discharge was obtained. For C$_3$F$_8$ concentration of 0.01%, utilized in this study, up to 0.02% of hydrogen could be introduced in the feed ([H$_2$]/[C$_3$F$_8$] ratio = 2) without any evidence of microdischarge formation.

For He–C$_3$F$_6$ GDBDs the effect of the excitation frequency was investigated from 15 to 30 kHz at 2.8 kV$_{p\text{-}p}$ and at 0.01% of C$_3$F$_6$. A linear growth of the discharge power and of the deposition rate from 19 to 34 nm min^{-1} was detected. In Fig. 19.7(a) the normalized FTIR absorption spectrum of film deposited at 25 kHz is reported. The main spectral feature is the band between 900 and 1400 cm^{-1} due to the overlap of CF$_x$ stretching vibration modes in the forms of CF (1350 cm^{-1}), CF$_2$ (1225 and 1190 cm^{-1}), and CF$_3$ (980 cm^{-1}) [44–46,48,64–66]. The broad and weak band between 1550 and 1900 cm^{-1} can be ascribed to the C=C and C=O stretching modes [44–46,48,64–67]. The F/C ratio detected by XPS is 1.5 and it is not affected by frequency variation, while, in agreement with FTIR results, a slight increase of the oxygen concentration is observed at lower frequencies. The static WCA is always 106–108° at all frequencies. The curve-fitting of the high-resolution C 1s region allows one to obtain interesting information on the structure of the organic matrix. As for Teflon-like thin films deposited by low-pressure PECVD, this region is quite complex. Figure 19.7(b) shows the C 1s peak best fit for the coating deposited at 25 kHz. Five components, related to CF$_3$ (294.5 ± 0.2 eV), CF$_2$ (292.5 ± 0.2 eV) [56], CF (290.1 ± 0.3 eV), C–CF, CF=C and CO (288.0 ± 0.2 eV), and C–C (285.0 ± 0.3 eV), were utilized for the fitting [42,47,48,66,68]. The weak shoulder in the high binding energy region, centered at approximately 296.5 eV, could be due to a shake-up component of unsaturated carbons.

SEM analyses allowed one to observe that at 15 kHz a rough surface was obtained, with some evidences of powder and globules formation. In contrast, at high frequency smooth surfaces were deposited, without significant evidence of powder formation.

The GDBD was characterized by intense emissions from helium, fluorine atoms, and CF$_2$ in the spectral range 240–350 nm (A^{1}B$_1$ – X^{1}A$_1$ system) [62,63]. A continuum, assigned by d'Agostino *et al.* [40,41] to CF$_2^+$, centered at approximately 290 nm, is also present. Several emissions from impurities, such as nitrogen, oxygen, and water, were observed along with CO$^+$ (first negative system) signals [62,63].

The effect of feed composition was investigated in He–C$_3$F$_8$–H$_2$ GDBDs, by varying the hydrogen-to-monomer ratio in the feed from 0 to 2, at 25 kHz and 2.8 kV$_{p\text{-}p}$.

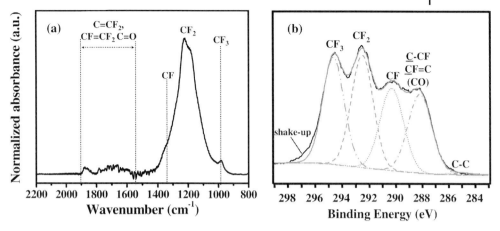

Fig. 19.7 Normalized FT-IR spectrum and best fit of C1s signal of a fluorocarbon film deposited in He–C_3F_6 GDBD at 25 kHz ($V_a = 2.8$ kV$_{p-p}$, [C_3F_6] = 0.01%).

Figure 19.8 shows that the deposition rate trend obtained by increasing the hydrogen concentration in the feed was characterized by a maximum. In good agreement with the results obtained by low-pressure plasmas fed with saturated fluorocarbons and hydrogen [40,42,46], the deposition rate is extremely low without hydrogen, then it increases to the maximum at 12 nm min^{-1} for [H_2]/[C_3F_8] ratio of 1, and it decreases to approximately 2 nm min^{-1} for [H_2]/[C_3F_8] = 2.

The infrared spectra of deposits show a shift at higher wavenumber and the broadening of the overall CF_x stretching band as a function of the hydrogen

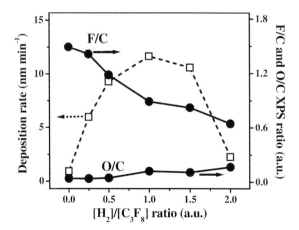

Fig. 19.8 Deposition rate and XPS F/C and O/C ratios of film deposited at different values of the [H_2]/[C_3F_8] feed ratio.

concentration in the feed, to indicate a reduction of the fluorine content of the coating and the formation of a more disordered and crosslinked polymeric network [46]. By increasing the hydrogen content in the feed gas, the broad band of C=O and C=C groups between 1600 and 1850 cm^{-1} increases and the OH stretching at 3400 cm^{-1} appears. When the [H$_2$]/[C$_3$F$_8$] ratio in the feed exceeds 0.5 also the CH$_x$ stretching band at 2950 cm^{-1} is detected.

By increasing the [H$_2$]/[C$_3$F$_8$] ratio in the feed gas from 0 to 2, the XPS F/C and O/C ratios pass from 1.5 to 0.7 and from 0.04 to 0.17, respectively (Fig. 19.8). In agreement with what reported for low-pressure PECVD, the evolution of XPS C 1s high-resolution signal shows that by simply changing the feed composition it is possible to obtain coating with different composition and distribution of CF$_x$ groups. The best fit of C 1s signal can be performed with the same five components already described for the fitting of thin films deposited in GDBDs fed with He–C$_3$F$_6$. By increasing the hydrogen content, the component at (288.0 ± 0.4) eV, correlated to crosslinking degree, increased, while the components at higher BE are reduced.

He–C$_3$F$_8$–H$_2$-fed glow dielectric barrier discharges were characterized by the same spectral features already observed in He–C$_3$F$_6$-fed GDBD. However, as hydrogen was added to the feed gas, the CH emission (A$^2\Delta$–X$^2\Pi$) at 4300 Å appears, while the C$_2$ Swan band completely disappears [63].

19.4
Conclusion

In the present contribution it has been shown that it is possible to deposit thin hydrocarbon films over a wide range of glow dielectric barrier discharge conditions using helium–ethylene-containing feeds. The observed oxygen uptake can be ascribed to contaminations in the deposition chamber and/or to post-deposition reactions with atmospheric oxygen or water. The chemical characteristics of the coatings, their oxygen content in particular, depend on the ethylene concentration in the feed gas. The disappearance of oxygenated species emissions in the spectra of ethylene-containing GDBDs could be due to the well-known high reactivity of hydrocarbons with oxygenated species, such as OH radicals [69,70]. Reactions of ethylene and other hydrocarbon fragments with oxygen-containing species could lead to the formation of volatile non-polymerizing species and/or oxygenated film precursors, in part responsible of the oxygen uptake of the coatings.

Concerning the deposition of fluorocarbon films it was observed that, as expected, the composition of the reactive gas plays a key role in the deposition process. The concentration limit that allows to operate in a glow regime is quite low if compared to ethylene-containing DBD. This behavior can be ascribed to the higher molecular weight of fluorocarbon precursors with respect to ethylene, and to the presence of an electronegative atom such as fluorine. In spite of this concentration limit, which strongly narrows the GDBD operational window, thin film deposition processes were performed with an adequate process control. In particular, deposition rates up to 35 nm min^{-1} and a F/C ratio of 1.5 were obtained with hexafluoropropene as monomer, while with perfluoropropane the deposition rate was extremely low and increased with the addition of hydrogen in the feed. The increase of hydrogen concentration allowed to vary the F/C ratio and the crosslinking degree of the coating as observed for low-pressure plasmas [40,42,46]. For both He–C_3F_6- and He–C_3F_8–H_2-fed GDBDs interesting results were obtained from optical emission spectroscopy investigations.

References

1 Kogelschatz, U., Eliasson, B. and Egli, W. (1997) *J. Phys. IV France*, **7**, C4–C47.

2 Kogelschatz, U. (2003) *Plasma Chem. Plasma Process.*, **23** (1), 1.

3 Kanazawa, S., Kogoma, M., Moriwaki, T. and Okazaki, S. (1988) *J. Phys. D: Appl. Phys.*, **21** (5), 838.

4 Yokoyama, T., Kogoma, M., Moriwaki, T. and Okazaki, S. (1990) *J. Phys. D: Appl. Phys.*, **23**, 1125.

5 Yokoyama, T., Kogoma, M., Kanazawa, S., Moriwaki, T. and Okazaki, S. (1990) *J. Phys. D: Appl. Phys.*, **23** (3), 374.

6 Massines, F., Rabehi, A., Decomps, P., Ben Gadri, R., Segur, P. and Mayoux, C., (1998) *J. Appl. Phys.*, **83**, 2950.

7 Massines, F. and Gouda, G. (1998) *J. Phys. D: Appl. Phys.*, **31**, 3411.

8 Gherardi, N., Gouda, G., Gat, E., Ricard, A. and Massines, F. (2000) *Plasma Sources Sci. Technol.*, **9**, 340.

9 Massines, F., Segur, P., Gherardi, N., Khamphan, C. and Ricard, A. (2003) *Surf. Coat. Technol.*, **174–175**, 8.

10 Seeböck, R., Esrom, H., Carbonnier, M. and Romand, M. (2000) *Plasmas Polym.*, **5** (2), 103.

11 Seeböck, R., Esrom, H., Charbonnier, M., Romand, M. and Kogelschatz, U. (2001) *Surf. Coat. Technol.*, **142–144**, 455.

12 Borcia, G., Anderson, C.A. and Brown, N.M.D. (2004) *Appl. Surf. Sci.*, **221**, 203.

13 Borcia, G., Anderson, C.A. and Brown, N.M.D. (2004) *Appl. Surf. Sci.*, **225**, 186.

14 Miralaï, S.F., Monette, E., Bartnikas, R., Czeremuszkin, G., Latrèche, M. and Wertheimer, M.R. (2000) *Plasmas Polym.*, **5** (2), 63.

15 Guimond, S., Radu, I., Czeremuszkin, G., Carlsson, D.J. and Wertheimer, M.R. (2002) *Plasmas Polym.*, **7** (1), 71.

16 Massines, F., Messaoudi, R. and Mayoux, C. (1998) *Plasmas Polym.*, **3** (1), 43.

17 Massines, F., Gouda, G., Gherardi, N., Duran, M. and Croquesel, E. (2001) *Plasmas Polym.*, **6** (1/2), 35.

18 Fridman, A., Chirokov, A. and Gutsol, A. (2005) *J. Phys. D: Appl. Phys.*, **38**, R1.

19 Gherardi, N., Martin, S. and Massines, F. (2000) *J. Phys. D: Appl. Phys.*, **33**, L104.

20 Martin, S., Massines, F., Gherardi, N. and Jimenez, C. (2004) *Surf. Coat. Technol.*, **177–178**, 693.

21 Fanelli, F., Fracassi, F. and d'Agostino, R. (2005) *Plasma Process. Polym.*, **2**, 688.

22 Sonnenfeld, A., Tun, T.M., Zajčková, L., Kozlov, K.V., Wagner, H.-E., Behnke, J.F. and Hippler, R. (2001) *Plasmas Polym.*, **6** (4), 237.

23 Zhu, X.D., Arefi-Khonsari, F., Petit-Etienne, C. and Tatoulian, M. (2005) *Plasma Process. Polym.*, **2**, 407.

24 Courdec, P. and Catherine, Y. (1987) *Thin Solid Films*, **146**, 93.

25 Mutsukura, N. and Miyatani, K. (1995) *Diam. Relat. Mater.*, **4**, 342.

26 Novikov, N.V., Voronkin, M.A., Dub, S.N., Lupich, I.N., Malogolovets, V.G., Maslyuk, B.A. and Podzayarey, G.A. (1997) *Diam. Relat. Mater*, **6**, 574.

27 Kim, B.K. and Grotjohn, T.A. (2000) *Diam. Relat. Mater.*, **9**, 37.

28 Hong, J., Goullet, A. and Turban, G. (2000) *Thin Solid Films*, **364**, 144.

29 Robertson, J. (2002) *Mater. Sci. Eng.*, **R37**, 129.

30 Liu, D., Yu, S., Liu, Y., Ren, C., Zhang, J. and Ma, T. (2002) *Thin Solid Films*, **414**, 163.

31 Niinomi, M., Kobayashi, H., Bell, A.T. and Shen, M. (1973) *J. Appl. Phys.*, **44**, 10.

32 Kobayashi, H., Bell, A.T. and Shen, M. (1974) *Macromolecules*, **7**, 3.

33 Donohoe, K.G. and Whydeven, T. (1979) *J. Appl. Polym. Sci.*, **23**, 2591.

34 Goossens, O., Dekempeneer, E., Vangeneugden, D., Van deLeest, R. and Leys, C. (2001) *Surf. Coat. Technol.*, **142–144**, 474.

35 Paulussen, S., Rego, R., Goossens, O., Vangeneugden, D. and Rose, K. (2005) *J. Phys. D: Appl. Phys.*, **38**, 568.

36 Girard-Lauriault, P.-L., Mwale, F., Iordanova, M., Demers, C., Desjardins, P. and Wertheimer, M.R. (2005) *Plasma Process. Polym.*, **2**, 263.

37 Liu, D., Benstetter, G., Liu, Y., Yang, X., Yu, S. and Ma, T. (2003) *New Diam. Front. Carb. Technol.*, **13** (4),191.

38 Klages, C.-P., Eichler, M. and Thyen, R. (2003) *New Diam. Front. Carb. Technol.*, **13** (4), 175.

39 Bugaev, S.P., Korotaev, A.D., Oskomov, K.V. and Sochugov, N.S. (1997) *Surf. Coat. Technol.*, **96**, 123.

40 d'Agostino, R., Cramarossa, F., Fracassi, F. and Illuzzi, F. (1990) in *Plasma Deposition, Treatment and Etching of Polymers* (ed. R. d'Agostino), Academic Press, 95.

41 d'Agostino, R. (1997) in *Plasma Processing of Polymers* (eds R. d'Agostino, P. Favia and F. Fracassi,), *NATO ASI Series, E: Appl. Sci.* **346**, Kluwer Academic, 3.

42 Favia, P., Perez-Luna, V.H., Boland, T., Castner, D.G. and Ratner, B.D. (1996) *Plasma Polym.*, **1**, 299.

43 Kim, H.Y. and Yasuda, H.K. (1997) *J. Vac. Sci. Technol. A*, **15** (4), 1837.

44 Mackie, N.M., Castner, D.G. and Fisher, E.R. (1998) *Langmuir*, **14**, 1227.

45 Butoi, C.I., Mackie, N.M., Gamble, L.J., Castner, D.G., Barnd, J., Miller, A.M. and Fisher, E.R. (2000) *Chem. Mater.*, **12**, 2014.

46 Mackie, N.M., Dalleska, N.F., Castner, D.G. and Fisher, E.R. (1997) *Chem. Mater.*, **9**, 349.

47 Cicala, G., Milella, A., Palumbo, F., Favia, P. and d'Agostino, R. (2003) *Diam. Relat. Mater.*, **12**, 2020.

48 Chen, R., Gorelik, V. and Silverstein, M.S. (1995) *J. Appl. Polym. Sci.*, **56**, 615.

49 Chen, R. and Silverstein, M.S. (1996) *J. Appl. Polym. Sci. A: Polym. Chem.*, **34**, 207.

50 Sandrin, L., Silverstein, M.S. and Sacher, E. (2001) *Polymer*, **42**, 3761.

51 Sawada, Y. and Kogoma, M. (1997) *Powder Technol.*, **90**, 245.

52 Prat, R., Koh, Y.J., Babukutty, Y., Kogoma, M., Okasaki, S. and Kodama, M. (2000) *Polymer*, **41**, 7360.

53 Thyen, R., Weber, A. and Klages, C.-P. (1997) *Surf. Coat. Technol.*, **97**, 426.

54 Vinogradov, I.P., Dinkelmann, A. and Lunk, A. (2004) *J. Phys. D: Appl. Phys.*, **37**, 3000.

55 Vinogradov, I.P. and Lunk, A. (2005) *Plasma Process. Polym.*, **2**, 201.

56 Beamson, G. and Briggs, D. (1992) *High Resolution XPS of Organic Polymers*, J. Wiley.

57 Dischler, B., Bubenzer, A. and Koidl, P. (1983) *Solid State Commun.*, **48**, 2.

58 Rinstein, J., Stief, R.F., Ley, L. and Beyer, W. (1998) *J. Appl. Phys.*, **84**, 7.

59 Bourée, J.E., Godet, C., Etemadi, R. and Drévillon, B. (1996) *Synth. Mater.*, **76**, 191.

60 Retzko, I., Friedrich, J.F., Lippitz, A. and Unger, W.E.S. (2001) *J. Elect. Spectro. Relat. Phenom.*, **121**, 111.

61 Kulikovsky, V., Vorlicek, V., Bohac, P., Kurdyumov, A. and Jastrabik, L. (2004) *Thin Solid Films*, **447–448**, 223.

62 Striganov, A.R. and Sventiskii, N.S. (1968) *Tables of Spectral Lines of Neutral and Ionized Atoms*, IFI/Plenum, New York/Washington.

63 Pearse, R.W.B. and Gaydon, A.G. (1976) *The Identification of Molecular*

Spectra, 4th edn , Chapman and
Hall, London.

64 Martinu, L., Biederman, H. and
Nedbal, J. (1986) *Thin Solid Films*,
136, 11.

65 Durrant, S.F., Ranger, E.C., daCruz,
N.C., Castro, S.G. and Bica de
Moraes, M. (1996) *Surf. Coat.
Technol.*, **86–87**, 443.

66 Seth, J. and Babu, S.V. (1993) *Thin
Solid Films*, **230**, 90.

67 Geigenback, H. and Hinze, D. (1959)
Phys. Stat. Sol. A, **81**, 1045.

68 Cioffi, N., Losito, I., Torsi, L., Farella,
I., Valentini, A., Sabbatini, L.,
Zambonin, P.G. and Bleve-Zacheo, T.
(2002) *Chem. Mater.*, **14**, 804.

69 Davis, D.D., Fischer, S., Schiff, R.,
Watson, R.T. and Bollinger, W. (1975)
J. Chem. Phys., **63**, 5.

70 Howard, C.J. (1976) *J. Chem. Phys.*,
65, 11.

20
Remark on Production of Atmospheric Pressure Non-thermal Plasmas for Modern Applications
R. Itatani

20.1
Introduction

Ionized gases or plasmas are chemically active media in which exotic chemical reactions take place because activated particles such as electrons, excited atoms or molecules, and ions are present among neutral atoms or molecules and they undergo reactions in thermal equilibrium or in non-equilibrium depending on the operation conditions. The most well-known industrial applications at atmospheric pressure are the arc as thermal equilibrium plasma and the ozonizer as non-equilibrium plasma using volume reaction in air.

Recently developed applications are not volume reaction in plasmas but surface reaction of plasmas with solid. For these applications such plasmas as chemically active but thermally low are preferable, because base material should not be changed and only the surface is modified by plasma. This is the reason why non-thermal plasmas are preferable for modern applications of plasmas.

Studies on non-thermal plasmas at atmospheric pressure covered from corona discharge to silent discharge. Corona discharge appears only where an electric field is not uniform, that is, electrodes of thin wire, needle, and edge of a plane. Non-thermal plasma at atmospheric pressure in uniform electrode configuration is the silent discharge and the most typical one is an ozonizer. This type of discharge has an insulator to prevent arcing and is the prototype of present dielectric barrier discharge. However, discharges in ozonizers show always non-uniform plasma and consist of multi-filament columns, so that the use of these types of discharges for surface treating is limited. On the other hand, non-thermal plasmas of large area at high pressure are already commonly used in TEA lasers, such as CO_2 lasers and XeCl lasers as examples from before 1988. Their plasmas are uniform and not filamentary with metal electrodes. After the report of a Sofia University research group in 1988 [1], diffuse-type dielectric barrier discharges (diffuse DBDs) have attracted the attention of scientists and engineers, because these uniform atmospheric pressure non-thermal plasmas are expected to have large potential in a wide range of applications.

Advanced Plasma Technology. Edited by Riccardo d'Agostino, Pietro Favia, Yoshinobu Kawai, Hideo Ikegami, Noriyoshi Sato, and Farzaneh Arefi-Khonsari
Copyright © 2008 WILEY-VCH Verlag GmbH & Co. KGaA, Weinheim
ISBN: 978-3-527-40591-6

As is well known, although gas temperature is much lower than electron temperature at low pressure, both temperatures become almost the same at around above half atmospheric pressure in a continuous discharge, so that to generate non-thermal plasmas at high pressure is not so straightforward as in the low-pressure case and requires some consideration. Here, we discuss factors to be considered to generate atmospheric pressure non-thermal plasmas for modern applications.

Generally, there are three methods. The first is to remove the heat of gas in plasmas, the second is not to give the gas energy, and the third is to use a separate active region of the plasmas.

20.2
Why Atmospheric Pressure Non-thermal Plasmas Are Attractive

For industries, it is very important that equipments are simple, do not need strong casings against pressure, allow high throughput, are able to treat large-size objects, and allow the introduction of continuous processes.

In the case of treating something with a large surface area like textile and paper, it takes a long time to evacuate because of the huge amount of adsorbed material and moisture. Time to processing will be reduced remarkably if such materials can be treated at atmospheric pressure without any evacuation process.

There are certain applications that cannot be carried out in a reduced pressure environment. Ambient atmosphere is essential for biological applications, otherwise one cannot keep alive the target to be treated. In the case that a plasma is applied to living things and the human body, operations should be done in open air. Moist materials are also processed in normal or higher pressure condition. In such a way, there are many fields for atmospheric pressure non-thermal plasmas to be applied. Applications of atmospheric pressure non-thermal plasmas (APNTP) are tabulated in Table 20.1. Typical discharge forms of APNTP are summarized in Table 20.2.

Tab. 20.1 Application of atmospheric pressure non-thermal plasmas.

Improvement of surface condition wettability, adhesive ability
Acceleration test facility
Sterilization
DNA introduction
Cleaning of surfaces: ashing, removing undesired materials, reduction of oxide
Air purification
Air pollutant reduction
Deposition of functional thin films such as TiO_2, etc.
Coating of protection film
Modification of bacteria and seeds

Tab. 20.2 Typical discharge forms for APNTP.

Townsend discharge
Corona
Glow
DBD filamentary
DBD uniform
Torch with/without dielectric barriers
RF/microwave torch with/without electrodes

20.3
Origin of Activities of Plasmas

As is well known, active species in plasmas are electrons, ions, and excited neutral particles. At atmospheric pressure, the mean free path between particles is very short and then active species are produced only in the active space where ionization occurs. However, outside of the active space, charged particles of opposite sign recombine quickly with each other and excited species are de-excited through radiation and collision. In these loss processes, each loss channel has a different loss rate or different time constant of loss. Generally, loss of excited particles able to undergo optical transitions is faster and loss of charged particles is much faster than de-excitation of metastable particles because coulomb force acts over longer ranges than forces between neutral particles.

Metastable states influence gas discharge phenomena because of long life and act as the energy reservoir within and in the vicinity of plasmas. Metastable particles themselves have long life. In addition to this, resonant transfer of the energy of one metastable particle to another particle in the ground state is so efficient in the same kind of gas that metastable particles can exist outside of the plasma and are conveyed out with gas flow.

Photons of resonant line, i.e. emission from the lowest excited level, are trapped and conveyed from inside to outside plasmas through the resonant transfer. In the case that this line is of ultraviolet energy, then photoemissions from walls act as positive feedback to the discharge.

20.4
Limits of Similarity Law of Gas Discharge

The similarity law tells us invariants exist, by keeping which similar, almost similar discharge can be obtained. Table 20.3 lists such invariants and how the values change.

Therefore, it is easily understood that to generate glow discharge similar to low and medium pressure is very difficult in the atmospheric pressure range. The cathode fall is invariant, so that the energy input density is proportional to current density, which becomes of the order of $100 \, A \, cm^{-2}$. This current density cannot be sustained not only in steady state without cooling but also even in continuous periodical discharge.

Tab. 20.3 Invariants of the similarity law and values resulting from them.

Invariant	Values at atmospheric pressure compare to 1 torr	Value at 1 torr ($\mu A \, cm^{-2}$)
pd	Plasma size $1/760 = 1.3 \times 10^{-3}$	
ω/p	Time variation should be 760 times faster	
j/p^2	Current density 5.8×10^5 times larger	Several to 400 (normal glow)
Ej/p^3	Power density 4.4×10^8 times larger	

Diffuse DBDs reported to date have of the order of $1 \, mA \, cm^{-2}$ as the maximum current density.

Similarity law informs only on charged particles and does not inform on neutral particles in metastable states and in excited states. Particles in metastable states have long mean free paths or long lives in spite of high pressure due to resonant transference with particles of the same kind in the ground state. Photons of the resonance line are also trapped between excited particles in the lowest state and particles in the ground state. These effects are more important at high pressure than at low pressure because the mean free path is shorter and degree of ionization is less at high pressure and vice versa at low pressure; accordingly the current density or charged density at high pressure is much less than that from the similarity law as discussed above.

20.5
Reduction of Gas Temperature

Input energy to plasma is transferred to electrons and ions at first through acceleration by an electric field. In the next stage energies on charged particles are transferred to neutral particles by collisions and finally the energies of neutral particles go to walls or spread out.

At atmospheric pressure, collisions are normally so frequent that all the temperatures become the same and the heat energy flows from the plasma to the wall surrounding the plasma. However, in the special case mentioned below, there exist very narrow slots where thermal equilibrium conditions are broken and non-thermal situations established.

For simplicity, consider a binary collision where each mass is m_1 and m_2. Energy transfer ratio is $2m_1 m_2/(m_1 + m_2)^2$, so that the fraction of energy transfer from an electron to a massive particle is $2m/M$ and that between particles of the same mass is around $1/2$. That means the energy transferring channel is much narrower between an electron and heavy particles, such as ions or molecules, than that between an ion and a molecule.

On the basis of the transport equations of energy for neutrals, one can write

$$\frac{\partial}{\partial t} N < \varepsilon >_N + \vec{\nabla} \vec{q}_N = +nNv\sigma_\varepsilon < \varepsilon > \tag{20.1}$$

Then, energy gain of neutrals is

$$nNv\sigma(2m/M)3kT_e/2 \tag{20.2}$$

where σ is the cross-section of momentum transfer between electrons and neutrals.

In steady state, energy loss is through diffusion, and using the relation

$$\vec{q} = -(3NkD/2)\vec{\nabla}T_g$$

this loss is written as

$$-(3NkD/2)\nabla^2 T_g \tag{20.3}$$

By putting $-\nabla^2 \rightarrow 1/\Lambda^2$ and equating Eqs. (20.1) and (20.2), then

$$T_g/T_e = n\sigma v(2m/M)(\Lambda^2/D) \tag{20.4}$$

Equation (20.4) tells us that a shorter diffusion length provides a lower gas temperature.

In transient (pulse) state, by putting the first term of left-hand side equal to the right-hand side, then

$$T_g/T_e = n\sigma v(2m/M)\tau \tag{20.5}$$

Equation (20.5) tells us that a shorter time provides a lower gas temperature.

When gas flow is introduced into a plasma, the residence time of the gas in the plasma becomes short, then the gas temperature becomes lower than without gas flow.

A magnetic field can be introduced to give motions to a plasma by the $J \times B$ force. This method is applied to electric circuit breakers in order to cool down the arc plasma generated when electric current is interrupted. This method also makes it possible to sweep out charged particles from the active region of a plasma and makes it possible to control freely relative motion of the plasma and gas flow.

20.6
Examples of Realization of the Above Discussion

For the steady-state plasma, it is essential to cool the gas temperature by reducing the diffusion length of gas in order to enhance heat loss of gas in the plasma. Cooling down of electrodes is inevitable and sometimes the wall also. The first example is a narrow gap discharge of 0.1 mm of a metal electrode generated by 2.45 GHz [2]. The second example is known as packed bed discharges consisting of many narrow current channels on the surfaces of dielectric particles or powder between electrodes.

To reduce plasma-on time automatically, there are two methods, one is to use serially connected capacitors and the other is to use dielectric material on electrodes equivalent to capacitors distributed on the electrodes [3]. The action is the same. Both can stop discharge automatically by charge-up of capacitor due to current flow by the discharge. By applying alternating voltage, pulse current flows in an alternating manner.

The most active method to reduce plasma-on time is to apply pulse voltage of intermittent and short period between electrodes with/without dielectrics. The presence of dielectrics on an electrode does not matter. In this case, sharp voltage rise of narrow pulses is more preferable to get the plasma in more non-thermal equilibrium.

Cooling down by gas flow is realized as a plasma torch [4,5] and the temperature becomes lower by separating more the distance of the work place from the plasma. To give a plasma motion is equivalent to injecting cooler gas to the plasma or to let it touch a wall, so that many variations will be proposed and tested for various applications.

20.7
Large-Area Plasma Production

Discharges having dielectrics on their electrodes, called dielectric barrier discharges (DBDs), have been much studied to understand and to improve ozone production. However, this type of discharge consists of many of filamentary discharge channels between electrodes [6], so that DBDs are not considered for applications for modification of material surfaces.

Evidence that glow discharge exists even for gas pressure higher than 1 atm is well known [7]. Studies concerning industrial application began after 1988, as reported by Kogoma *et al.* [1]. Uniform DBD (APGD) is most attractive from the viewpoint of not only industrial application but of physics too, because many of recent applications are surface modifications, and these require plasmas of large area and of uniformity in density and temperature. The uniform DBD was obtained with He as working gas at the first stage. However, at present, there have been many studies to generate uniform DBDs with various kinds of gas, and the physics of DBDs is becoming clearer [8–16].

20.8
Summery of Evidence To Date to Obtain Uniform DBDs

One type is a glow-type configuration having cathode glow at low pressure and observed with rare gases such as He, Ne, Ar, and mixture of a small amount of additives with these gases.

Another type is the Townsend discharge type and glow is observed close to the anode with nitrogen as a working gas. In both cases metastable particles have an important roll in making the discharge easy for the next half cycle [3].

As a power source, sinusoidal voltage of frequency higher than several kilohertz is preferable and trapping of ions in the discharge region is effective to make discharge voltage low [11].

Rectangular or pulse voltage with fast rise time is preferable to make plasmas of large area with nitrogen and air. The shape of electrodes also generally has the effect to produce a plasma. Sharp edges can expand the range of the conditions on gas. Electric power sources change the characteristics of the discharge, and the plasma changes from uniform to filamentary on increasing input power [9].

There are many factors that affect diffuse and filamentary DBD conditions such as properties of dielectric material, gas and gas mixture, configuration of electrodes, condition of surface of dielectric barriers, relationship among gap size, thickness of dielectrics, kind of gas, waveform of applied power, and so on, to optimize performance.

20.9
Consideration to Realize Uniform Plasmas of Large Area

Evidence that diffuse DBDs appear in some limited conditions should be considered following the general theory of gas discharge

Consider a plasma of large area as a large number of small discharge tubes connected in parallel. In such a case parallel operation is impossible with negative resistance tubes; in contrast, it is possible with positive resistance tubes, according to the circuit theory.

However, parallel operation is possible in the case that all of the electrodes are in one envelope in certain condition, that is, as the normal glow discharge. This suggests that charged particles or metastable ones and photons come from nearby region where discharge takes place and induce a discharge.

Positive resistance characteristics are found in the Townsend discharge region and abnormal glow region. The former shows a brighter region near the anode and the latter has it near the cathode [11].

With needle-type electrodes, a corona appears as the first step of discharge and the discharge current increases with increasing voltage, that is, showing a positive resistance.

In contrast, with plane parallel electrodes, corona discharge does not appear and direct spark discharge takes place. Then, negative resistance characteristics are always shown and, usually, uniform plasma is not expected except for very short gaps.

20.10
Factors to be Considered to Realize Uniformity of DBD Plasma

Electric charge stays and is impossible to move on the surface of a dielectric material. Therefore, the surface of dielectrics facing a plasma has the memory of the point where discharge occurs at a previous cycle, and at the following reverse cycle the voltage due to the charge at the point acts additively. So that, once discharge takes place at some points, then discharge occurs successively in the following cycles at the same points on the dielectrics. To realize diffuse DBD, it is necessary to eliminate this localized memory effect of dielectrics placed on electrodes.

In order to reduce this memory effect of the surface of dielectrics, motion of charge in the lateral direction to the discharge is important. This is realized by enhancing diffusion of charged particles and metastable particles in the lateral direction and by controlling surface conductivity of the barrier. Both act as to provide uniformly active species in front of dielectrics and to make surface potential distribution uniform.

Another approach is to use the ion trapping effect to suppress ion loss from the plasma region by selection of the frequency of the applied voltage [14].

Introduction of optimum amount of cross flow of working gas along barrier surface is also effective to spread active species on the surface.

To ensure uniform starting of the discharge, the most powerful method is to provide enough electrons uniformly in front of the electrodes or on the surface of barriers, or to make a thin plasma in the discharge space [17,18]. UV irradiation to get photoemission of electrons and auxiliary weak discharge to feed charged particles are' already applied as common methods in TEA gas lasers.

20.11
Remote Plasmas

Gas flow carries active species produced in an active plasma to the surface to be processed where the density of active species as well as the temperature of the plasma become lower but still active due to metastable particles.

In order to cool the gas temperature, a certain amount of gas flow is necessary. In such a case, discharges with electrodes are much more stable than electrodeless discharges.

When cooling a plasma by mixing the cold gas of the same species, both a temperature drop and less decrease of density of metastable particles are expected. For simplicity a cold gas of density N_{cold} and temperature T_{cold} is mixed at $x = 0$ into the plasma having hot gas of temperature T_{hot} and of density N_{hot}. The energy transport equation is written in steady state as

$$\vec{\nabla}\vec{q}_{\text{hot}} = -N_{\text{hot}}N_{\text{cold}}\sigma_{\text{h-c}}v_{\text{h-c}}(1/2)(3/2)(T_{\text{hot}} - T_{\text{cold}}) = -(3N_{\text{hot}}kD/2)\nabla^2 T_{\text{hot}} \tag{20.6}$$

and its solution is

$$T_{\text{hot}} = [T_{\text{hot}}(0) - T_\infty]\exp\left(-\sqrt{\frac{(N_{\text{cold}} + N_{\text{hot}})\sigma v}{2kD}}x\right) + T_\infty \tag{20.7}$$

where T_∞ is the gas temperature far from the plasma region.

By comparing the middle term of Eq. (20.6) with Eq. (20.2), one can see that energy transfer between neutrals is much larger than that between electrons and neutrals, so that remote plasma systems provide low-temperature gas reactors with low flux density but relatively high-energy reaction, which may be due to metastable atoms of rare gas.

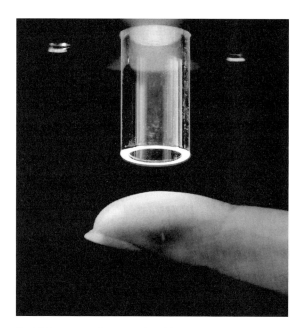

Fig. 20.1 Remote plasma produced by 2.45 GHz microwave, friendly to the human skin. (Courtesy of Adtec Plasma Technology Co. Ltd.)

In remote plasmas, only long-lived excited particles work and to get the lower gas temperature causes the lower density of active species. It may be difficult to eliminate this contradiction. However, by using the gas having metastable state of long life, remote plasmas are quite useful tools to treat delicate targets such as human bodies, organic materials, substrates of integrated circuits, and so on.

For surface cleaning, inductively coupled radiofrequency torches or microwave torches are more suitable than plasmas with electrodes, to avoid flashover on the surfaces on which some delicate constructions exist.

Recently, remote plasmas have been applied to the field relating to medical use and there is one chapter in this book describing 13.56 MHz needle-type plasmas. Furthermore, a plasma torch of 2.45 GHz microwave is also planned to be applied for sterilization of human skin. Figure 20.1 shows how a microwave remote plasma is cool and friendly to human skin. This kind of plasma also shows sterilization and disinfection effects. However, these are still in the development stage, and why, how, and by what the remote plasma acts is not yet clear perfectly.

20.12
Conclusion

General remarks on factors to generate APNTP have been presented on the basis of the theoretical point of view. Observed facts of APGD are easily deduced and well explained qualitatively with discharge theory previously established for high-pressure conditions. However, not enough yet is known to understand quantitatively and to design a reactor suitable for a specific application.

References

1 Kanazawa, S., Kogoma, M., Moriwaki, T. and Okazaki, S. (1988) *J. Phys. D: Appl. Phys.*, **21**, 838.

2 Kono, A., Sugiyama, T., Goto, T., Furuhashi, H. and Uchida, Y. (2001) *Jpn. J. Appl. Phys.*, **40**, L238.

3 Kogelschatz, U. (2002) *IEEE Trans. Plasma Sci.*, **30**, 1400.

4 Koinuma, H., Inomata, K., Ohkubo, H., Hashimoto, T., Shiraishi, T., Miyanaga, A. and Hayashi, S. (1992) *Appl. Phys. Lett.*, **60**, 816.

5 Hubicka, Z., Cada, M., Sicha, M., Churpita, A., Pokorny, P., Soukup, L. and Jastrabik, L. (2002) *Plasma Sources Sci. Technol.*, **11**, 195.

6 Kogelschatz, U. (2003) Dielectric-barrier discharges: their history, discharge physics, and industrial applications, in *Plasma Chemistry and Plasma Processing, Springer, Netherland*, **23**, pp. 1–46.

7 von Engel, A., Seeliger, R. and Steenbeck, M. (1933) *J. Phys.*, **85**, 144.

8 Yokoyama, T., Kogoma, M., Kanazawa, S., Moriwaki, T. and Okazaki, S. (1990) *J. Phys. D: Appl. Phys.*, **23**, 374.

9 Yokoyama, T., Kogoma, M., Moriwaki, T. and Okazaki, S. (1990) *J. Phys. D: Appl. Phys.*, **23**, 1125.

10 Gherardi, N., Gouda, G., Gat, E., Ricard, A. and Massines, F. (2000) *Plasma Sources Sci. Technol.*, **9**, 340.

11 Naude, N., Cambronne, J.-P., Gherardi, N. and Massines, F. (2005) *J. Phys. D: Appl. Phys.*, **38**, 530.

12 Massines, F., Rabahi, A., Decomps, P., Gadri, R.B., Segur, P. and Mayoux, C. (1998) *J. Appl. Phys.*, **83**, 2950.

13 Golubovskii, Yu.B., Maiorov, V.A., Behnke, JF., Tepper, J. and Lindmayer, M. (2004) *J. Phys. D: Appl. Phys.*, **37**, 1346.

14 Liu, C., Tsai, P.P. and Roth, J.R. (1993) *Proc. 20th IEEE Int. Conf. on Plasma Science, Vancouver, Canada, 7–9 June.*

15 Roth, J.R., Rahel, J., Xin, Dai, and Sherman, D.M. (2005) *J. Phys. D: Appl. Phys.*, **38**, 555.

16 Rahel, J. and Sherman, D.M. (2005) *J. Phys. D: Appl. Phys.*, **38**, 547.

17 Palmer, A.J. *Appl. Phys. Lett.*, (1974) **25**, 138.

18 Karnyushin, V.N. *et al.* (1978) *Sov. J. Quantum Electron*, 319.

21
Present Status and Future of Color Plasma Displays
T. Shinoda

The plasma display market is growing rapidly. The market value reached $10 billion in 2005. Especially the plasma television (plasma TV) market is strong in Japan and the market size became larger than the industrial market in the world. In these circumstances color plasma display (PDP) technology is one of the key issues to support the growing market. This chapter discusses the advance in the PDP technologies from the initial stage of the PDP development to the future.

21.1
Introduction

In the late 1960s and early 1970s researches in color PDPs had been started in many laboratories with AC and DC types. The DC type had been studied intensely for flat panel high-definition television (HDTV) in NHK and a 40-inch-diagonal prototype color plasma TV was shown in 1993 [1]. For the AC type, researches with opposed discharge technologies had not been successful and surface discharge technology became the mainstream for color PDP development in the early 1980s and researched at Fujitsu and Hiroshima University [2,3]. Color PDP technologies needed a long time and tough researches to be successful. Figure 21.1 shows the development of color PDPs. Until 1992, research for practical full color PDPs was not successful and when a 21-inch PDP had been put into practical use the modern PDP era opened in 1993 [4]. The target market of color PDPs was mainly public information displays for business application. After the development of the 21-inch PDP the TV technologies for large areas, such as 40-inch-diagonal, have been investigated. At the same time, improvements in luminous efficiency, process technologies, materials, and production equipment have been developed. The first major investment for large-area PDPs was decided and mass production of 42-inch-diagonal WVGA systems was started in 1996. The growth of the plasma TV market started by putting a 42-inch-diagonal display into the market with surface discharge technology. The early market, however, was for business use and not for consumer television use. NEC, Mitsubishi, Pioneer, Matsushita, and Hitachi in Japan, and LG

Advanced Plasma Technology. Edited by Riccardo d'Agostino, Pietro Favia, Yoshinobu Kawai, Hideo Ikegami, Noriyoshi Sato, and Farzaneh Arefi-Khonsari
Copyright © 2008 WILEY-VCH Verlag GmbH & Co. KGaA, Weinheim
ISBN: 978-3-527-40591-6

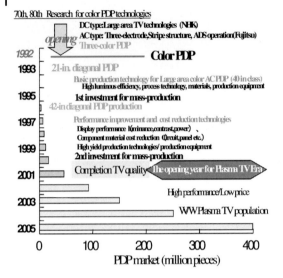

Fig. 21.1 Color PDP development.

electronics and Samsung in Korea have joined the PDP leagues and competition started. From 1996 to 2000 the performance for TV was largely improved, especially for peak luminance, contrast, and power consumption and the display quality has reached levels comparable to that of cathode ray tubes (CRTs). The cost of the component materials, such as driving circuits, glass materials, and phosphor materials, have been decreasing and the production yield has largely improved resulting in the reduction of PDP costs. The second investment led to the quantity of production increasing at a growth rate of about 100% each year. Production reached 1.6 million sets in 2003. In particular, a 32-inch plasma TV has been put onto the market with a price of less than 60 000 yen in Japan and the plasma TV market has grown rapidly. The year 2001 was called the opening year of the plasma TV era. And the total size of the PDP market has reached more than $1 billion: $1.6 billion in 2001 and $10 billion in 2005. The growth rate is expected to continue until 2008, the size of the market having reached 500 million sets per year in 2005.

The production of color PDPs has also increased. In 1996, a 42-inch WVGA PDP was first put onto the market, and larger and smaller WXGA and HDTV families have been added, and currently from 32-inch to 61-inch. The market will expand to three directions. The first is the market for smaller size from 30- to 40-inch-diagonals with about 1 million pixels that replace the conventional TV. In this market lower price and lower power consumption displays will be needed. The second is the market for larger sizes from 40- to 60-inch-diagonals with more than 2 million pixels that are used for home core displays in the broadband networking arena. In this market, high-resolution displays will be needed. The third is the market for the largest size of 40- to 80-inch-diagonals with about 2 million pixels used for home theaters. In this market larger displays will be needed. From these requirements, the technologies for power consumption, higher resolution, larger size, and lower price should be developed in the future. Progress has been supported with the development of technologies concerning both PDP engineering and manufacturing processes.

21.2
Development of Color PDP Technologies

Color PDP technologies are based on the three-electrode surface discharge structure. The application of surface discharge technology to color PDPs was reported first by Takashima et al. for segment-type color PDPs in 1973 [5]. Dick also reported the technology of matrix-type monochrome PDPs in 1974 [6]. The author started to study color PDPs with surface discharge technologies in 1979 [2] and finally reached the three-electrode structure as shown in Table 21.1 [7,8]. Figure 21.2 summarizes the technical issues raised when the author started to study color PDPs. The issues were how to realize color PDPs, with long operating life, high luminance, high resolution, and full color operation. Developing a new panel structure has solved the first four issues. And developing a new driving technology has solved the last one. Finally, the three-electrode PDP structure has been completed as shown in Fig. 21.3 and a practical 21-inch-diagonal color PDP has been developed with these technologies as shown in Fig. 21.4. The current large-area plasma displays have been developed from 40- to 60-inch-diagonal PDPs based on the developed structures.

21.2.1
Panel Structure

Figure 21.5 shows the principle of the structure of the color surface discharge technologies made by the author in 1979. Two kinds of stripe electrodes isolated by a dielectric thin film were placed orthogonally on a single substrate. These were covered by a thin dielectric and an MgO layer. Phosphors were deposited on the other substrate. A gas mixture, which emits much UV radiation by discharge, was introduced between the substrates. The structure has improved the major issue of phosphor degradation by discharge that prevented color PDPs for practical use. A new problem of change in driving voltage was, however, raised due to the degradation of the MgO surface by ion bombardment at the cross point of the electrodes. Then the three-electrode structure was introduced to solve the problem in 1984 as shown in the Fig. 21.6. Each discharge cell has two kinds of cells, address cell and display cell. At the address cell, two kinds of electrodes, address and scan electrodes, cross and at the display cell two kinds of display electrodes, scan and sustain electrodes, are placed in parallel on the single substrate. There are two functions to complete a display: cell selecting and display. The address cell is used only for selecting the display cells and then display discharges are maintained at display cells. In the usual display system, discharges are initiated only 480 times per second in the address cells and 60 000 times per second in display cells. Then the operating life of the three-electrode structure was extended 100 times more than the

Tab. 21.1 Development of structures for color PDPs.

Year	Researchers	Electrodes	Substrates with electrode	Display type	Phosphors	Electrodes	Ribs	Comments
1973	Takashima [5]	Two	Single	Transmitting	Green	Segment	Glass sheet	
1974	Dick [6]	Two	Single	–	Non	Matrix	Non	
1979	Shinoda [2]	Two	Single	Transmitting	RGB	Matrix	Non	
1984	Shinoda [7]	Three	Single	Transmitting	RGB	Matrix	Stripe + mesh	
1985	Dick [8]	Three	Double	–	Non	Matrix	Stripe	
1989	Shinoda [9]	Three	Double	Reflective	RG	Matrix	Stripe + mesh	Production
1992	Shinoda [10]	Three	Double	Reflective	RGB	Matrix	Stripe	Production
1999	Komaki [11]	Three	Double	Reflective	RGB	Matrix	Waffle	Production
1999	Toyoda [12]	Three	Double	Reflective	RGB	Matrix	Meander	Production

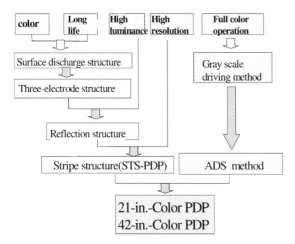

Fig. 21.2 Technical points to develop practical color PDPs.

two-electrode structure that has 60 480 times discharges in a cell. Finally, the address electrode and display electrodes were placed on different substrates to reduce the capacitance in the address cell. The luminance level was not sufficient because a transmitting-type structure was used. Introducing a reflecting-type structure has solved this.

With these technologies, color PDPs have become practical. The structure has continued to develop to realize high luminous efficiency. Figure 21.7 compares the development of the panel structures from the first practically available one to the future. The first challenge for color PDP production was a three-color PDP for financial

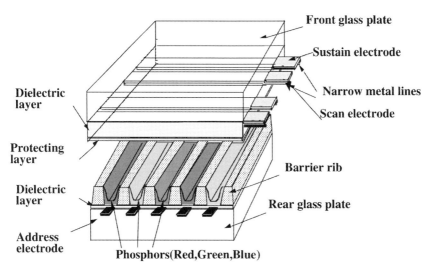

Fig. 21.3 Three-electrode PDP with stripe phosphor and barrier rib.

Fig. 21.4 The first practical full-color 21-inch-diagonal PDP (1992) of surface discharge color PDP (1979).

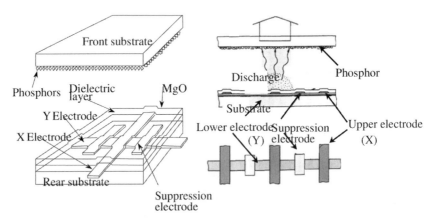

Fig. 21.5 First experimental panel structure of surface discharge color PDP (1979).

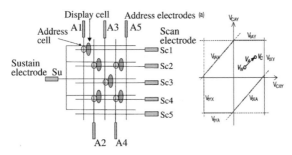

Fig. 21.6 Electrode configuration of three-electrode PDP (1984).

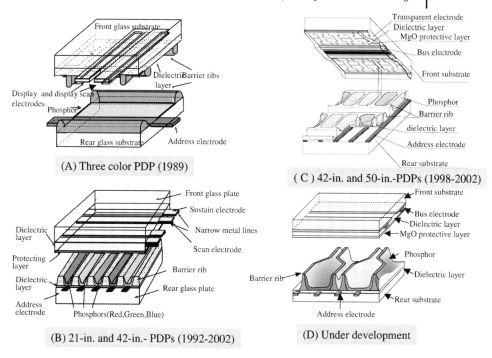

Fig. 21.7 Development of panel structure for three-electrode PDPs.

display with the structure of Fig. 21.7(a) in 1989 [9]. The first practical structure was complex in which the address electrode was exposed to the gas and not covered by phosphors. The first full-color PDP was a 21-inch-diagonal PDP with the structure of Fig. 21.7(b) in 1993 [10]. The resolution was four times that of the three-color PDP and it was not possible to fabricate with the conventional technologies, so simple stripe rib and phosphor structure was introduced. It was designed such that the address electrode was covered by phosphors to realize a high resolution. These developments succeeded to complete the color PDP technologies and fabrication process in the first stage. Recently, to improve the power consumption and display quality, new approaches have been investigated. A waffle rib structure has been introduced as shown in Fig. 21.7(c) [11]. This improved the luminous efficiency of $1.8\,\mathrm{lm\,W^{-1}}$ in practical PDPs. Although the structure was much more complex compared to the stripe structure, process improvement such as the introduction of high strain point glass and sandblasting made it possible. The highest luminous efficiency has been reported by the structure of Fig. 21.7(d) [12]. This is called the Delta-cell structure.

21.2.2
Driving Technologies

For AC PDPs it was thought that high-speed addressing was difficult due to controlling the wall charge for operation. Sufficient accumulation of wall charge needs a sufficiently wide operating pulse resulting in slow addressing speed and thus it is not possible to realize the 256 gray scale of HDTV specification on AC PDPs.

The author has proposed the ADS (address period, display period separation) method. In this method the address period and display period are completely separated [13]. In the address period wall charges are formed in all the cells over the panel. And in the subsequent display period the discharges are initiated and sustained in only the displays cells with accumulated wall charges by applying sustain pulses. This largely increased the addressing speed and then a 256 gray scale has become possible for AC PDPs. The reset step in the address period is also very important for operation of the AC PDP. The initial particles and wall charges have been reset and set up to realize a stable operation of the cells. A ramp waveform has been applied to reduce the luminance while the reset resulting in high contrast ratio [14].

21.3
Latest Research and Development

Thirty- to 60-inch PDPs have been put into practical use using the three-electrode surface discharge technologies explained [12,13]. These are supported by the following researches on improvement of the performances and reduction in the cost. Next the latest results of researches on panel structure, luminous efficiency improvement, analysis of PDP discharge, and driving method will be discussed.

21.3.1
Analysis of Discharge in PDPs

More than 10 years has passed since the author unveiled the three-electrode structure in 1984 and the ADS method in 1992. Both technologies became the general standard technologies of AC PDPs and color PDPs have been produced based on the technologies. The theory of driving has not been completed.

Sakita et al. have tried to analyze the addressing process in the ADS for the three-electrode PDP by introducing a new concept of Vt close curve [14]. The Vt close curve has been defined as the line connected among the threshold voltage of weak discharge in PDP cells. When we define the vertical axis as the cell voltage between address and scan electrodes (V_{CAX}) and then the horizontal axis as one between sustain and scan electrodes (V_{CXY}), the hexagonal Vt close curve can be defined in the plane as shown in Fig. 21.8. The cell voltage is defined as the inside voltage in the discharge gap which is obtained by the externally applied voltage and wall voltage by accumulated wall charge on the dielectric layer.

The lamp-reset process in the reset step has easily been analyzed by using the Vt close curve and it has been possible to analyze the ideal reset condition.

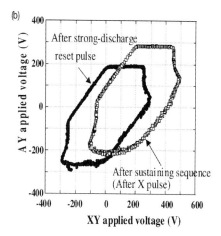

Fig. 21.8 Vt close curve.

21.3.2
High Luminance and High Luminous Efficiency

To realize high luminance and high luminous efficiency is another important issue to be researched. Currently the luminous efficiency is almost $1\,lm\,W^{-1}$ and the 42-inch PDP consumes about 250 W typically. It is desired to operate PDPs with about 100 W to penetrate widely in the world market. Lowering the power consumption is also valuable to realize thin and lightweight units and to reduce costs. The next target of luminous efficiency is $3\,lm\,W^{-1}$. A new structure with delta arrangement of colors (Delta structure) was developed and $3\,lm\,W^{-1}$ luminous efficiency obtained in experiments [15]. Figure 21.9 shows the panel structure. The figure does not show

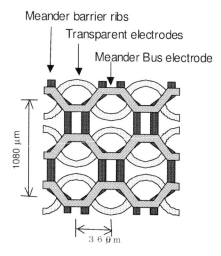

Fig. 21.9 Cell structure of Delta PDP.

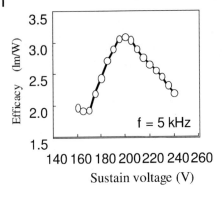

Fig. 21.10 Luminous efficiency of Delta PDP.

address electrodes formed between meander barrier ribs on the back substrate. In the Delta structure, barrier ribs are meandered, so that the non-discharge gap area can be small and the discharge gap can be large. Bus electrodes are also meandered along the meander barrier ribs so that the bus electrodes do not disturb the light emission. The transparent electrodes have an arc shape. The sustain gap width is narrow at the center of a cell, becoming wider towards the edge of a cell. Therefore the disruption of the discharge by the barrier ribs, that is, the discharge is concentrated in the center of a cell. Figure 21.10 shows the dependence of luminous efficiency on the sustain voltage. The sustain frequency is 5 kHz. The luminous efficiency has a strong dependence on the sustain voltage. The efficiency begins to increase sharply at 170 V and reaches $3.1 \, \text{lm} \, \text{W}^{-1}$ at 200 V.

21.3.3
ALIS Structure

A PDP structure named ALIS (alternate lighting of surface method) which increases luminance and reduces driving circuits has been proposed [16]. This has two advantages of high resolution and high luminance. Small size HDTV PDPs, such as a 32-inch PDP, have been realized by this technology.

Figure 21.11 shows the principle. This is essentially the three-electrode PDP and the panel structure is almost as same as the conventional one. The big difference between this and the conventional one is the arrangement of the display electrodes.

In the conventional structure the gap between display electrodes in the discharge site is narrower than the gap between display electrodes in the non-discharge site in the vertical direction. This gives a memory margin by isolating the discharge in adjacent cells. In contrast, in the ALIS structure the gaps between the display electrodes are all the same. Then ALIS is operated in an interlaced way as for the CRT to isolation between adjacent cells. That is, the display image is constructed with the mixing of two fields made by each odd or even display line. A resolution twice that of the conventional structure with the same display lines was realized with this technology. The back plate is as same as the conventional one.

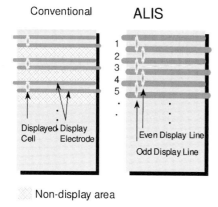

Conventional ALIS

Non-display area

Fig. 21.11 The principle of ALIS PDP.

Therefore this structure has advantages in terms of cost because the production process and the scale of driving circuits are almost same as the conventional ones.

21.4
Conclusion

Both monochrome and color PDP histories and color PDP technologies have been discussed. The three-electrode PDP and ADS driving method have made color PDPs a success in practical use. After the development of 21- and 42-inch PDPs, the research and development of color PDPs became active and research on the basic two technologies has been developing and improving continuously, as discussed in this chapter. The latest research results on analysis, panel structure, luminous efficiency, and driving method make us confident that color PDPs are being pushed forward to ideal flat panel display devices.

References

1 Yamamoto, T., Kuriyama, T., Seki, M., Katoh, T., Takei, T., Kawai, T., Murakami, H. and Shimada, K. (1993) A 40-inch-diagonal HDTV plasma display. *SID Digest*, **93**, 165–168.

2 Shinoda, T., Yoshikawa, K., Miyashita, Y. and Sei, H. (1980) Surface discharge color AC-plasma display panels. *late news in Biennial Display Research Conference.*

3 Uchiike, H. (1986) Mechanisms of 3-phase driving operation in surface-discharge ac-plasma display panels. *Int. Display Research Conf.*, 358–361.

4 Yoshikawa, K., Kanazawa, Y., Wakitani, M., Shinoda, T. and Ohtsuka, A. (1992) full color AC plasma display with 256 gray scale, Japan Display '92, pp. 605–608.

5 Takashima, K., Nakayama, N., Shirouchi, Y., Iemori, T. and Yamamato, H. (1973) Surface discharge type plasma display panel. *SID Digest*, 76–77.

6 Dick, G.W. (1974) Single substrate AC plasma display. *SID Int. Symp., Dig.*, 124–125.

7 Shinoda, T. and Niinuma, A. (1984) Logically addressable surface discharge ac plasma display panels with a new write electrode. *SID Digest*, **84**, 172–175.

8 Dick, G.W. (1985) Three-electrode per pel AC plasma display panel, *Int. Display Research Conf.*, 45–50.

9 Shinoda, T., Wakitani, M., Nanto, T., Yoshikawa, K., Otsuka, A. and Hirose, T. (1993) Development of technologies in large-area color ac plasma displays. *SID Digest*, **93**, 161–164.

10 Shinoda, T., Wakitani, M., Nanto, T., Awaji, N. and Kanagu, S. (2000) Development panel structure for a high resolution 21-in.-dagonal color plasma display panel. *IEEE Trans. ED*, **47** (1), 77–81.

11 Komaki, T., Taniguchi, H. and Amemiya, K. (1999) High luminance AC-PDPs with waffle-structured barrier ribs, IDW '99. pp. 587–590.

12 Toyoda, O., Kosaka, T., Namiki, F., Tokai, A., Inoue, H. and Betsui, K. (1999) A high-performance delta arrangement cell with meander barrier ribs. *IDW*, **99**, 599–602.

13 Shinoda, T., Wakitani, M. and Yoshikawa, K. (1998) High level gray scale for AC plasma display panels using address-display period-separated sub-field method. *Trans. IEICE*, **J81-C-2** (3), 349–355. (in Japanese).

14 Weber, L.F. (1998) Plasma display device challenges, Asia Display '98. pp. 15–27.

15 Sakita, K., Takayama, K., Awamoto, K. and Hashimoto, Y. (2001) High-speed address driving waveform analysis using wall voltage transfer function for three terminals and vt close curve in three-electrode surface-discharge AC-PDPs, SID'01, **32**, pp. 1022–1025.

16 Hashimoto, Y., Seo, Y., Toyoda, O., Betsui, K., Kosaka, T. and Namiki, F. (2001) High-luminance and highly luminous-efficient AC-PDP with delta cell structure, SID'01, **32**, pp. 1328–1331.

17 Kanazawa, Y., Ueda, T., Kuroki, S., Kariya, K. and Hirose, T. (1999) High-resolution interlaced addressing for plasma displays, SID'99 Digest, pp. 154–157.

18 Kishi, T., Sakamoto, T., Tomio, S., Kariya, K. and Hirose, T. (2001) A new driving technology for PDPs with cost effective sustain circuit, SID'01, **32**, pp. 1236–1239.

22
Characteristics of PDP Plasmas
H. Ikegami

Pictures on a plasma display panel (PDP) consist of millions of light spots, each of which is produced by a tiny barrier discharge between transparent electrodes fabricated in a lattice-like structure. The plasma facing surface of the electrode is coated by a non-conductive material, MgO. Characteristics of such PDP plasmas are studied as to their spatial structure, power consumption, voltage–current characteristics, and basic plasma parameters.

22.1
Introduction

The PDP has a history of 40 years of development before reaching today's commercial use for flat color TV. The PDP plasma carries not only fundamental issues of discharge physics, but also exhibits various interesting features of atmospheric, barrier, glow discharges.

Taking an XGA-grade color PDP for a flat 42-inch (92 cm × 52 cm) TV as an example, the PDP consists of $1024 \times 768 \times 3$ (red, green, blue) $\approx 2.4 \times 10^6$ discharge cells, each of which has an opening of approximately 0.3×0.6 mm^2 on the front face. The lattice-like discharge structure is sandwiched between a front and rear glass plate with a separation of 0.15 mm as shown in Fig. 22.1. One discharge cell volume is approximately 0.027 mm^3.

For the discharge electrodes, indium tin oxide (ITO) of 0.25 mm width (denoted "Display Electrode" in Fig. 22.1) is plated in an array arrangement with a separation of 0.1 mm on the inner surface of the front glass plate, then coated with a dielectric layer, and covered by MgO as a protective layer, which is the innermost surface of the front glass panel to be exposed to PDP plasmas.

The inner surface of the rear glass plate is separated from the front glass by barrier ribs approximately 0.15 mm high. The array of address electrodes and barrier ribs is formed perpendicularly to the display electrode on the rear plate. Then the rear panel surface is coated by phosphors to produce (RGB) colors. The schematic of Fig. 22.1

Advanced Plasma Technology. Edited by Riccardo d'Agostino, Pietro Favia, Yoshinobu Kawai, Hideo Ikegami, Noriyoshi Sato, and Farzaneh Arefi-Khonsari
Copyright © 2008 WILEY-VCH Verlag GmbH & Co. KGaA, Weinheim
ISBN: 978-3-527-40591-6

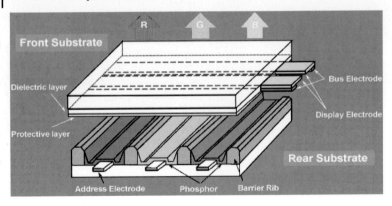

Fig. 22.1 Structure of plasma display panel (PDP).

shows a unit picture cell (pixel) composed of three discharge cells for the (RGB) color unit.

Discharges in each cell are so-called barrier discharges sustained by alternative voltage pulses applied between the ITO display electrodes, whose surface is coated by an insulating layer of MgO for protection against plasmas. Discharge gases are a mixture of noble gases, mostly He, with several percent of Xe under a pressure of approximately 600 torr, or higher. Color light spots of PDP are from those phosphors coated on the rear glass plate excited by 147 nm UV emission from Xe atoms in the He–Xe gas plasmas.

22.2
PDP Operation

Repetitive pulse discharges with microcurrent channels in the atmospheric barrier discharges are a key feature of the PDP plasma. A simplified model of the PDP operation is described in the following.

In order to address when/where to initiate a discharge, or to write in, a firing voltage V_f is applied between the address electrode and one of the display electrodes in the cell. The triggered plasmas build up a wall voltage V_w due to electron accumulation (*wall charge*) on the insulator surface of the anodic electrode. Consequently the effective voltage between the electrode drops by V_w and the discharge stops if $(V_f - V_w)$ gets too small to maintain the discharge. Once the wall charge has set, the charged-up anodic electrode works in the next reversed voltage pulse as a cathode, and the discharge can be sustained by the effective voltage $|V_s + V_w|$ between the display electrodes without any help of the address electrode. During this period, the other electrode gets the wall charge, and pulse discharges are thus alternatively sustained. The *sustain voltage* V_s is smaller than V_f, but $|V_s + V_w| \geq V_f$. It must be noted that the wall charge keeps on increasing after the discharge stops as long as the external positive pulse and the plasma remains in the cell. To quench the discharge, a shorter or a lower voltage pulse is sent so that poor wall charges are produced to give $|V_S + V_w| \leq V_f$ and the discharge discontinues.

The discharge current rises at the front edge of the imposed pulse voltage and reaches a maximum, then starts decreasing as the wall charge builds up. The discharge current ceases in 0.5 µs with a half-width of approximately 0.25 µs. As the closest spot to the counter-electrode on the anodic surface edge is occupied by the electrons, the discharge microspot moves along the anode surface looking for a virgin area, so the discharge pulse width, as well as the sustain voltage, depend very much on the shape and size of the electrode.

Now a rough estimation can be made of the electric power consumption with an XGA-grade 42-inch PDP operation. In one discharge cell of PDP, when alternative pulses of ±150 V with a pulse width of 3 µs are applied, each barrier discharge generates a peaked discharge current pulse of 300 µA with a half-width of approximately 0.25 µs at the front pulse edge. The energy dissipated by one discharge pulse is thus estimated to be 1×10^{-8} J.

With a PDP color TV to generate a gray scale of 256 ($=2^8$) steps, eight sub-fields are needed in one frame field of 1/60 Hz ($=16.7$ ms) duration. When the 256 gray scale steps are combined with three color cells of (RGB), approximately 17 million colors are generated.

Let us suppose an average of 500 discharge pulses within one frame field in every cell. Assuming a working spot rate of 40% in the 2.4×10^6 discharge spots in the whole screen, one can estimate the electric power consumption with this PDP to be approximately 300 W under a moderate operating condition. There will be additional power dissipation in the driving circuitry, accordingly the overall power consumption with the PDP will be estimated at above 300 W.

22.3
PDP Plasma Structure

Although the PDP discharge in each cell is a so-called barrier discharge driven by alternate pulses, its voltage–current characteristics indicate that each PDP discharge falls in the category of the normal glow discharge mode. As shown in Fig. 22.2 there are two regions of strong light intensity in the normal glow discharge: one is the negative glow and the other is the positive column. An interesting feature is that as the separation between the anode and the cathode decreases, the positive column disappears, but the negative glow region remains, which is considered to be the main body of the PDP plasma. With an applied voltage of 200 V or $E/p = 16.7$ V cm^{-1} torr^{-1}, the PDP discharge current density is 0.4 A cm^{-2}, which indicates that the PDP discharge belongs to the normal glow discharge mode.

It is shown [1] in Fig. 22.3 that by changing the anode–cathode separation from 3 mm down to 0.5 mm, eventually the negative glow remains without changing its size, although the positive column has disappeared. In this case the electrode is a square of 1 mm × 1 mm and the experiment is performed by repetitive pulse discharges with neon gas at 100 torr.

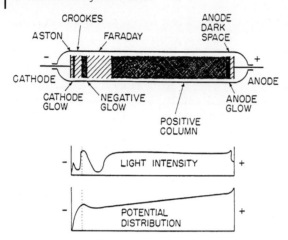

Fig. 22.2 Normal glow discharge plasma column.

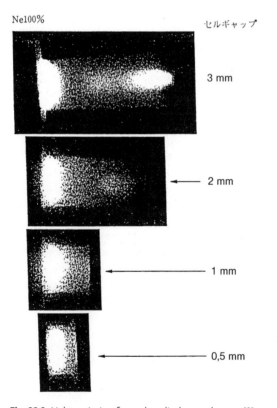

Fig. 22.3 Light emission from glow discharge plasmas [1].

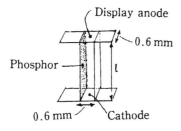

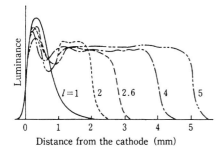

Fig. 22.4 Schematic of the experimental setup and the luminance along the plasma column vs. electrode separation [2,3].

The NHK group in Japan made extensive researches [2,3] concerning the negative glow with a variable electrode separation as a simulation of their DC PDP using a mixture gas of He and Xe (2%) at 100 torr with a fixed DC discharge current of 200 μA. As shown in Fig. 22.4, both electrodes are a square of 0.6 mm × 0.6 mm and their separation is variable from 5 mm to 1 mm. Plasma column is confined between two glass plates, one of which is coated by phosphors.

As the anode and cathode approach, the positive column goes away and the negative glow region of the glow discharge plasma remains without changing its size and luminance.

Since the AC PDP employs alternative pulse discharges between the electrodes placed on the same plane in parallel, the plasma build-up behavior will be more complicated. The discharge will start at the electrode edge separated by 0.1 mm, which is good for initiating the discharge, but too close to generate a negative glow. Because of the wall charge, the discharge spot on the anode backs away looking for the virgin area and the effective separation of the electrode will increase with increasing currents. During the period, the PDP discharge maintains as a negative glow.

22.4
Plasma Density and Electron Temperature

There are almost no measurements of plasma density and electron temperatures of PDP plasmas, because of the difficulty in that the plasma is sub-millimeter in size produced in a transient manner under nearly atmospheric pressure. For the ordinary probe technique, no dependable probe theory has been developed for atmospheric plasmas. Also, the PDP plasma is too small to avoid serious perturbations.

PDP plasma densities can be estimated with the use of the mobility data at atmospheric pressure [4] with the assumption that the PDP plasma is the negative glow plasma from where ions fall onto the cathode surface.

There, the helium ion currents to the cathode are given by the equation

$$i = Sne\mu E$$

where i is the discharge current of $300\,\mu$A, S is the cathodic area ($0.25\,\text{mm} \times 0.33\,\text{mm}$), e is the electric charge, μ is the mobility of helium ions in helium gas at 1 atm pressure (given by $26 \times 10^{-4}\,\text{m}^2\,\text{V}^{-1}\,\text{s}^{-1}$) [4], and E is the electric field in the cathode fall given by $2 \times 10^6\,\text{V}\,\text{m}^{-1}$. Putting these numbers into the above equation, we will get $n = 4 \times 10^{12}\,\text{cm}^{-3}$, which will give a PDP plasma density of a reasonable order of magnitude.

For the electron temperature, assuming that the electron temperature in the discharge plasma is nearly uniform, we will use the Schottky–Tonks–Langmuir theory for a positive column with the use of the universal curve [5] for calculating the electron temperatures Te/u_i (KeV^{-1}) as a function of cpR (torr cm) as shown in Fig. 22.5.

The constant c specific to helium gas is given [5] by $c = 3.9 \times 10^{-3}$ and the ionization potential of the helium atom $u_i = 24.56\,\text{eV}$. For the helium discharge of PDP plasma,

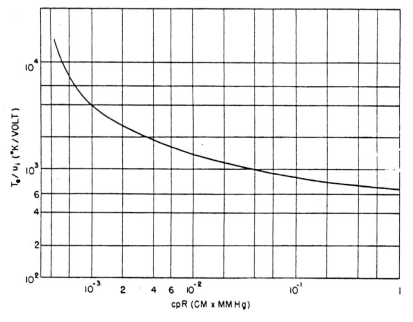

Fig. 22.5 Universal curve for calculating electron temperatures in a positive column as a function of *cpR*.

$p = 600 \, \text{torr}, \; R^{-1} = (0.025)^{-1} + (0.003)^{-1} = (0.0136)^{-1} \, \text{cm}^{-1}, \; cpR = 3.2 \times 10^{-2} \, \text{torr cm}.$
With this value of cpR, the electron temperature is given from the curve in Fig. 22.5 as
$T_e/u_i = 10^3 \, \text{K eV}^{-1}$ or $T_e = 2.1 \, \text{eV}$.

Although the obtained electron temperature is a reasonable for the atmospheric glow discharge plasma, it must be noted that the sustaining condition of the negative glow is quite different from that of the positive column which the Schottky–Tonks–Langmuir theory handles, because the negative glow plasma is maintained by the electrons from the cathode gaining their energy through the cathode fall.

22.5
Remarks

The characteristics of PDP plasmas have been discussed. One major issue for today's PDP consists in the improvement of power consumption, which will require combined efforts regarding (1) optimized gas mixture with its component and operating pressure, (2) electrode and cell shaping, and (3) plasma density and electron temperature control for higher efficiency UV emission.

References

1 IEEJ Technical Report 688, p. 14, Fig. 2.20 (in Japanese).

2 Matsuzaki, and Kamegawa. (1979) *NHK Technology Research*, **31**, Ser. No. 159.

3 NHK Display Devices research group, (1983) *Advances in Image Pickup and Display*, **6**, 112.

4 Brown, S.C. (1961) *Basic Data of Plasma Physics*, MIT Press, Cambridge, MA, 78.

5 von Engel, A. (1965) *Ionized Gases*, Clarendon, Oxford.

23
Recent Progress in Plasma Spray Processing
M. Kambara, H. Huang, and T. Yoshida

23.1
Introduction

The origin of "thermal spray processing" may go back to the coating method of the early 1900s, in which various powders were simply sprayed into combustion flames. The development of "plasma spray processing" was triggered, 50 years later, by the invention of the DC plasma torch [1]. Since then, this process has evolved into a variety of spray technologies, practically intending complete dissolution and vaporization of the injected materials, and is nowadays recognized as one of the most important and basic coating technologies. In the future, a real breakthrough may be achieved by a radical advancement of each conventional spray technique and also by a transverse coupling of various techniques developed for specific engineering applications. In this context, we will give an overview on the current status of "plasma spray processing" from its key process element, and the future directions of this technology are foreseen as coating, powder synthesis and environment-friendly technologies.

23.2
Key Elements in Thermal Plasma Spray Technology

In thermal spray processing, raw materials are sprayed either in the powder, solution or vapor form and subjected to physical, chemical, or both reactions in the plasma. According to the different interactions of sprayed materials with plasma, this technique can be classified into three elemental processes as summarized in Fig. 23.1: (i) plasma spray melting in which rather coarse powders are injected in the plasma and melted completely to form droplets; (ii) plasma spray synthesis in which various vapor mixtures are created via decomposition and/or chemical reactions of gas or liquid raw materials; and (iii) plasma spray evaporation in which fine powders are injected and decomposed completely to form vapor mixture and/or atom clusters [2,3]. In any elemental processing, the fundamental role of the plasma is to provide an "extreme high temperature" environment and its "flow". The thrust of "plasma spray processing" is therefore an effective use of "high-temperature gas

Advanced Plasma Technology. Edited by Riccardo d'Agostino, Pietro Favia, Yoshinobu Kawai, Hideo Ikegami, Noriyoshi Sato, and Farzaneh Arefi-Khonsari
Copyright © 2008 WILEY-VCH Verlag GmbH & Co. KGaA, Weinheim
ISBN: 978-3-527-40591-6

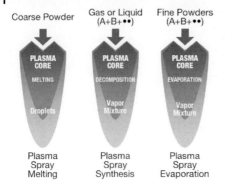

Fig. 23.1 Schematic of the elemental processes in various
plasma spraying technologies.

flow" that is not achievable by combustion flames. In some cases, the discharge
generation part itself is utilized. However, the "flow" should be controlled if the
unique characteristics of the thermal plasma are to be used effectively. Further
potential of thermal plasma will be exerted not simply from the role as a heating
medium but from the use of the physical and chemical phenomena during heating/
cooling in the plasma or plasma/substrate boundary. In this respect, the fundamental
approach to thermal plasma spray processing is essentially different from that
generally considered in low-pressure plasma processing. It is, however, difficult to
control the plasma flow magnetically except the electrically conducting path in the
plasma. Alternatively, the plasma could be controlled based on fluid dynamics. In
addition, power units, plasma torch and plasma flame all have to be taken as important
"cogs" in the "plasma system" as they are cooperatively interacting with each other
from the plasma control point of view. In other words, individual components of the
plasma system have to be developed considering the overall performance as the system
as a whole, which is particularly critical when scaling up the plasma facilities.

In application of thermal plasma to industry, another attention has to be paid to the
material amounts requested by industry, with respect to the capacity that specific
plasma technology can accommodate. As will be described in the following section,
plasma powder spraying is regarded as a core technology in the spray processing,
because this process is supposed to treat $1\,\mathrm{kg\,h^{-1}}$ at most. Therefore, a treatment of
volatile organic compounds (VOCs) at a rate as fast as $100\,\mathrm{kg\,h^{-1}}$ is regarded as an
innovative progress of this technology [4]. This in turn indicates that a 100 kW class
thermal plasma system can treat powder, liquid or vapor at 1, 10, and $100\,\mathrm{kg\,h^{-1}}$,
respectively, if it is of economical worth.

23.3
Thermal Plasma Spraying for Coating Technologies

Incorporation of additional functions of coatings to bulk base materials is becoming
more important in view of the current technological trends described by nanotech-

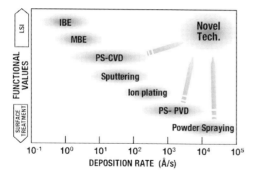

Fig. 23.2 Various coating technologies classified with their functional values and processing rates [3].

nology. Figure 23.2 summarizes typical thin-film technologies as a function of production speed and values achieved. The above elemental steps (i)–(iii) in the plasma spraying can be redefined, by the interaction with substrate or formation of precursors in the plasma/substrate boundary as: (i) plasma powder spraying, (PPS) (ii) plasma spray chemical vapor deposition (PS-CVD), and (iii) plasma spray physical vapor deposition (spray PVD), respectively. Plasma powder spraying can be classified as a leading coating technique due to its extremely fast production rate and affordability. In contrast, CVD and evaporation gives films with added functions despite slow deposition rates. The future technology, therefore, may be the ones which possess both high-rate deposition capability and high functional values, with "viable", "sustainable", and "affordable" aspects. Taking account of these, the arrows shown in Fig. 23.2 will be the directions that the current techniques should evolve in as the next generation of film/coating technologies.

23.3.1
Plasma Powder Spraying

"Affordability" or "cost-efficiency" has made plasma powder spraying (PPS) an attractive coating technique, together with its tailor-made coatings to cope with the problems of wear, corrosion, and thermal degradation arising in various structural applications. The major advantage of the use of plasma over combustion flame is its high temperature, reaching up to 15 000 K, which enables one to melt and deposit almost all refractory materials. In recent years, plasma powder spraying has been further applied to broader areas such as healthcare and semiconductor applications.

According to the scientific and engineering requirements, the research of this processing may be classified reasonably into three stages in the following order.

- First stage. Experimental and theoretical studies to analyze a series of processes that includes thermal history and trajectory of single particles injected, heated and accelerated in a plasma jet, and also their collision, deformation and solidification on the substrate.

- Second stage. Systematic study of plasma/surface interactions during formation of the sprayed coating and effects of various spraying parameters on the structure, composition, internal stress, and adhesive strength of individual layers as a result of the above-mentioned individual phenomena.
- Third stage. Evaluation and modification of the overall spraying process, from the effectiveness and coherency point of views, designed specifically for advanced application such as composite materials including FRM, multi-layered material and graded materials.

However, the reality is that most of the researches were shifted to the second and third stages with no notable research results gained in the first stage over the past 30 years. In addition, the research in the last two stages is only with the limited parameters controllable in the black box of the spraying apparatus. Even so, steady advancements have been somehow accomplished from the engineering point of view. There are essentially considerable "tough" distributions in controlling this technique, such as velocity distribution of the injected powders, non-uniform flow field of the plasma, and the corresponding inhomogeneous thermal history of the particles. Therefore, the basics of spraying have been somewhat put aside with no significant attentions paid. However, the coating is a result of convolution of such distributions, and the lack of basic and detailed insight in the first stage hinders critically the real breakthrough of this technology. Having realized the situation, efforts have been made to understanding of the basic phenomena of individual droplet dynamics with an aim to the control of distributions by both measurement and physical modeling approaches [5].

Figure 23.3 illustrates the range of residence times and impact velocities of droplets in different types of plasmas. It clearly shows that plasma spraying parameters vary strongly with the facilities used to generate plasmas. In the case of DC plasma, however, the intrinsic properties of DC discharge fluctuate periodically by nature of the order of milliseconds and this time scale is of the same order as the time needed to heat and melt the particles in the plasma jet. This means that the control of thermal history of particles is essentially difficult even if distributions related to the injected powders could be controlled. In contrast, radiofrequency (RF) plasma allows narrowing the undesired distributions in a rather large plasma volume of several tens of millimeters in diameter. Owing to its long residence time of 5–10 ms, it is possible to spray and melt larger particles and thereby achieve high material yields of 80%. However, once the velocity reduces to as slow as $20\,\mathrm{m\,s^{-1}}$, the particle tends to resolidify before it impacts on the substrate. In this respect, the "hybrid plasma" has a potential in that the plasma velocity is controlled in the range of $40–70\,\mathrm{m\,s^{-1}}$ by adjusting the power input of the central DC plasma jet, which permits the spraying of most refractory ceramics under atmospheric pressures.

Figure 23.4 shows the distributions of the velocity and size of the droplets in different plasma spraying techniques. The latter is defined as the range of particle sizes at which the powder can be fully melted without evaporation in the corresponding

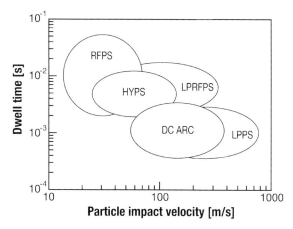

Fig. 23.3 Dwell times and velocities of droplet in different types of plasmas (RFPS: radiofrequency plasma spraying; HYPS: hybrid plasma spraying; APS: atmospheric plasma spraying; LPRFPS: low-pressure radiofrequency plasma spraying; LPPS: low-pressure plasma spraying).

plasma. Various plasma spraying processes are roughly divided into "large powder spraying at low velocity" and "small powder spraying at high velocity" regimes, which may correspond to the new and old techniques, respectively. Even so, an interesting example is that the adhesive strength between solidified Ti droplets (splats) and stainless substrate was improved from 40 to 200 MPa when the size of the feedstock Ti powder increased from 30 to 120 μm [3].

Many detailed analyses are still required to identify the effects of velocity and size on deformation and solidification of droplets. Nevertheless, the droplets are usually

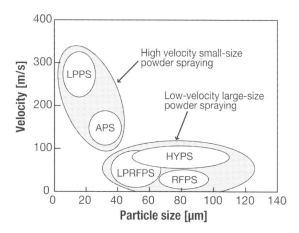

Fig. 23.4 Typical ranges of powder sizes and velocities achieved in various spraying techniques (abbreviations as in Fig. 23.3).

observed to deform in a similar manner, i.e. the flattening degree, the ratio of the splat diameter to the diameter of a droplet, of around 3–4, regardless of different droplet sizes. Therefore, it can be said that both parameters affect solidification more significantly than they do the deformation process. In particular, since the thermal resistance between the impinging droplet and substrate is affected strongly by velocity, size and surface condition of the substrate, the microstructure of the splat can vary from unidirectional structure to equiaxed grains, depending on the heat transfer characterized either by conduction towards the substrate or by radiation from droplet surface. This is further subjected to the status of the droplet, including undercooled or superheated, through the possible impact-induced nucleation event, and additionally to the partitioning of elements, i.e. formation of alloy phase, in the case of spraying alloys. In the case of deposition on materials with small thermal conductivity like ceramics, on the other hand, the heat extraction process is altered during deposition with an increase in the coating thickness, due to the reduced heat transfer toward the bottom solidified layers, which consequently affects the solidification and deformation behavior of the subsequent droplet.

23.3.2
Plasma Spray CVD

In thermal plasma chemical vapor deposition (PS-CVD), gaseous or liquid reactants are injected and completely dissociated to atomic elements as precursors. In most cases, as the substrate temperature is significantly lower than that of the tail flame, the rate-limiting step of deposition is considered to be in the plasma/substrate boundary. Therefore, unlike surface reaction observed in conventional CVD, the prominent feature of PS-CVD is transport of high radical flux. For some cases, cluster deposition is expected as an additional important feature, especially when extremely high-rate deposition is attained [6]. For example, SiC coatings have been successfully deposited by PS-CVD from $SiCl_4 + CH_4$ gas mixture at a rate of $1\,mm\,h^{-1}$ with a deposition efficiency of over 70%, due possibly to CH_x radical formation and also to the thermophoretic effect in the plasma/substrate boundary [7]. By using hybrid plasma, microcrystalline Si with a defect density of $7.2 \times 10^{16}\,cm^{-3}$ was produced from $SiH_4 + Ar$ plasma at a ultrafast rate of over $1000\,nm\,s^{-1}$, which is about 2000 times faster than that achieved by conventional CVD [8]. Computer simulations have suggested the effective roles of clusters with a size of $\sim$1 nm and of thermophoresis in the plasma/substrate boundary to achieve such ultrafast depositions [9]. Furthermore, the deposition of complex oxides such as optical materials was also attained by this technique, supporting the possibility of high-rate deposition assisted by hot cluster formations [10].

In the case of diamond synthesis, it is plausible that molecules such as CH_3 and C_2H_x are the precursors and the deposition rate is controlled by extraction of hydrogen near the surface. As a result, despite the deposition efficiency of only several percent, high deposition rates faster than $10\,\mu m\,h^{-1}$ are realized by the high flux of atomic hydrogen [11]. Commercially, large-area diamond films with a diameter of 6 inches and thickness of 1–2 mm are already produced by a 300 kW

DC plasma system. Another interesting example is that thick (over 20 μm) cBN films were obtained by using BF_3–Ar–H_2 gas mixture plasma under 50 torr with the PS-CVD method, although the thickness of the cBN films hardly reaches 1 μm by conventional low-pressure (several millitorr) deposition [12].

23.3.3
Plasma Spray PVD

In the thermal plasma physical vapor deposition (PS-PVD) method, atoms or clusters are utilized as precursors through evaporation of injected fine powders. Therefore, the most important feature in this method is the ability to vaporize powders completely (generally as small as 10 μm in diameter) and continuously in the plasma. Along with the merits of the analogous "flash evaporation" technique that facilitates deposition of compounds with totally different vapor pressures, the advantages of PS-PVD processing are its flexibility and the minimum formation of byproducts. In addition, higher vapor fluxes can be achieved by an increase in the powders to be injected, with increasing the plasma power inputs. However, the sizes of the powders have to be carefully selected according to their residence time in the plasma, as rather large particles may remain in the liquid phase without being evaporated [13]. Furthermore, a high rate of carrier gas is in general required to inject smoothly finer particles, intending complete evaporation within the plasma flow. This gas velocity, however, needs to be carefully controlled since the plasma is markedly disturbed and becomes unstable by extremely fast carrier gas velocities, deteriorating inevitably the quality of the coatings [14].

This method has been employed to deposit a variety of materials, including 123 oxide superconductors. In particular, an interesting result observed is that hot cluster deposition facilitates the epitaxial growth of the 123 phase although such peritectic phase formation from incongruent melt is in general difficult to control in the case of the powder spraying technique [15]. Similarly, PS-PVD has successfully demonstrated the deposition of thick SiC films at rates well above 300 nm s^{-1}, using ultrafine SiC powder as a starting material. Such SiC has shown enhanced thermoelectric properties by N-doping [16]. This process is often compared to the laser ablation process. European makers expect that PS-PVD will become advantageous over laser ablation due to its "large-area" and "ultrafast" characteristics, once ultrafine powders and precisely controllable powder feeding system both become conveniently accessible.

23.3.4
Thermal Barrier Coatings

The development of novel thermal barrier coatings (TBCs) by twin hybrid plasma spraying is one of the best examples of integrating all the potentials of different plasma spraying techniques, that is, composite coating production with the PPS, PS-PVD and PS-CVD processes.

The use of TBCs together with the development of superalloys and cooling systems enable continuous improvements in the efficiency and durability of gas turbine engines. Practically, a thickness of over several hundreds of micrometers is required for TBCs, so that deposition rate and cost-efficiency have become the major criteria in choosing the proper manufacturing method. Over the last decade, various processes have been adopted for TBC deposition and are divided into two main types: one is a vapor deposition technique, such as electron-beam PVD, plasma-enhanced CVD and laser CVD; the other is molten particle deposition typically represented by atmospheric plasma powder spraying. In short, TBCs made by the former are characterized by columnar microstructures while splat structures are commonly obtained by the latter. Due to this structural difference, the thermal and mechanical properties of such TBCs vary considerably. The vapor-deposited TBCs have the merits of excellent strain tolerance, excellent surface finish, good erosion resistance, and relatively long lifetime. However, these merits are gained at the expense of rather high thermal conductivity and high production cost. On the other hand, molten particle-deposited TBCs exhibit the lowest thermal conductivity, high processing efficiency, and low cost, but are less durable. Thus, the coatings with all the above advantages as added values will be the next generation of TBCs. In this respect, thermal plasma spraying is an attractive, promising, and natural choice for novel coating technologies as it inherently possesses the potential to achieve all the characteristics of both molten particle spraying and vapor deposition.

In a hybrid plasma, the large recirculation eddy present in conventional RF plasmas is extinguished by the high-velocity channel of the arc jet. The central channel of the plasma is also heated by the high-temperature jet, and the DC plasma jet can act as a lasting igniter and a strong propulsor to enable the axial injection of powder, which may improve the uniformity of the powder thermal history compared to the radial feeding method. A more homogeneous plasma flow than that of conventional RF plasma is another merit of the hybrid plasma, as is found in the temperature and axial velocity fields in a 100 kW hybrid plasma estimated by computer simulation shown in Fig. 23.5. Such high power capacity of the twin hybrid plasma spraying (THPS) system enables the full melting and/or evaporation of yttria-stabilized zirconia (YSZ) with high melting temperature and with unique chemical and physical properties for TBC topcoats. As powders are used for both PPS and PS-PVD processes, composite coatings can be formed by one-torch deposition due to inevitable distributions of powder size, velocity, and plasma conditions, i.e. the combination of melting and evaporation process of the injected powders. However, real engineered composites should be obtained by a precise and active control of the film structures, and this was made possible with a novel 300 kW THPS system shown in Fig. 23.6(a). This system is equipped with two hybrid plasma torches in a chamber, and substrates on a rotary substrate holder are exposed by turns to the two plasma flames which are capable of PS-PVD, PS-CVD, and PPS processes. That is, an alternate layering of the intended structures with any materials can be realized by assigning different processes separately to each of the two hybrid plasma torches. Figure 23.6(b) shows a photograph of the as-deposited YSZ coating by this cyclic deposition technique. It is seen that YSZ coating was performed on the 8 substrates simultaneously, exhibiting uniform top appearance over the 5×5 cm^2 substrate.

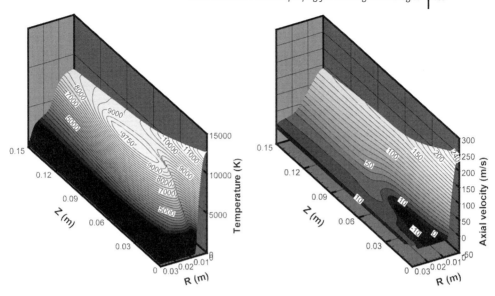

Fig. 23.5 Temperature and axial velocity field of a 100 kW hybrid plasma. The central high temperature and high velocity channel is caused by the DC jet.

The critical parameters to achieve a specific structure by the use of a single-torch system are at first identified. In the case of PPS, empirically, a better adhesion strength is achieved by rather large powders if full melting of the constituent is maintained. As shown in Fig. 23.7, both computer simulation and experiments on large YSZ (with size of 63–88 μm) powder spraying confirms that a flattening degree as large as 5 was reached when the droplet size is near 100 μm in a 100 kW hybrid plasma [17]. The good agreements between the numeric simulation results and the single splat deformation experiments in turn confirm that the viscosity and thermal resistance values obtained from *in situ* measurement are reasonable [18].

On the other hand, in the PS-PVD process, the powder feeding rate is found to be the first important parameter that affects the evaporation degree of powders. Figure 23.8 shows the variation of the deposition rate and coating structures with the powder feeding rate and distance from the torch [19]. At a relatively high powder feeding rate of 4 g min^{-1}, the coating was composed mostly of splats as a result of complete melting of the powders and deposition at the liquid state. In contrast, when the powder feeding rate was lowered to 1 g min^{-1}, only the aggregation of nanoparticles are observed, which is the typical structure of vapor deposition by plasma spray PVD. This variation in microstructure demonstrates clearly that the increase in the powder feeding rates increases the volume of the solids residing in the plasma which decreases readily the plasma temperature so as to be not high enough to evaporate the powders. On this basis, when the powder feeding rate was controlled to 2 g min^{-1}, a peculiar layered composite structure was achieved, in which the sprayed splats are enweaved between the PVD structure layers. Such tendency in the microstructure

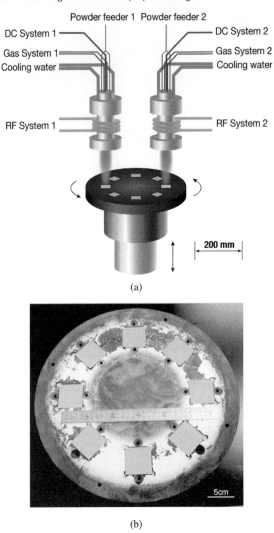

Powder feeder 1 Powder feeder 2

(a)

(b)

Fig. 23.6 (a) Schematic of a 300 kW twin hybrid plasma spray system and (b) photograph of the as-deposited YSZ coatings on a rotating holder.

was observed at any distance from the plasma torch, although the deposition rates decrease on increasing the distance. A more interesting result is that the PVD structure is achieved at a quarter of the deposition rate for the sprayed films, when reducing the feeding rate to one-fourth of that for the spraying. This is clear evidence of quite a good deposition efficiency of PS-PVD.

A denser microstructure was also achieved in a thick PS-PVD YSZ coating with a thickness of around 550 μm deposited on water-cooled stationary substrate at ultrahigh speed of over 150 μm min^{-1}. It is interesting to note that interlaced

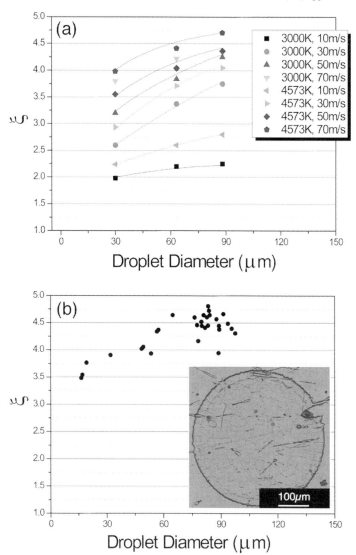

Fig. 23.7 Effect of droplet diameter on the flattening degree (substrate temperature = 723 K): (a) numerical simulation and (b) experimental results [22].

structures were revealed in such dense YSZ coating after etching with 50% HF for 10 min. From transmission electron microscopy (TEM) observations, a similar interlaced twin variants structure was observed clearly and identified as t' phase. More detailed analyses are required to understand the formation mechanism of such structures [20]. Nevertheless, an important result is that the coating with interlaced structure has shown much higher reflectivities against the infrared than the powder

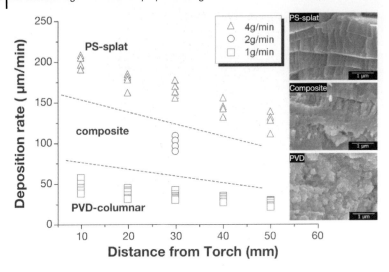

Fig. 23.8 Variations of deposition rates and the coating structures with the powder feeding rates at different torch to substrate distance [22].

sprayed coating, due probably to the layering structure at several micrometer intervals, parallel to the coating surface [21]. This suggests that the coating can reflect the infrared light effectively and therefore suppress radiation heating of underlying base materials, particularly when exposed at elevated temperatures. This is an additional advantage of YSZ as TBCs and also of the use of PS-PVD processing.

Based on the knowledge of PS-PVD by single-torch spraying, combinations of these two processes were performed by twin-torch deposition. In twin-torch depositions, Al_2O_3 and YSZ were first injected for the sake of a clear identification of layers by large contrast in scanning electron microscopy (SEM) observation, and then the depositions for YSZ/YSZ composite coating were performed afterwards. From the one-turn thickness of the PS-PVD YSZ layer, which is evident by the Al_2O_3 and YSZ layering, the thickness of PS-PVD YSZ measured is from 100 to 200 nm, which corresponds to the ultrafast deposition rate of 30–60 μm min^{-1} (10 times faster than that by EB-PVD) by considering the residence time of the substrate in the plasma flame of each turn. This growth rate agrees well with the single-torch deposition experiments. Composite coating with the other combination of PS-PVD Al_2O_3 and PPS YSZ was also produced similarly. Figure 23.9(a) shows the well-flattened YSZ splats (white) in the vapor-deposited Al_2O_3 matrix (black) [22]. Understanding the individual droplet deformation and solidification in the middle of the coating is as important for structure control purposes as the first splat formation behavior at the splat/substrate interface mainly for adhesion control purposes. This technique facilitates the direct observation of such droplet deformation in the coating with ease. The relationship between the thickness of Al_2O_3 PVD layer and powder feeding rate of Al_2O_3 was measured as shown in Fig. 23.9(b). Although a rather large distribution in the PVD layer thickness is observed at high feeding rate, the average

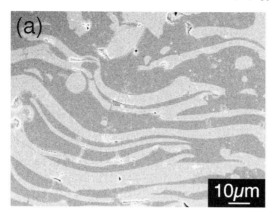

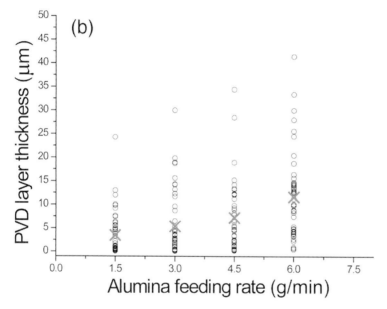

Fig. 23.9 (a) FE-SEM micrograph of PS-PVD Al_2O_3 and PPS YSZ composite coating and (b) the effect of powder feeding rate of Al_2O_3 on the thickness of PVD Al_2O_3 layers.

values of the thicknesses, indicated by the crosses, increase rather linearly with the feeding rate, which is readily explained by a small amount of PPS splats at higher Al_2O_3 powder feeding rate for PVD structure. Nevertheless, this is direct evidence that the PS-PVD layer thickness is controlled by the powder feeding rate. Also due to the discontinuous nature of the sprayed layer, the layer periodicity in the deposited coating is not significantly affected by the substrate rotation speed. Figure 23.10 shows the peculiar layered structure attained in the YSZ/YSZ composite coating deposited by small-particle PPS and PS-PVD. As is described earlier, the deformation

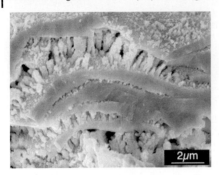

Fig. 23.10 FE-SEM image of the cross-section of layered YSZ coating deposited by PPS and PS-PVD using twin torch deposition [22].

and solidification of the droplet must be altered during deposition of the thermally low conducting materials due to the suppressed heat conduction toward the previously solidified underlying droplets. This effect must be pronounced for YSZ, resulting in quite unique composite structures and also in the large columnar grains within the splat. This altered heat transfer was confirmed by the decreasing tendency of the temperature monitored at the bottom of the substrate during deposition despite fixed plasma conditions. Such porous YSZ composite was deposited at high rate of $>3\,mm\,h^{-1}$ and achieved significantly reduced thermal conductivities of $0.7\,W\,m^{-1}\,K^{-1}$.

23.4
Thermal Plasma Spraying for Powder Metallurgical Engineering

23.4.1
Thermal Plasma Spheroidization

This process utilizes the same elemental process that is observed in thermal plasma powder spraying. That is, the injected and melted powders form into spherical droplets due to surface tension and retain this shape after solidification. Powders with fused, crashed or agglomerated shape can be used as the feedstock. Rods or wires can also be used, which are atomized in the plasma and form spherical droplets similarly. The only requirement is that the material should have a melting point well below its vaporization or decomposition temperature. A residence time of several milliseconds is usually needed, depending on the thermodynamic and mechanical properties of the source powders. Therefore, RF plasma is generally adopted for its relatively low plasma speed and no contaminations from electrodes. This process may be used simply to produce a spherical particle and also to contribute to smooth powder feeding for a variety of applications such as photocopying. It can also be applied to purification of the powder by physical or chemical processes. The former is achieved by evaporation of impurity elements that have high vapor pressure and the latter is normally by reactions at the droplet surface. Recently, this process has been employed in the semiconductor industry as the spherical Si preparation method.

23.4.2
Plasma Spray CVD

Plasma spray CVD is used for ultrafine powder synthesis with gaseous or liquid sources as is used in the conventional thermal CVD and metalorganic CVD. This process is categorized as chemical vapor deposition in terms of the overall product mass balance during chemical reactions. However, the elemental reactions involved in this process are far different from other CVD processes. That is, in PS-CVD, the atomized feedstock is decomposed to atoms or ions in the plasma, and subject to the non-equilibrium chemical reactions at the plasma/environment gas boundary. As a result, it is usually difficult to attain the desired single-phase material if the reactive gases are fed into the plasma all together. For example, in the case of ultrafine Si_3N_4 powder synthesis via $3SiCl_4 + 4NH_3 \rightarrow Si_3N_4 + 12HCl$ reaction, the $SiCl_4$ and NH_3 molecules are fully dissociated into atoms in the plasma and N atoms tend to form N_2 molecules around 5000 K. As a result, the actual reactants are mainly Si, N_2, H, and Cl in the boundary layer. However, as the N_2 molecule is very stable and less reactive, the efficiency of the nitriding process is limited to about 10%. As an effective counter-measure, injection of the additional NH_3 gas into the plasma tail flame helps to generate active NH_2 or NH radicals and thereby improve the reaction efficiency. For other materials systems such as oxides and carbides, such problems are not so evident and the synthesis process becomes much simpler. In fact, the latter was commercialized in the 1970s and the laboratory research including the nitrides fine powder synthesis was completed by the 1980s [23].

23.4.3
Plasma Spray PVD

Similar to spray -CVD, spray -PVD was originally developed for synthesis of ultrafine powders. Among important parameters such as plasma power input and the intrinsic powder properties, the size of the powder is one of the most critical parameters in this process. In order to fully evaporate the injected powders within the residence time in RF plasma, the size of the powder should be generally in the range 1–10 μm. Conversely, if an appropriate sized powder is prepared, evaporation of large amounts of powders is possible at rates around $1\,kg\,h^{-1}$ with a 100 kW RF plasma.

In the PS-PVD process, particles are formed effectively at the tail flame with no byproduct formation. This adds another advantage to the PS-PVD as a suitable processing route to obtain high-purity, single-phase and ultrafine particles. Especially, with a precise control of the cooling process, it is also possible to prepare complex ultrafine particles composed of high melting temperature and non-equilibrium phases even if different powders are atomized and mixed together in the plasma. Moreover, in the case of materials with largely different vapor pressures, peculiar fine particles can be synthesized, e.g. high-melting metal ultrafine particles with low-melting metal surface or semiconducting particles with metal embedding support. Other special examples are the synthesis of fullerenes and platelet carbon fine particles by injection of soot in the plasma [24]. Furthermore, the "reactive plasma spray PVD", incorporating chemical reactions into the ultrafine particle formation, is also promising for synthesis of oxides, carbides, and nitride fine particles [25]. However, the powder feeding system technically restricts the potential of this method, that is, vapor concentration fluctuates with the feeding rate which is not yet equipped with the required long-time stability at present.

23.5
Thermal Plasma Spraying for Waste Treatments

Over the last 10 years, there has been no marked increase in the total volume of wastes [26]. However, general municipal wastes have increased and the ratio of incineration is reaching 80% of all the waste treatments in the last few years, which makes detoxification of fly ashes including dioxins an imperative task to tackle urgently. In addition, due to the limited area for landfilling, reduction of volume and recovery of byproducts are encouraged and promoted. The countermeasures are adsorption/ collector filters, combustion methods, electron beam irradiations, and various plasma treatments including low-energy waste treatments by non-thermal plasma spraying are also proposed for VOCs and are already introduced for practical use [27–31].

The extreme high-temperature environment attained by thermal plasma is also an advantage over other conventionally used heating techniques, as this enables complete melting and/or evaporation without being limited by disposal of materials. Thermal plasmas are therefore used for melt-reduction in volume of fly ashes and low-level radioactive wastes and also for decomposition of highly concentrated VOCs and polychlorinated biphenyls (PCBs) [32–34]. In general, DC plasmas are employed at megawatt level outputs for such treatments and the current technical issues in this process are to improve fusion efficiency and to prolong the lifetime of plasma torches [33]. The engineering countermeasure is applied by rotating torch system to suppress local heat-up and evaporation thereby maintaining uniform melting of the waste at around its melting temperature. Such high enthalpy can also accept large amounts of waste injection, i.e. an injection of liquid and solid wastes as well as gaseous materials. Another approach is that for chlorofluorocarbons (CFCs) decomposition with RF plasma torches, in which VOCs can be transformed into plasma at a rate of $1 \, \mathrm{kg \, min^{-1}}$ with $500 \, \mathrm{L \, min^{-1}}$ vapor moisture and reduced effectively below the detection limit of dioxins [4].

Despite these technical developments, the critical problem arising in waste treatments is usually the operation cost. In this aspect, at present, one should admit that plasma spray processing is not advantageous compared to other competing techniques. Even so, this can become a valuable process if "extreme high temperature" at "atmospheric operation", available only with thermal plasma, are to be incorporated effectively in the treatment. These may be the simultaneous recovery of reusable elements via evaporation separation during detoxification of the melt fly ashes concentrated with volatile heavy metal elements, and also the complete decomposition of highly concentrated VOCs. Furthermore, other functionalized treatments are reductional segregation with hydrogen plasmas, high-rate oxidation with oxygen plasmas, and a combination of higher power outputs with RF plasma with other long-lasting technologies [34].

From a legislation perspective, the operational cost of such waste treatments can vary with the effluent control of the byproducts such as dioxin. That is, as waste control becomes more tightened with the increased concerns of society, the thermal plasma treatments could become more advantageous in terms of both waste volumes and materials to be treated. "Detoxification" and "volume reduction" are a global concerns, although there is a slight difference in the ratio of incineration/landfilling of wastes between countries [26]. Therefore, the development of waste treatment systems with thermal plasma may expand the market of waste treatments and reduce consequently the operational costs, thereby promoting the reduction in hazardous substances globally.

23.6
Concluding Remarks and Prospects

Various aspects of plasma spray processing have been briefly reviewed as a next-generation versatile technology. Many of industries working with the conventional thermal plasmas are requiring at least one order of magnitude increase in the throughput of any product. To meet this, it is necessary to establish plasma systems with at least 1 MW level output powers as a basic standard. In fact, plasma heating generators with a maximum output power of megawatt level (Japan) and of several tens of megawatts (South Africa) are considered for practical use.

Even so, a comprehensive design of the system incorporated with various characteristics of thermal plasmas will be critical to cope with an apparent increase in time and financial resources. In the case of plasma powder spraying, for example, complete melting of and coating with larger particles at higher power inputs will be the primary target from the economical perspective. To exploit the ultrafast deposition characteristics more effectively, thin-film production with less than 1 μm in thickness by finer particle spraying and high uniformity in thickness over a wide area can also be the directions from the additional functionality perspective. In fact, the latter has already been applied successfully to the production of electrostatic chucks in the semiconductor industry, which is a fascinating trend of this technology [35].

Meanwhile, plasma spraying at medium pressure of around several torr can be a unique approach to expand the potential of plasma spray processing. Effectiveness of this processing may be in that various precursors can be created stably through melting, evaporation, and chemical reactions in the plasma, promoting the functionalities of the products. In addition, owing to the characteristics of medium pressure plasma, such precursors are transported/treated in rather high-density plasma flow, which increases the throughput and thereby retains the potential of plasma spraying as an affordable process technique. Because of such characteristics different from those of low-pressure and thermal plasmas, this plasma process may be distinguished by calling it "mesoplasma" spraying. A recent example is the deposition of epitaxial silicon films at high rates via mesoplasma CVD [36]. As such, a variety of developments in plasma spraying technology are much expected as a key technology in the future.

References

1 Berndt, C.C. (2001) in The origins of thermal spray literature, Proc. *Thermal Spray* 2001: New Surfaces for a New Millennium, (eds C.C. Berndt, K.A. Khor, and E.F. Lugscheider), ASM International, Ohio, 1351.

2 Yoshida, T. (1990) The future of thermal plasma processing. *Mater. Trans. JIM*, **31**, 1.

3 Yoshida, T. (1994) The future of thermal plasma processing for coating. *Pure Appl. Chem.*, **66**, 1223.

4 Decomposition of organic halides by radio frequency ICP plasma, Patent, JP2732472 (1998).

5 Thermal-spray processing of materials, (2000) MRS Bull, 25 July, 12.

6 Yamaguchi, N., Sasajima, Y., Terashima, Y. and Yoshida, T. (1999) Molecular dynamics study of cluster deposition in thermal plasma flash evaporation. *Thin Solid Films*, **345**, 34.

7 Murakami, H., Yoshida, T. and Akashi, K. (1988) High rate thermal plasma CVD of SiC. *Adv. Ceram. Mater.*, **3**, 423.

8 Chae, Y.K., Ohone, H., Eguchi, K. and Yoshida, T. (2001) Ultrafast deposition of microcrystalline Si by thermal plasma chemical vapor deposition. *J. Appl. Phys.*, **89**, 8311.

9 Han, P. and Yoshida, T. (2001) Ultrafast deposition of microcrystalline Si by thermal plasma chemical vapor deposition. *J. Appl. Phys.*, **89**, 8311.

10 Kulinich, S.A., Shibata, J., Yamamoto, H., Shimada, Y.,

Terashima, K. and Yoshida, T., (2001) Highly c-axis oriented $LiNb_{0.5}Ta_{0.5}O_3$ thin films on Si fabricated by thermal plasma spray CVD. *Appl. Surf. Sci.*, **182**, 150.

11 Yin, H.Q., Eguchi, K. and Yoshida, T. (1995) Diamond deposition on tungsten wires by cyclic thermal plasma chemical vapor deposition. *J. Appl. Phys.*, **78**, 3540.

12 Matsumoto, S. and Zhang, W. (2000) High-rate deposition of high-quality, thick cubic boron nitride films by bias-assisted DC jet plasma chemical vapor deposition. *Jpn. J. Appl. Phys.*, **39**, L442.

13 Yoshida, T. and Akashi, K. (1977) Particle heating in a radio-frequency plasma torch. *J. Appl. Phys.*, **48**, 2252.

14 Fauchais, P. (2004) Understanding plasma spraying. *J. Phys. D: Appl. Phys.*, **37**, R86.

15 Terashima, K., Eguchi, K., Yoshida, T. and Akashi, K. (1988) Preparation of superconducting Y-Ba-Cu-O films by a reactive plasma evaporation method. *Appl. Phys. Lett.*, **52**, 1274.

16 Wang, X.H., Yamamoto, A., Eguchi, K., Obara, H. and Yoshida, T. (2003) Thermoelectric properties of SiC thick films deposited by thermal plasma physical vapor deposition. *Sci. Technol. Adv. Mater.*, **4**, 167.

17 Huang, H., Eguchi, K. and Yoshida, T. (2006) High power hybrid plasma spraying of large yttria stabilized zirconia powder. *J. Therm. Spray Technol.*, **15**, 72.

18 Shinoda, K., Kojima, Y. and Yoshida, T. (2005) In-situ measurement

system for deformation and solidification phenomena of plasma-sprayed zirconia droplets. *J. Therm. Spray Technol.*, **14**, 511.

19 Huang, H., Eguchi, K. and Yoshida, T. (2003) Novel structured yttria stabilized zirconia coatings fabricated by hybrid thermal plasma spraying. *Sci. Technol. Adv. Mater.*, **4**, 617.

20 Heuer, A., Chaim, R. and Lanteri, V. (1987) The displacive cubic–tetragonal transformation in ZrO_2 alloys. *Acta Metall.*, **35**, 666.

21 Ma, T., Kambara, M., Huang, H. and Yoshida, T.Effect of microstructure on reflectance of thermal barrier coatings. (in preparation).

22 Eguchi, K., Huang, H., Kambara, M. and Yoshida, T. (2005) Ultrafast deposition of YSZ composite for thermal barrier coatings by twin hybrid plasma spraying technique. *J. Jpn. Inst. Metals*, **69**, 17.

23 Lee, H.J., Eguchi, K. and Yoshida, T. (1990) Preparation of ultrafine silicon-nitride and silicon nitride and silicon carbide mixed powders in a hybrid plasma. *J. Am. Ceram. Soc.*, **73**, 3356.

24 Yoshie, K., Kasuya, S., Eguchi, K. and Yoshida, T. (1992) Novel method for C60 synthesis: a thermal plasma at atmospheric pressure. *Appl. Phys. Lett.*, **61**, 2782.

25 Yoshida, T., Kawasaki, A., Nakagawa, K. and Akashi, K. (1979) The synthesis of ultrafine titanium nitride in an RF plasma. *J. Mater. Sci.*, **14**, 1624.

26 http://www.env.go.jp/doc/toukei/contents/index.html.

27 Yamamoto, T., Ramanathan, K., Lawless, P.A., Ensor, D.S., Newsome, J.R., Plaks, N. and Ramsey, T.H. (1992) Control of volatile organic compounds by an ac energized

ferroelectric pellet reactor and a pulsed corona reactor. *IEEE Trans. Ind. Appl.*, **28**, 528.

28 Masuda, S. and Nakao, H. (1999) Control of NOx by positive and negative pulsed corona discharges. *IEEE Trans. Ind. Appl.*, **26**, 374.

29 Chang, J.S. (2000) Recent development of gaseous pollution control technologies based on non-thermal plasmas. *Oyo Butsuri*, **69**, 268.

30 Urashima, K. and Chang, J.S. (2000) Removal of volatile organic compounds from air streams and industrial flue gases by non-thermal plasma technology. *IEEE Trans. Dielect. Elect. Insul.*, **7**, 602.

31 Miziolek, A.W., Daniel, R.G., Skaggs, R.R., Rosocha, L.A., Chang, J.S. and Herron, J.T. (1999) Non-thermal plasma processing and chemical conversion of halons: reactor considerations and preliminary results, *Halon Options Technical Working Conf.* 501.

32 Inaba, T. and Iwao, T. (2000) Treatment of waste by dc Arc discharge plasmas. *IEEE Trans. Dielect. Elect. Insul.*, **7**, 684.

33 Hayashi, A. (2000) Prolongation of lifetime of electrode in plasma torch for incineration ash melting process. *J. Plasma Fusion Res.*, **76**, 742.

34 Sakano, M., Tanaka, M. and Watanabe, T. (2001) Application of radio-frequency thermal plasmas to treatment of fly ash. *Thin Solid Films*, **386**, 189.

35 Ishiguro, C. (2005) Handotai Sasaeru Kuroko-tachi, *Nikkei Business*, 17 Jan. 100.

36 Kambara, M., Yagi, H., Sawayanagi, M. and Yoshida, T. (2006) High rate epitaxy of silicon thick films by medium pressure plasma CVD. *J. Appl. Phys.*, **99**, 074901.

24
Electrohydraulic Discharge Direct Plasma Water Treatment Processes

J.-S. Chang, S. Dickson, Y. Guo, K. Urashima, and M.B. Emelko

24.1
Introduction

The application of plasma technologies for the treatment of drinking and waste water is not new. Three categories of plasma treatment technologies exist: remote, indirect, and direct. Remote plasma technologies involve plasma generation in a location away from the medium to be treated (e.g. ozone). Indirect plasma technologies generate plasma near to, but not directly within, the medium to be treated (e.g. UV, electron beam). More recently, direct plasma technologies (i.e. electrohydraulic discharge) have been developed that generate plasma directly within the medium to be treated thereby increasing treatment efficiency [1,2]. Three types of electrohydraulic discharge systems (pulsed corona electrohydraulic discharge (PCED), pulsed arc electrohydraulic discharge (PAED), and pulsed power electrohydraulic discharge (PPED)) have been employed in numerous environmental applications, including the removal of foreign objects (e.g. rust, zebra mussels) [3], disinfection [4,5a,5b], chemical oxidation [6–9], and the decontamination of sludges [10]. This chapter will compare the characteristics of the pulsed corona, pulsed arc and pulsed power electrohydraulic discharge systems, discuss the treatment mechanisms generated by electrohydraulic discharge technologies, review the literature investigating the application of these technologies to water treatment, and discuss the issues and research needs associated with these technologies.

24.2
Characteristics of Electrohydraulic Discharge Systems

The types of electrohydraulic discharge systems differ in several operational characteristics, as summarized in Table 24.1, due to their different configurations as well as the different amounts of energy injected by each type of system. The PCED system employs discharges in the range of 1 J per pulse, while the PAED and PPED systems use discharges in the range of 1 kJ per pulse and larger. The pulsed corona system operates at a frequency of 10^2 to 10^3 Hz with the peak current below 100 A and the

Advanced Plasma Technology. Edited by Riccardo d'Agostino, Pietro Favia, Yoshinobu Kawai, Hideo Ikegami, Noriyoshi Sato, and Farzaneh Arefi-Khonsari
Copyright © 2008 WILEY-VCH Verlag GmbH & Co. KGaA, Weinheim
ISBN: 978-3-527-40591-6

Tab. 24.1 Characteristics of Electrohydraulic Discharge Systems.

Property	Pulsed corona (PCED)	Pulsed arc (PAED)	Pulsed power (PPED)
Operating frequency [Hz]	10^2–10^3	10^{-2}–10^2	10^{-3}–10^1
Current [A]	10^1–10^2	10^3–10^4	10^2–10^5
Voltage [V]	10^4–10^6	10^3–10^4	10^5–10^7
Voltage rise [s]	10^{-7}–10^{-9}	10^{-5}–10^{-6}	10^{-7}–10^{-9}
Pressure wave generation	Weak	Strong	Strong
UV generation	Weak	Strong	Weak

voltage rise occurring on the order of nanoseconds. A streamer-like corona is generated within the liquid to be treated, weak shock waves are formed, and a moderate number of bubbles are observed [11,12]. This system also generates weak UV radiation [13] and forms radicals and reactive species in the narrow region near the discharge electrodes.

PAED employs the rapid discharge of stored electrical charge across a pair of submerged electrodes to generate electrohydraulic discharges forming a local plasma region. The PAED system operates at a frequency of 10^{-2}–10^2 Hz with the peak current above 10^3 A and the voltage rise occurring on the order of microseconds [14,15]. An arc channel generates strong shock waves with a cavitation zone [15,16] containing plasma bubbles [17] and transient supercritical water conditions [18]. This system generates strong UV radiation and high radical densities, which have been observed to be short-lived in the cavitation zone [15,19]. Pulsed spark electrohydraulic discharge (PSED) system characteristics are similar to those of PCED, with a few characteristics falling between those of PCED and PAED.

The PPED system operates at a frequency of 10^{-3} to 10^1 Hz with the peak current in the range of 10^2–10^5 A [7]. The voltage rise occurs on the order of nanoseconds [7]. This type of system generates strong shock waves and weak UV radiation.

24.3
Treatment Mechanisms Generated by Electrohydraulic Discharge

Traditional water and wastewater treatment technologies can be broadly classified into three categories: biological, chemical, and physical processes. Biological methods are typically used to treat both municipal and industrial wastewaters, and are generally not successful at degrading many toxic organic compounds. Chemical processes typically involve the addition of a chemical to the system to initiate a transformation; some chemicals commonly added, such as chlorine, are often associated with negative effects on both human and environmental health. Regulations governing the use of many of these chemicals are becoming increasingly stringent. Physical processes do not involve chemical transformation, and therefore generally serve to remove contaminants from the bulk fluid phase through concentrating them in a liquid or sludge phase. Many drinking water systems are challenged with difficult to treat target compounds such as organics (e.g. NDMA) and pathogens (e.g. *Cryptosporidium*, viruses, etc.). To be effectively removed from drinking water supplies, many of these recalcitrant compounds require specialized and often target-specific treatment technologies. The necessity of multiple treatment technologies can be quite costly, especially for small systems. In order to meet the challenges presented by continuously emerging contaminants and increasingly stringent regulations, new and promising treatment technologies, such as electro-hydraulic discharge, must be developed.

Depending on the technology, the treatment mechanisms generated by plasma technologies include (1) high electric fields; (2) radical reactions (e.g. ozone, hydro-gen peroxide); (3) UV irradiation; (4) thermal reactions; (5) pressure waves; (6) electronic and ionic reactions; and (7) electromagnetic pulses (EMP). In general, both electron and ion densities are proportional to the discharge current, while UV intensity, radical densities, and the strength of the pressure waves generated are proportional to the discharge power. Direct plasma technologies (i.e. electrohydraulic discharge) have the potential to be more efficient than either indirect or remote plasma technologies as they capitalize, to some degree, on all of these mechanisms due to the direct application [1,2]. Figure 24.1 illustrates the treatment mechanisms initiated by PAED.

Because electrohydraulic discharge systems do exploit all of the treatment mechanisms generated by the plasma reaction, both chemical and physical, these technologies have the ability to effectively treat a broader range of contaminants than other conventional and emerging technologies [1]. Preliminary research has indicated that PAED offers advantages over indirect plasma methods in that it

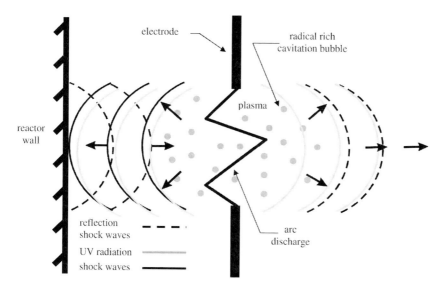

Fig. 24.1 Water treatment mechanisms initiated by PAED.
After [6].

Tab. 24.2 Comparison of Plasma and Conventional Water Treatment Processes [1].

Target compounds	Cl/ClO$_2$	Ozone	Electron beam	PCED	PAED	UV-C
Microorganisms	adequate	good	adequate	good	good	good
Algae	none	partial	none	partial	good	adequate
Urine components	adequate	good	good	good	good	none
VOCs	none	adequate	good	good	adequate	none
Inorganics	none	partial	partial	adequate	adequate	none

can provide comparable or superior treatment of microorganisms, algae, volatile organics, nitrogenous municipal waste compounds, and some inorganics [1,2,5,6,20]; these findings are qualitatively summarized in Table 24.2 [1]. Moreover, these benefits are available concurrently from one technology as opposed to a series of treatment technologies. In addition, preliminary investigations conducted with a limited number of target compounds have indicated that effective water treatment with PAED utilizes less than 50% of the kilowatt-hours required by other plasma technologies (e.g. UV) for equivalent levels of treatment [5].

24.4
Treatment of Chemical Contaminants by Electrohydraulic Discharge

Several studies have demonstrated that electrohydraulic discharge can effectively treat aqueous chemical contaminants such as atrazine [8], paraquinone [8], 4-chlorophenol [7], 3,4-dichloroanaline [7], phenol [21–24], dyes [25], urine compounds [2], methyl *tert*-butyl ether (MTBE) [26], and 2,4,6-trinitrotoluene [7,9]. Several investigations employing pulsed streamer corona discharge (PCED) have been conducted [21–23,25]; however, it has been observed that organic compound treatment by PCED requires the addition of activated carbon, a photocatalyst, or the superimposition of glow corona above the liquid surface [1,27,28]. Willberg *et al.* [7] investigated the removal of 4-chlorophenol (4-CP), 3,4-dichloroanaline (3,4-DCA), and 2,4,6-trinitrotoluene (TNT) by PPED. A 4.0 L vessel was employed in batch mode as the reaction chamber, and connected to a PPED power supply with a discharge energy of 7 kJ per pulse. Figure 24.2 shows that a 35% conversion of 4-CP was obtained after a cumulative energy input of 69.4 kWh m^{-3} (250 kJ L^{-1}) at an applied voltage of 10.2 kV. Figure 24.2 also shows the production of *p*-benzoquinone and chloride as a function of cumulative power input. Stoichiometrically, the chloride production can be attributed entirely to the 4-CP degradation; however, only 45% of the 4-CP degraded is present for in the form of *p*-benzoquinone. Experiments investigating the degradation of TNT through a combination of PPED and ozone were also conducted. Greater than 99% TNT degradation was achieved after a cumulative input energy of 126.4 kWh m^{-3} (455 kJ L^{-1}). This conversion rate is significantly larger than that accomplished by PPED alone on the 4-CP; however, it is unclear if the increased conversion can

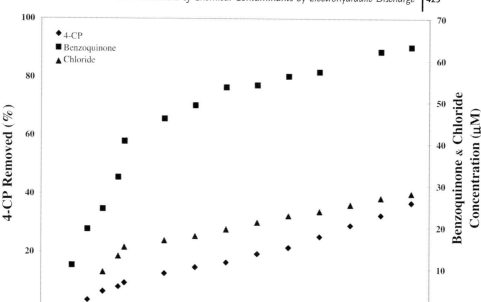

Fig. 24.2 4-Chlorophenol (4-CP) removal by PPED as a function of cumulative energy input. $V = 10.2\,kV$, $E = 7\,kJ$ per pulse, [4-CP] $= 200\,\mu M$, pH $= 5$. After [7].

be attributed to ozone addition due to the fact that the reaction pathways for 4-CP and TNT are not the same. The authors concluded that their results demonstrated the potential application of the electrohydraulic discharge process to the treatment of hazardous wastes.

Lang *et al.* [9] investigated the kinetics of TNT removal by PPED as a function of aqueous phase ozone concentration, pH, discharge energy, and water gap distance. A 4.0 L vessel was employed in batch mode as the reaction chamber, and connected to the PPED power supply. It was observed that the rate of TNT degradation increased with increasing aqueous phase ozone concentrations (to $150\,\mu M$) (Fig. 24.3), increasing pH (from 3.0 to 7.9) increasing discharge energy (from 5.5 to 9 kJ), and decreasing water gap distance (from 6 to 10 mm). Several intermediate degradation products (nitrate, 2,4,6-trinitrobenzoic acid (TNBA)) were detected over the course of these experiments; however, effective mineralization of over 90% of the initial TNT present was achieved.

Karpel Vel Leitner *et al.* [8] identified various reaction mechanisms initiated by PAED by treating various molecules with known degradation pathways. Their experiments employed a 5 L reaction chamber operated in batch mode connected to a spark gap type power supply (0.5 kJ per pulse) with rod-to-rod electrodes in water. They employed both maleic and fumaric acids, which are known to photoisomerize

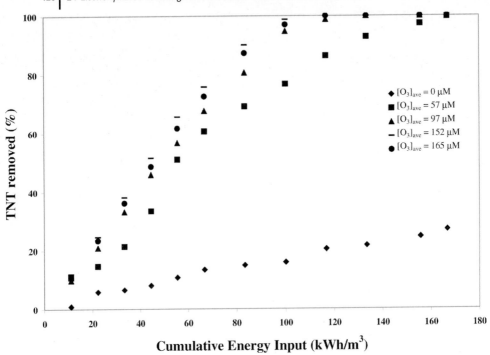

Fig. 24.3 The effect of ozone concentration on 2,4,6-
trinitrotoluene (TNT) removal by PPED as a function of
cumulative energy input. $V = 10.2$ kV, $E = 7$kJ per pulse,
[TNT] $= 170\,\mu$M, pH $= 4.7$, water gap $= 8$ mm. After [9].

by UV radiation, to confirm the presence of UV radiation in PAED systems. Approximately 35% of the maleate and fumarate ions removed in their experiments were converted to fumarate and maleate ions respectively. It was therefore concluded that PAED does induce photochemical reactions, but that photolysis is not the only phenomenon responsible for the abatement of these molecules. Karpel Vel Leitner *et al.* [8] also treated solutions of nitrate ions to demonstrate reduction reactions. They found that 75–86% of the nitrate ions removed were converted to nitrites, and concluded that reducing species played a significant role in the transformation, as photolysis alone could not account for the observed nitrite production. Karpel Vel Leitner *et al.* [8] investigated the presence of oxidizing species on aqueous solutions of hydroquinone. Although the hydroquinone was oxidized, the study was not able to identify the mechanism responsible for the oxidation reaction. The authors concluded that oxidizing species are present, but that further investigation is required to identify and quantify these species.

The final objective of the Karpel Vel Leitner *et al.* [8] study was to investigate the effect of various additives, initial concentration, and operating conditions on the degradation of atrazine by PAED. Several additives were considered, including:

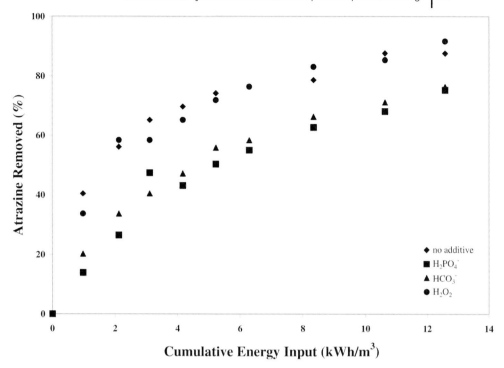

Fig. 24.4 The effect of selected additives on atrazine removal by
PAED as a function of cumulative energy input. $V = 3.4$ kV,
$E = 0.5$ kJ per pulse, $[\text{atrazine}]_o = 0.5\ \mu\text{mol L}^{-1}$. No additive
pH $= 6.9$; 5 mmol L^{-1} H$_2$PO$_4^-$ pH $= 6.5$; 5 mmol L^{-1} HCO$_3^-$
pH $= 8.5$; 0.5 mmil L^{-1} H$_2$O$_2$ pH $= 6.9$. After [8].

5 mmol L^{-1} of bicarbonate, 5 mmol L^{-1} of dihydrogenphosphate, and 500 μmol L^{-1}
of hydrogen peroxide; PAED alone, with no additives, constituted the reference case.
Figure 24.4 shows that hydrogen peroxide had no significant effect on degradation of
atrazine by PAED, and that both bicarbonate and dihydrogenphosphate addition
inhibited the removal of atrazine by PAED. Initial atrazine concentrations were also
found to be a factor in removal by PAED, with the percentage of atrazine removal
increasing with decreasing initial concentrations [8]. Increasing the gap distance
between the electrodes in the aqueous solution from 1.5 to 4 mm increased removal
rates from approximately 50% to greater than 95%, with a cumulative input energy of
approximately 12.5 kWh m^{-3} (45 kJ L^{-1}) in each case. Additionally, this increase in
gap distance resulted in a decrease in the formation of the byproduct deethylatrazine
(DEA) [8]. The increased efficiency with larger gap distances was attributed to an
increase in the size of the spurs induced by the plasma.

Angeloni *et al.* [29] investigated the effects of initial solution pH (6–8.5), charging
voltage (2–3 kV), detention time (5–30 min) (i.e. cumulative input energy), and water–
arc-electrode gap (1–3 mm) on the removal of 10 mg L^{-1} of MTBE from an aqueous

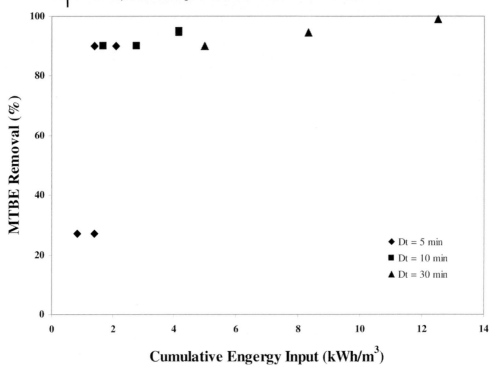

Fig. 24.5 MTBE removal as a function of cumulative energy input by PAED. Water–arc-electrode gap = 1 mm; initial solution pH range: 6–8.5; charging voltage range: 2–3 kV; flow rate range: 0.1–0.6 L min^{-1}; detention time range: 0–30 min. After [29].

solution by PAED. A spark gap-type power supply (0.3 kJ per pulse) with rod-to-rod type electrodes in the aqueous solution was employed together with a 3 L flow-through reactor. Figure 24.5 shows that greater than 99% removal was observed with a cumulative input energy of 12.5 kWh m^{-3} (45 kJ L^{-1}) (which corresponded to a detention time of 30 min), an initial solution pH of 7, a charging voltage of 3 kV, and a water–arc-electrode gap of 1 mm. The experimental investigation indicated that the initial solution pH did not have a significant effect on MTBE removal; however, the MTBE decomposition increased with increasing charging voltage and increasing detention times. Figure 24.6 shows that MTBE decomposition decreased with increasing water–arc gap distances; this was attributed to the fact that PAED discharges weaken with increasing water–arc gap distances. UV photolysis was not considered to be a mechanism of MTBE decomposition in these experiments; however the author indicated that further research was required to determine which mechanism(s) were responsible for MTBE decomposition. No significant liquid MTBE decomposition by-products were observed; however the author concluded that further research was required to examine potential by-product formation.

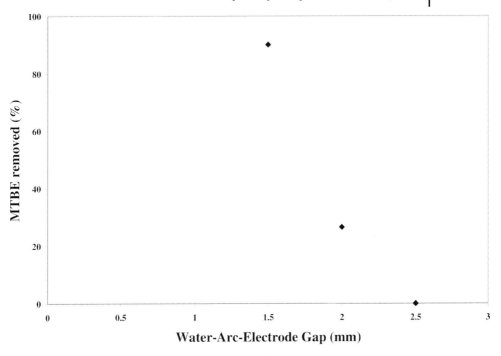

Fig. 24.6 Effect of the water–arc-electrode gap on removal of
MTBE. Initial solution pH $= 7$; charging voltage $= 3$ kV; flow
rate $= 0.1$ L min^{-1}. After [29].

24.5
Disinfection of Pathogenic Contaminants by PAED

The relative ease of disinfection of *E. coli*, as compared to protozoan pathogens such
as *Cryptosporidium parvum*, is commonly recognized. *E. coli* disinfection has been
achieved with several treatment technologies, including UV irradiation. Many strains
of *E. coli* have demonstrated an ability to repair after irradiation with low- and
medium-pressure UV. *E. coli* inactivation by electrohydraulic discharge was reported
by Ching *et al.* [20], who employed PPED with discharges of 7 kJ per pulse at 5.5 kV.
   Subsequently, Emelko *et al.* [5a,5b] investigated the disinfection of 0.7 L of
4.0×10^7 cfu mL^{-1} *E. coli* and *B. subtilis* suspensions in 0.01 M PBS at pH $= 7.4$ by
PAED with 0.3 kJ per pulse at 2.2 kV and a water gap of 1 mm. Figure 24.7 shows the
$\log_{10}$ inactivation of *E. coli* cells and *B. subtilis* in PBS as a function of the cumulative
power input per liter of solution treated. This figure clearly indicates average *E. coli*
inactivation of 2.6-, 3.3-, and 3.6-log after the respective application of 5.6, 13.9, and
25 kWh m^{-3} (20, 50, and 90 kJ L^{-1}) (corresponding to detention times of approximately 0.8, 1.5, and 5.8 min, respectively). Slightly larger removals were obtained for
inactivation of *B. subtilis*. A logit plot of these data was not constructed because of the

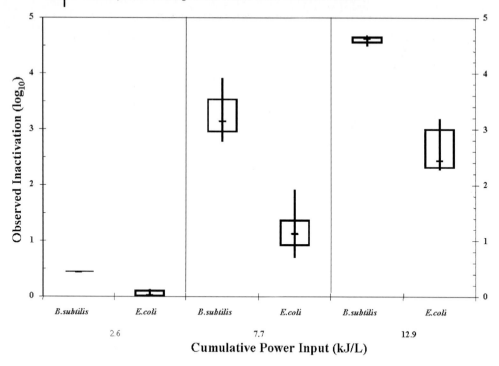

Fig. 24.7 Box and whisker plots of *E. coli* and *B. subtilis* (suspended in 0.01 M PBS) inactivation by PAED. $V = 2.2$ kV, $E = 0.3$ kJ per pulse, pH $= 7.4$, water gap $= 1$ mm. After [5b].

relatively small number of data points and the high level of disinfection achieved within the first data point 9.4 kWh m^{-3} (2.6 kJ L^{-1} power input).

It is noteworthy that the pattern of *E. coli* inactivation observed by Emelko *et al.* [5a,5b] is similar to that observed by Ching *et al.* [20] (Fig. 24.8), in that the rate of inactivation is initially high, and decreases as the cumulative power input increases. The observations made by Emelko *et al.* [5b] differ from those of Ching *et al.* [20], however, as the Emelko *et al.* [5b] experiments achieved a slightly higher level of *E. coli* inactivation for significantly lower cumulative energy inputs. This difference is likely due to the fact that the Emelko *et al.* [5b] experiments employed a PAED system, while Ching *et al.* [20] used a PPED system. While both systems fall under the category of direct plasma technologies, they operate slightly differently, and as a result the UV radiation generated by PPED is weaker than that generated by PAED [1].

24.6
Municipal Sludge Treatment

Conventional municipal wastewater treatment consists of preliminary processes (screening and grit removal), primary settling to remove heavy solids and floatable

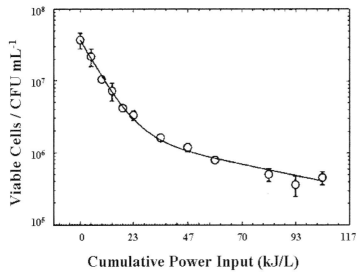

Fig. 24.8 *E. coli* (suspended in 0.01 M PBS) inactivation by PPED.
$V = 5.5$ kV, $E = 7$ kJ per pulse, pH $= 7.4$. After [20].

materials, and secondary biological aeration to metabolize and flocculate colloidal and dissolved organics. Waste sludge drawn from these unit operations is thickened and processed for ultimate disposal. Current options for ultimate disposal include land application, incineration, or landfill disposal. The characteristics of municipal sludge that affect its suitability for application to land include organic content (about 40% of total solids if measured as volatile solids), nutrients, pathogens, metals, and toxic organics [30]. The concentration of metals and toxic organics in municipal sludge vary widely. The typical metal concentrations in municipal sludge are shown in Table 24.3.

Because of the presence of toxic metals and organics, land application of municipal sludge is under greater scrutiny now. For large metropolitan areas, landfill disposal is getting more and more costly. Cost and air pollution concerns also limit the use of incineration as an option for ultimate disposal of municipal sludge. Due to the high organic content of many sludges, however, PPED may be used for sludge volume reduction (i.e. conversion of organics to hydrocarbon gases) [32].

Tab. 24.3 Typical Metal Content in Municipal Sludge (Dry Solids, mg kg^{-1}) [31].

Metal	Arsenic	Cadmium	Chromium	Cobalt	Copper	Iron	Lead
Range	1.1–230	1–3410	10–99 000	11.3–2490	84–17 000	1000–154 000	13–26 000
Median	10	10	500	30	800	17 000	500

Metal	Manganese	Mercury	Molybdenum	Nickel	Selenium	Tin	Zinc
Range	32–9780	0.6–56	0.1–214	2–5300	1.7–17.2	2.6–329	101–49 000
Median	260	6	4	80	5	14	1700

24.7
Concluding Remarks

This chapter conducted a review of the application of direct plasma technologies to the removal/inactivation of chemical and microbial contaminants in water. The findings indicate that electrohydraulic discharge technologies, and specifically PAED, generate a range of treatment mechanisms, both chemical and mechanical, suitable for removing/inactivating both chemical and microbial contaminants. Furthermore, the results of bench-scale experiments completed to date indicate that electrohydraulic discharge technologies have the potential to treat these contaminants as effectively, and more economically, than conventional treatment technologies. Although the results to date are very promising, many questions and issues must still be resolved before these technologies will become commonplace in treatment facilities. These issues include, but are not limited to: optimization of the reactor for a range of operating conditions, a more complete understanding of the reactions initiated by this technology and therefore the range and types of contaminants that could be treated effectively, potential additives and catalysts to the reactions initiated, possible hazardous byproduct formation, a more complete understanding of the economics of these systems as compared to more conventional treatment technologies, and the effectiveness of electrohydraulic discharge at various points within the conventional treatment process.

References

1 Chang, J.-S. (2001) *Sci. Technol. Adv. Mater.*, **2**, 571–576.
2 Chang, J.-S., Urashima, K., Uchida, Y. and Kaneda, T. (2002) Research Report of Tokyo Denki University, Tokyo, *Denki Daigaku Kogaku Kenku J.*, **50**, 1–12.
3 Bryden, A.D. (1995) US Patent 5,432,756.
4 Sato, M., Ohgiyama, T. and Clements, J.S. (1996) *IEEE Trans. Ind. Appl.*, **32**, 106–112.
5 Emelko, M.B., Dickson, S.E., Chang, J.-S. (2003) In *Proc. AWWA Water Quality Technology Conference*, Pittsburgh, PA: AWWA (a). Emelko, M.B., Dickson, S.E., Chang, J.-S. and Lee, L. (2004) In *Proc. AWWA Water Quality Technology Conference*, San Antonio, TX: AWWA (b).
6 Karpel Vel Leitner, N., Urashima, K., Bryden, A., Ramot, H., Touchard, G. and Chang, J.-S. (2001)

In *Proc. Third Int. Symp. on Non-Thermal Plasmas*, 23–27 April (eds J.-S.Chang and J.Kim), Kimm Press, Tajun, Korea, pp. 39–44.
7 Willberg, D.M., Lang, P.S., Hochemer, R.H., Kratel, A. and Hoffmann, M.R. (1996) *Environ. Sci. Technol.*, **30**, 2526–2534.
8 Karpel Vel Leitner, N., Syoen, G., Romat, H., Urashima, K. and Chang, J.-S. (2005) Water Res. (in press).
9 Lang, P.S., Ching, W.-K., Willberg, D.M. and Hoffmann, M.R., (1998) *Environ. Sci. Technol.*, **32** (20), 3142–3148.
10 Warren, D., Russel, J. and Siddon, T. (1996) In *Proceedings AOT-3*, Cincinnati, 27–29 October 1996.
11 Teslenko, V.S., Zhukov, A.J. and Mitrofanov, V.V. (1995) *Lett. ZhTF*, **21**, 20–26.

12 Jomni, F., Denat, A. and Aitken, F. (1996) *Proc. Conf. Record of ICDL '96.*

13 Hoffman, M.R. (1997) *Proc. 2nd Int. Environ. Appl. Adv. Oxid. Tech.*, EPRI Report CR-107581.

14 Robinson, J.W. (1973) *J. Appl Phys.*, **44**, 76.

15 Chang, J.-S., Looy, P.C., Urashima, K., Bryden, A.D. and Yoshimura, K. (1998) *Proc. Asia-Pacific AOT Workshop.*

16 Martin, E.A. (1958) *J. Appl. Phys.* **31**, 255.

17 Robinson, J.W., Ham, M. and Balaster, A.N. (1973) *J. Appl. Phys.*, **44**, 72–75.

18 Ben'Kovskii, V.G., Golubnichii, P.I. and Maslennikov, S.I. (1974) *Phys. Acoust.*, **20**, 14–15.

19 Jakob, L., Hashem, T.M., Burki, S., Guidny, N.M. and Braun, A.M. (1993) *Photochem. Photobiol. A: Chem.*, **7**, 97.

20 Ching, W.K., Colussi, A.J., Sun, H.J., Nealson, H. and Hoffmann, M.R. (2001) *Environ. Sci. Technol.*, **35** (20), 4139–4144.

21 Sharma, A.K., Locke, B.R., Arce, P. and Finney, W.C. (1993) *J. Hazardous Waste Hazardous Mater.*, **10**, 209–219.

22 Sun, B., Sato, M. and Clements, J.S. (1999) *J. Phys. D: Appl. Phys.*, **32**, 1.

23 Sun, B., Sato, M. and Clements, J.S. (2000) *Environ. Sci. Technol.*, **34**, 509.

24 Hoeben, W.F.L.M., van Veldhuizen, E.M., Rutgers, W.R. and Kroesen, G.M.W. (2000) *J. Phys. D: Appl. Phys.*, **32**, L133.

25 Sato, M., Yamada, Y. and Sugiarto, A.T. (2000) *Trans. Inst. Fluid Flow Machine*, **107**, 95.

26 Angeloni, D.M., Dickson, S.E., Chang, J.-S. and Emelko, M.B. (2004) *Proc. OWWA/OMWA Joint Annual Conference & Trade Show 2004*, OWWA/OMWA, Niagara Falls, ON.

27 Locke, B.R., Sato, M., Sunka, P., Hoffmann, M.R. and Chang, J.S. (2006) *Ind. Chem. Eng. Res.*, **45**, 882–905.

28 Grymonpre, D.R., Finney, W.C., Clark, R.J. and Locke, B.R. (2004) *Ind. Eng. Chem. Res.*, **43**, 1975.

29 Angeloni, D.M., Dickson, S.E., Emelko, M.B. and Chang, J.S. (2006) *Japanese J. Appl. Phys.* **45** (106), 8290–8293.

30 Metcalf & Eddy, Inc., (2003) *Wastewater Engineering, Treatment and Reuse*, McGraw-Hill, Boston, MA.

31 US EPA, (1984) Environmental Regulations and Technology, Use and Disposal of Municipal Wastewater Sludge, EPA/625/10-84-003, US Environmental Protection Agency.

32 Warren, D., Russel and Siddon.1996. *Presentation at 3rd Int. Conf. on Advanced Oxidation Technology, Cincinnati, OH, 27–29 October.*

25
Development and Physics Issues of an Advanced Space Propulsion

M. Inutake, A. Ando, H. Tobari, and K. Hattori

Electric propulsion (EP) is one of the most promising space propulsions due to its high specific impulse and enables a long-term space mission with less consumption of propellant. Various types of advanced space propulsion devices based on the plasma technology have been proposed and are under development in order to be utilized not only for a main engine of a small spacecraft but for a large-sized engine in long-term space missions.

A magneto-plasma-dynamic arcjet (MPDA) is one of the promising electric thrusters with a higher specific impulse and a relatively large thrust for space missions such as a manned interplanetary flight and an earth-impact asteroid de-orbiting. The MPDA plasma is accelerated axially by a self-induced $J \times B$ force. Thrust performance of the MPDA is expected to increase by applying a magnetic nozzle instead of a solid nozzle. In order to improve the flow characteristics of an MPDA and to get a much higher thruster performance, two methods have been investigated in the HITOP device (Tohoku University). One is to use a magnetic Laval nozzle in the vicinity of the MPDA muzzle for converting high ion thermal energy to axial flow energy. The other is the combination of ion heating by use of ICRF (ion cyclotron range of frequency) waves and ion acceleration in a divergent magnetic nozzle.

Plasma flow characteristics of an MPDA have been measured in various shapes of a magnetic field configuration by use of a spectrometer and Mach probes, and magnetic/electric probes. Ion-acoustic Mach number of the plasma flow in a uniform magnetic channel is limited to unity and increased up to almost 3 in a gradually divergent magnetic nozzle. A small Laval nozzle located near the MPDA muzzle could successfully convert the thermal energy to the flow energy. The results are compared with the one-dimensional isentropic flow model. It is found that the experimentally determined specific heat ratio is lower than the ideal value of 5/3 for a monatomic gas.

Ion heating in a fast-flowing plasma is proposed in the Variable Specific Impulse Magneto-plasma Rocket (VASIMR) project of NASA in order to control the plasma flow velocity by converting the thermal energy to the flow energy in a diverging nozzle. We have performed the ion heating experiment using the MPDA plasma. The fast-flowing plasma is successfully heated by use of an ICRF antenna in the magnetic

Advanced Plasma Technology. Edited by Riccardo d'Agostino, Pietro Favia, Yoshinobu Kawai, Hideo Ikegami, Noriyoshi Sato, and Farzaneh Arefi-Khonsari
Copyright © 2008 WILEY-VCH Verlag GmbH & Co. KGaA, Weinheim
ISBN: 978-3-527-40591-6

beach configuration. Dispersion relation of the waves excited by a helical antenna was obtained and agreed well with Doppler-shifted shear and compressional Alfven waves. Ion cyclotron resonance heating by the shear Alfven wave is observed for the first time in the fast-flowing plasma with a low collisionality.

25.1
Introduction

Since the earliest days of space explorations, there has always been a demand for improvements of space propulsion systems. Chemical propulsion brings us beyond the earth's gravity successfully. However, it takes longer period and larger costs for an interplanetary mission by using chemical propulsion. It is no use in an interstellar mission where the speed of a rocket is required to exceed at least 10% of the speed of light. No matter how large the rocket is, no matter how many stages it has, no one can achieve the speed required for an interplanetary mission by using a chemical propulsion system.

Electric propulsion (EP) is one of the most promising space propulsions due to its high specific impulse I_{sp} which is defined by a ratio between a thrust and a propellant weight flow rate [1]. An EP system enables a long-term space mission with a less consumption of the propellant. Electric power generated from solar panels or nuclear reactors is utilized to ionize a propellant gas and to accelerate the ionized propellant by electrothermal, electrostatic, or electromagnetic effects. It is exhausted downstream of the thruster with a velocity much higher than that in a chemical rocket. Recently, various types of advanced space propulsion devices have been proposed and are under development in order to be utilized not only for a main engine of a small spacecraft but for a large-sized one for long-term space missions, such as earth-impact asteroid de-orbiting and a manned Mars exploration [2].

For a manned interplanetary space thruster, both a higher specific impulse and a larger thrust are required to shorten the flight term. Though a fusion-plasma thruster is one of the ultimate solutions for the purpose, realization of a fusion reactor needs several more decades of research. A high-powered magneto-plasma-dynamic arcjet (MPDA) is one of the promising candidates for a manned Mars spacecraft. The MPDA plasma is accelerated axially by a self-induced $J \times B$ force. It has a high thrust density, which is in proportion to the square of the discharge current, and a thrust efficiency is improved in higher current operation. Thrust performance of an MPDA is expected to improve by applying an external magnetic nozzle instead of a solid nozzle [3,4]. An MPDA operation with a magnetic field will reduce erosion of electrodes and diffuse current attachment on the anode, which are crucial problems for a steady-state and high-current operation of the MPDA. However, the acceleration mechanism becomes very complicated due to the addition of $J_r \times B_z$ rotational, $J_\theta \times B_r$ Hall current, and magnetic-nozzle accelerations. It is important to clarify various effects of the external magnetic field and to obtain an optimum magnetic nozzle configuration in the vicinity of the MPDA exit and to improve thrust efficiency.

Recently, NASA proposed another advanced thruster for manned explorations to Mars. The propulsion device, named Variable Specific Impulse Magneto-plasma Rocket (VASIMR), is a high-power, radiofrequency-driven magneto-plasma rocket. It is proposed to control the ratio of specific impulse to thrust at constant power [5]. The rocket provides a helicon plasma source and a combined system of ion cyclotron heating and magnetic nozzle, where a flowing plasma is heated by ICRF (ion cyclotron range of frequency) heating and plasma thermal energy is converted to flow energy in the magnetic nozzle. The ICRF heating and magnetic nozzle effect in a fast-flowing plasma are two key issues for the development of the advanced space thruster.

Radiofrequency (RF) heating technologies of plasma ions have been developed in magnetically confined fusion researches. Ion heating condition in a fast-flowing plasma, however, would be different from that in a confined, stationary fusion plasma on several points such as the short transit time for ions to pass only once through a heating region and the variation of the resonance frequency due to the Doppler shift effect. Ion heating also would be unsuccessful in a dilute plasma where plenty of the background neutral gas will penetrate deeply into the core plasma and the heated ion energy will be quickly lost due to the fast charge-exchange loss. It is urgently required to develop a combined system of RF heating and a magnetic nozzle.

In Section 25.2 are presented basic concepts of rocket system and performance of electric propulsion. In Section 25.3, we present two kinds of basic experiments to improve the plasma performance for the advanced thruster. One is improvement of MPDA performance operated with an additional magnetic Laval nozzle. The other is ICRF ion heating experiments of a fast-flowing plasma exhausted from the MPDA and conversion of plasma thermal energy to flow energy in a diverging magnetic nozzle.

25.2
Performance of Rocket Propulsion Systems

A rocket is accelerated by ejecting propellant mass according to Newton's laws of dynamics. A spacecraft's equation of motion is directly derived from conservation of the total momentum of the spacecraft and its exhaust stream:

$$\frac{d}{dt}(Mu) = 0 \tag{25.1}$$

Then we obtain

$$M\frac{du}{dt} = -u_{ex}\frac{dM}{dt} \tag{25.2}$$

where M is the mass of the spacecraft, $\dot{u} = du/dt$ its acceleration vector, u_{ex} the velocity vector of the exhaust stream relative to the spacecraft, and $\dot{m} = dM/dt$ the rate of change of spacecraft mass due to propellant mass ejection. The right-hand

Payload mass   mass of rocket structure

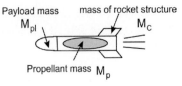

Propellant mass M_p

Fig. 25.1 Mass components of a rocket.

term of Eq. (25.2) is the thrust of the rocket, $T = \dot{m}u_{ex}$. The ratio of the thrust to the expelled propellant mass consumption rate measured in units of the sea-level weight, $\dot{m}g$, is called the specific impulse, $I_{sp} = \dot{m}u_{ex}/\dot{m}g = u_{ex}/g$.

When u_{ex} is constant during the rocket acceleration, the increment of the spacecraft velocity, ΔV, is obtained by integrating Eq. (25.1) as

$$\Delta V = \int_i^f dv = -\int_i^f u_{ex} \frac{dm}{m} = u_{ex} \ln \frac{m_i}{m_f} \tag{25.3}$$

where m_i and m_f are the total spacecraft mass at the initial (i) and final (f) stage of the acceleration period. This rocket equation was first derived in 1903 by K.E. Tsiolkovsky and shows that larger mass ratio of m_i/m_f is necessary to achieve larger ΔV for a constant u_{ex}.

Initial and final rocket masses consist of mass of payload M_{pl}, mass of rocket system M_C, and mass of propellant M_p as shown in Fig. 25.1:

$$m_i = M_{pl} + M_C + M_p \tag{25.4}$$

$$m_f = M_{pl} + M_C \tag{25.5}$$

Additionally, we define the ratio of rocket structure constant $\varepsilon = M_C/(M_p + M_C)$, and the ratio of payload $\Lambda = M_{pl}/M_i$. Then, Eq. (25.3) can be expressed as

$$\Delta V = u_{ex} \ln[\varepsilon + \Lambda(1 - \varepsilon)] \tag{25.6}$$

Consequently, Λ is derived as

$$\Lambda = \frac{1}{1 - \varepsilon} \exp\left(-\frac{\Delta V}{u_{ex}}\right) - \frac{\varepsilon}{1 - \varepsilon} \tag{25.7}$$

This equation can be extended to an N-stage propulsion system and Λ in the N-stage propulsion system is expressed as follows:

$$\Lambda = \left[\frac{\exp(-\Delta V/Nu_{ex}) - \varepsilon}{1 - \varepsilon}\right]^N \tag{25.8}$$

Figure 25.2 shows relations between Λ and the ratio of ΔV to u_{ex} in various rocket stages under the condition of $\varepsilon = 0.1$. As can be seen in the figure, $\Delta V/u_{ex}$ cannot

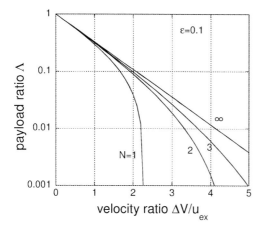

Fig. 25.2 Payload ratio Λ as a function of $\Delta V/u_{ex}$.

exceed 2.3 in a single-stage rocket system. Even in a multi-stage rocket system, it cannot reach 10. As a large ΔV is necessary for an interplanetary flight, propellant exhaust speeds in the range above $10^4\,\mathrm{m\,s^{-1}}$ are desirable. Unfortunately, exhaust speeds of a conventional chemical rocket with liquid or solid propellant are limited by the available energy of combustion reaction and are less than a few times $10^3\,\mathrm{m\,s^{-1}}$.

In EP systems electric power is utilized to ionize and accelerate the propellant gases. The ionized gas (plasma) can be accelerated in electric and magnetic fields and exhaust speed reaches more than $10^5\,\mathrm{m\,s^{-1}}$.

There are several methods in the EP system: electrothermal, electrostatic, electromagnetic, and combined systems. In electrothermal propulsion, propellant gas is electrically heated and then expanded through an adequately shaped solid nozzle to convert its thermal energy to exhaust stream energy. Resistojet and arcjet are categorized in electrothermal propulsion. The former heats the propellant in touch with an electrically heated chamber wall or heater coil, and the latter heats it by an electric arc plasma. The propellant gas can be selected more freely since the physical properties of the heating process are independent of any chemistry. Performance of the electrothermal thruster, however, is affected by "frozen" flow loss due to unrecovered energy which is frozen in the internal modes and the dissociation of molecules. The exhaust velocity of the flow is limited to $u_{ex} < \sqrt{2c_p T}$, where c_p is the specific heat at constant pressure per unit mass and T is propellant temperature heated upstream of the expanding nozzle. Since the power supply system is simple, these thrusters have been used in attitude control and orbit insertion of satellites.

An ion thruster is one example of the electrostatic thrusters [6]. Ions in ionized propellant are accelerated by an electric field generated between grid electrodes and subsequently neutralized by an equal flux of electrons extracted from another emitter. An extremely high exhaust velocity can be achieved in the ion thruster. However, ion flux extraction is limited by a shielding effect of the acceleration electric field due to space charge. The achievable current density can be calculated according to the Child–Langmuir law. When the acceleration voltage becomes large, exhaust velocity

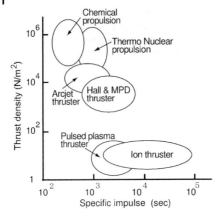

Fig. 25.3 Thrust density and specific impulse for various thrusters.

and current flux increase, resulting in larger thrust and specific impulse. However, the thrust efficiency (ratio of thrust to electric power) becomes worse and power system and insulation become complicated. System optimization considering thrust density, exhaust speed, efficiency, and power system should be performed for a given space mission.

An MPDA plasma is accelerated by an electromagnetic force [7]. It has coaxial electrodes, a central cathode and an annular anode. An MPDA plasma is formed between coaxial electrodes and discharge current flows through the plasma radially. Simultaneously, an azimuthal magnetic field is self-induced by the discharge current. Then the plasma is accelerated axially by a self-induced $J \times B$ force. It has a high thrust density, which is in proportion to square of the discharge current, and thrust efficiency is improved in a higher current operation.

There are some other electric propulsion systems, such as a Hall thruster and a pulsed plasma thruster (PPT). Performance of these thrusters is categorized by the thrust and specific impulse. Figure 25.3 shows typical operation regimes of the thrust and specific impulse for various thrusters.

25.3
Experimental Researches for an Advanced Space Thruster

25.3.1
Experimental Apparatus and Diagnostics

We have performed two kinds of basic experiments to control the plasma flow exhausted from an MPDA for the advanced thruster. Experiments are performed in the HITOP (high density Tohoku plasma) device of Tohoku University [8–13]. The device, shown in Fig. 25.4, consists of a large cylindrical vacuum chamber (diameter $D = 0.8$ m, length $L = 3.3$ m) with external magnetic coils. Various types of magnetic

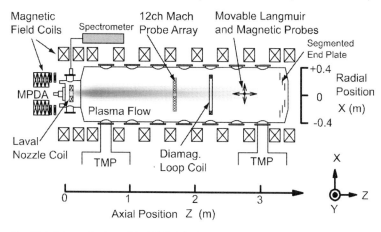

Fig. 25.4 Schematic view of the HITOP device.

field configuration can be formed by adjusting each coil current. A high-power, quasi-steady (1 ms) MPDA is installed at one end-port. It has a coaxial structure with a center tungsten rod cathode (10 mm in diameter) and an annular molybdenum anode (30 mm in inner diameter). A fast-acting gas valve can inject helium or argon gas quasi-steadily for 3 ms. Arc discharge is initiated when the gas flow rate becomes steady state. Quasi-steady (1 ms) discharge current I_d is supplied by a pulse-forming network (PFN) power supply with a maximum I_d of 10 kA. A typical discharge voltage is 200 V.

Several diagnostic instruments are installed on the HITOP device. Flow velocity U and ion temperature T_i in the region near the MPDA are measured by the Doppler shift and broadening of HeII line spectra ($\lambda = 468.58$ nm) and by Mach probes in the region far downstream of the MPDA.

The ion acoustic Mach number M_i is calculated as follows:

$$M_i = \frac{U_z}{C_s} = \frac{U_z}{\sqrt{k_B(\gamma_e T_e + \gamma_i T_i)/m_i}} \qquad (25.9)$$

The Mach probe has two plane surfaces, one of which faces upstream of the flow and the other faces perpendicularly to the axial flow. Effect of the magnetic field is negligible in the ion saturation current, because the ion Larmour radius is much larger than the size of the probe tip in the present experiments. The ion Mach number can be derived from the ratio of the two ion saturation current densities J_{para} and J_{perp}, which are collected in the parallel and perpendicular tips, respectively:

$$M_i = \kappa \frac{J_{para}}{J_{perp}} \qquad (25.10)$$

The above relation is calibrated with spectroscopic measurements [8]. Magnetic field components in the plasma flow are measured by a magnetic probe array [9]. Ion and electron temperatures far downstream of the MPDA are measured by an electrostatic energy analyzer and a fast voltage-scanning Langmuir probe, respectively. The plasma energy is measured using a diamagnetic loop coil.

25.3.2
Improvement of an MPDA Plasma Using a Magnetic Laval Nozzle

In the case of a uniform magnetic channel, it is found that both axial and rotational flow velocities of the MPDA plasma increase linearly with an increase of the discharge current, and at the same time ion temperature increases more steeply. This results in the limitation of the ion Mach number to less than 1 in the vicinity of the muzzle [10]. This indicates that there is an upper limit of the Mach number of nearly unity in the case of a uniform magnetic channel. This phenomenon is quite similar to the so-called choking where the mass flow rate attains the maximum at the condition that $M = 1$ in a conventional gas dynamics.

In the case of a long diverging magnetic nozzle which has a uniform field only near the MPDA muzzle and then a gradually diverging field, a supersonic plasma flow with ion Mach number up to 2.8 has been successfully obtained in the far downstream region of the MPDA, as shown in Fig. 25.5 [11]. The Mach number in the downstream region decreases with time elapsed after the discharge initiation. This is caused by a charge-exchange collision with neutral He atoms which are produced on the end wall through a surface recombination of the plasma flow and diffused upstream.

From multi-channel magnetic probe measurements, it is found that the diamagnetic effect is so strong that the externally applied uniform field is modified to an effectively converging nozzle [9]. This result suggests that when an electromagnetically accelerated supersonic plasma is injected into the effectively convergent nozzle, the flow should be decelerated or suddenly jump into a subsonic flow through a shock wave and would be re-accelerated through the effectively formed

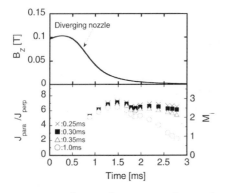

Fig. 25.5 Production of a supersonic flow in a long diverging nozzle. Time after the discharge initiation is shown. Mach number in the far downstream decreases gradually with the elapsed time due to charge-exchange collisions with surface-recombined neutral He atoms.

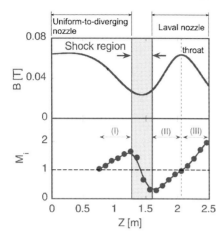

Fig. 25.6 Formation of a standing shock wave in front of a magnetic Laval nozzle and re-acceleration of the post-shock subsonic flow to a supersonic one through the Laval nozzle.

converging nozzle to a sonic flow at the nozzle throat and further to a supersonic flow if a diverging nozzle is provided. The flow remains a sonic flow if the uniform field is provided without any diverging field.

As shown in Fig. 25.6, this behavior was actually observed in experiments in the long diverging nozzle configuration. A supersonic flow produced in the long diverging nozzle (region (I) in Fig. 25.6) is intentionally injected into a Laval nozzle which is located in the far downstream region. A standing shock wave appears in front of the Laval nozzle (region (II)) and the subsonic flow after the shock wave is re-accelerated to a supersonic flow (region(III)) by passing through the Laval nozzle [12].

In order to closely investigate the Laval nozzle effect on the thermal energy conversion to a flow energy, a short magnetic Laval nozzle is installed near the MPDA muzzle as shown in Fig. 25.7. In Fig. 25.8 are shown flow characteristics with and without the Laval nozzle. It is shown clearly that the flow velocity and ion temperature downstream of the nozzle are higher and lower than those without the nozzle, respectively. On the other hand, variation of the flow parameters upstream of the nozzle is reversed. This shows that the subsonic flow upstream of the nozzle perceives the existence of the nozzle and self-adjusts so as to satisfy the sonic condition ($M = 1$) at the nozzle throat. It is also confirmed that total energy of the flow energy and the thermal energy is kept almost constant in the region of the Laval nozzle. The axial profile of the Mach number in the Laval nozzle is compared with that predicted by a one-dimensional isentropic flow model and in a qualitative agreement with each other. Quantitatively, the Mach number measured downstream of the nozzle throat is lower than the predicted one. This would be due to an under-expansion, since the observed cross-section of the plasma flow downstream of the throat does not coincide with that of the vacuum magnetic channel. This may be suppressed by increasing the magnetic field strength of the nozzle and/or by

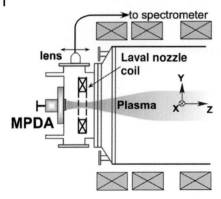

Fig. 25.7 A short magnetic Laval nozzle installed near the MPDA muzzle.

increasing the characteristic length scale of the magnetic nozzle. It is noted that an appropriate converging ratio of the Laval nozzle should be selected according to the Mach number of the plasma flow ejected from the MPDA.

25.3.3
RF Heating of a High Mach Number Plasma Flow

RF ion heating in a fast-flowing plasma is quite different from that in a confined, stationary fusion plasma as described in Section 25.1. Experiments are performed in

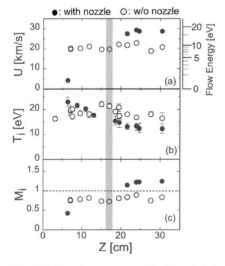

Fig. 25.8 Flow characteristics with (filled circles) and without (open circles) the Laval nozzle. A subsonic flow is converted into a supersonic one passing through the magnetic Laval nozzle.

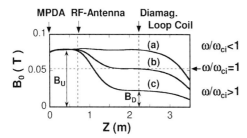

Fig. 25.9 Axial profiles of the magnetic field configuration with three types of the diverging nozzle ratio. The wave excited in the region where $\omega/\omega_{ci} < 1$ propagates downstream into the region of (a) $\omega/\omega_{ci} < 1$, (b) $\omega/\omega_{ci} = 1$, and (c) $\omega/\omega_{ci} > 1$. The antenna is located at $Z = 0.63$ m and the diamagnetic loop coil at $Z = 2.23$ m.

various types of the long diverging magnetic field configurations, as shown in Fig. 25.9.

The upstream magnetic field B_U is kept constant and the downstream one B_D is varied to form a magnetic beach configuration. Various types of antenna such as Rogowski-type and loop-type antennas with an azimuthal mode number $m = 0$, and double loop antennas with $m = \pm 1$ (Nagoya type-III antenna) and $m = \pm 2$ have been tried to inductively excite waves in the ICRF [13]. Helical antennas with right- or left-handed windings and $m = \pm 1$, as shown in Fig. 25.10, are adopted in the present experiments. The antenna current is driven by a FET-inverter power supply.

At first, dispersion relations of excited waves are obtained in a uniform field. Shear and compressional Alfvén waves are identified, as shown in Fig. 25.11, by comparing the experimental results with theoretical curves in which the Doppler shift of a fast flowing plasma is taken into account.

Figure 25.12 shows typical waveforms of the discharge current I_d, the antenna current I_{RF}, and an observed diamagnetic coil signal $W_\perp$. The diamagnetic coil signal apparently increases during the RF excitation.

To confirm the ion heating we measured spatial profiles of ion and electron temperature and density. The ion temperature T_i increases from 3.9 to 6.3 eV and also

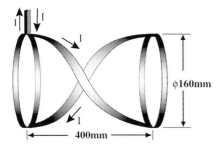

Fig. 25.10 Helical antenna with an azimuthal mode number $m = \pm 1$.

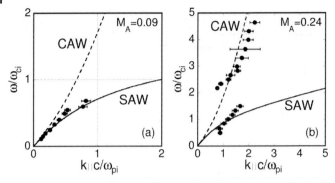

Fig. 25.11 Dispersion relations of the excited waves propagating downstream in a fast-flowing plasma with a Alfvén Mach number M_A. (a) He plasma and (b) Ar plasma. RF waves are excited by an $m = \pm 1$ helical antenna with right-handed winding. Shear Alfvén wave (SAW) and compressional Alfvén wave (CAW) are identified. Theoretical dispersion curves are calculated by taking into account the effect of the Doppler shift due to the fast flow.

electron temperature T_e from 1 to 1.5 eV. The electron density is $1 \times 10^{19}\,\mathrm{m}^{-3}$ and slightly decreases during the RF excitation.

Figure 25.13 shows dependence of the ratio $\Delta W_\perp / W_\perp$ on the magnetic field in the downstream region B_U for three different plasma densities. As is shown in the figure, a clear indication of the ion cyclotron resonance is observed in the case with a lower plasma density, where the cyclotron frequency f_{RF} is much larger than the ion–ion collision frequency ν_{ii}. The optimum B_D where the ratio $\Delta W_\perp / W_\perp$ becomes

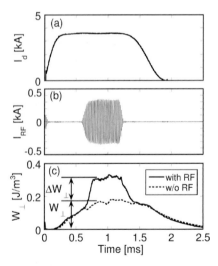

Fig. 25.12 Temporal evolutions of (a) discharge current I_d, (b) antenna current I_{RF}, and (c) diamagnetic coil signal $W_\perp$. Helium plasma. $B_Z = 0.7\,\mathrm{kGs}$ (uniform); $f_{RF} = 80\,\mathrm{kHz}$ ($\omega/\omega_{ci} = 0.3$).

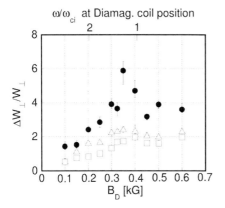

Fig. 25.13 Ratio of $\Delta W_\perp / W_\perp$ as a function of the magnetic field B_D. Helium plasma. $B_U = 0.7\,\mathrm{kGs}$; $f_{RF} = 160\,\mathrm{kHz}$. ($\bullet$) $f_{RF}/\nu_{ii} = 8.4$ ($n = 0.52 \times 10^{18}\,\mathrm{m}^{-3}$), ($\square$) $f_{RF}/\nu_{ii} = 2.4$ ($n = 1.9 \times 10^{18}\,\mathrm{m}^{-3}$), ($\square$) $f_{RF}/\nu_{ii} = 1.6$ ($n = 2.7 \times 10^{18}\,\mathrm{m}^{-3}$).

maximum is slightly shifted to a lower value than the cyclotron resonance. This is caused by the Doppler effect due to the fast-flowing plasma. The increased perpendicular ion energy is converted to a parallel one according to the adiabatic invariant in a collisionless plasma sufficiently expanded along a diverging magnetic nozzle.

25.4
Summary

Performance of an MPDA thruster is improved by using a magnetic Laval nozzle and by RF heating. A subsonic flow near the MPDA muzzle is converted to a supersonic flow through the conversion of the thermal energy to the flow energy. The plasma flow self-adjusts so as to satisfy the sonic condition at the throat of the magnetic nozzle. RF wave heating of a fast-flowing plasma is performed using a helical ICRF antenna in various diverging magnetic nozzles. Both shear and compressional Alfvén waves are excited and identified by comparing with theoretical curves which take into account the Doppler effect. Ion cyclotron resonance heating by the shear Alfvén wave is observed for the first time in the fast-flowing plasma with a low collisionality.

Acknowledgments

This work was supported in part by a Grant-in-Aid for Scientific Research from Japan Society for the Promotion of Science. Part of this work was carried out under the Cooperative Research Project Program of the Research Institute of Electrical Communication, Tohoku University.

References

1 Jahn, R.G. (1968) *Physics of Electric Propulsion*, Mcgraw-Hill.
2 Frisbee, R.H. (2003) *J. Propul. Power*, **19**, 1129.
3 Sasoh, A. and Arakawa, Y. (1995) *J. Propul. Power*, **11**, 351.
4 Tahara, H., Kagaya, Y. and Yoshikawa, T. (1997) *J. Propul. Power*, **13**, 651.
5 Chang Diaz, F.R., Squire, J.P., Bengston, R.D., baity, F.W. and Carter, M.D. (2000) *AIAA Paper*, **2000–3756**, 1.
6 Kuninaka, H. (1998) *J. Propul. and Power*, **14**, 1022.
7 Toki, K., Shimizu, Y. and Kuriki, K. (1997) **IEPC** *(Int. Electric Propulsion Conf.)*, **97–120**, 1.
8 Ando, A., Watanabe, T.S., Watanabe, T.K., Tobari, H., Hattori, K. and Inutake, M. (2005) *J. Plasma Fusion Res.*, **81**, 451.

9 Tobari, H., Sato, R., Hattori, K., Ando, A. and Inutake, M. (2003) *Adv. Appl. Plasma Sci.*, **4**, 133.
10 Ando, A., Ashino, M., Sagi, Y., Inutake, M., Hattori, K., Yoshinuma, M., Imasaki, A., Tobari, H. and Yagai, T. (2001) *J. Plasma Fusion Res.*, **4**, 373.
11 Inutake, M., Ando, A., Hattori, K., Tobari, H. and Yagai, T. (2002) *J. Plasma Fusion Res.*, **78**, 1352.
12 Inutake, M., Ando, A., Hattori, K., Yoshinuma, M. Imasaki, A., Yagai, T., Tobari, H., Murakami, F. and Ashino, M. (2000) *Proc. Int. Conf. on Plasma Physics*, Quebec, Vol. 1, 148.
13 Inutake, M., Ando, A., Hattori, K., Yagai, T., Tobari, H., and Kumagai, R., Miyazaki, H. and Fujimura, S. (2003) *Trans. Fusion Technol.*, **43**, 118.

Index

Advanced Plasma Technology. Edited by Riccardo d'Agostino, Pietro Favia, Yoshinobu Kawai, Hideo Ikegami, Noriyoshi Sato, and Farzaneh Arefi-Khonsari
Copyright © 2008 WILEY-VCH Verlag GmbH & Co. KGaA, Weinheim
ISBN: 978-3-527-40591-6